T0141811

Smart Innovation, Systems and Technologies

Volume 69

Series editors

Robert James Howlett, Bournemouth University and KES International,
Shoreham-by-sea, UK
e-mail: rjhowlett@kesinternational.org

Lakhmi C. Jain, University of Canberra, Canberra, Australia;
Bournemouth University, UK;
KES International, UK
e-mails: jainlc2002@yahoo.co.uk; Lakhmi.Jain@canberra.edu.au

About this Series

The Smart Innovation, Systems and Technologies book series encompasses the topics of knowledge, intelligence, innovation and sustainability. The aim of the series is to make available a platform for the publication of books on all aspects of single and multi-disciplinary research on these themes in order to make the latest results available in a readily-accessible form. Volumes on interdisciplinary research combining two or more of these areas is particularly sought.

The series covers systems and paradigms that employ knowledge and intelligence in a broad sense. Its scope is systems having embedded knowledge and intelligence, which may be applied to the solution of world problems in industry, the environment and the community. It also focusses on the knowledge-transfer methodologies and innovation strategies employed to make this happen effectively. The combination of intelligent systems tools and a broad range of applications introduces a need for a synergy of disciplines from science, technology, business and the humanities. The series will include conference proceedings, edited collections, monographs, handbooks, reference books, and other relevant types of book in areas of science and technology where smart systems and technologies can offer innovative solutions.

High quality content is an essential feature for all book proposals accepted for the series. It is expected that editors of all accepted volumes will ensure that contributions are subjected to an appropriate level of reviewing process and adhere to KES quality principles.

More information about this series at http://www.springer.com/series/8767

Anna Esposito · Marcos Faudez-Zanuy
Francesco Carlo Morabito
Eros Pasero
Editors

Multidisciplinary Approaches to Neural Computing

 Springer

Editors
Anna Esposito
Dipartimento di Psicologia
Università della Campania "Luigi
  Vanvitelli"
Caserta
Italy

and

International Institute for Advanced
  Scientific Studies (IIASS)
Vietri sul Mare
Italy

Marcos Faudez-Zanuy
Fundació Tecnocampus
Pompeu Fabra University
Mataro
Spain

Francesco Carlo Morabito
Department of Civil, Environmental,
  Energy, and Material Engineering
Mediterranea University of Reggio Calabria
Reggio Calabria
Italy

Eros Pasero
Dipartimento di Elettronica e
  Telecomunicazioni
Politecnico di Torino, Laboratorio di
  Neuronica
Torino
Italy

ISSN 2190-3018   ISSN 2190-3026   (electronic)
Smart Innovation, Systems and Technologies
ISBN 978-3-319-86031-2   ISBN 978-3-319-56904-8   (eBook)
DOI 10.1007/978-3-319-56904-8

Preface

Multidisciplinary Approaches to Neural Computing belongs to a book series dedicated to recent advances in computational and theoretical issues of artificial intelligence methods. In this volume, particular attention is given to the dynamics of signal exchanges and the role played by artificial neural networks in giving meanings to aspects of information processing and social information processing in order to approximate noncomputable functions in the Turing exception of the term. The content of the book is organized in sections and each edition affords and discusses new ANN topics on the basis of their contributions in integrating algorithms and procedures for the processing of dynamic signals, in anticipation of the implementation of intelligent avatars, interactive dialog systems, and reliable complex autonomous systems for facilitation learning, decision making, aging, and improve the common well-being.

Each edition of the book is related to a long running International Conference, WIRN (International Workshop on Neural Networks, currently at the 28th edition). After the conference, the topics of major interest are selected and researchers proposing these topics are invited to contribute to the book.

The current edition is composed of the following topics:

1. Introduction
2. Algorithms
3. ANN Applications
4. Industrial Applications of Computational Intelligence Approaches
5. Social and Biometric Data for Applications in Human-Machine Interactions

Given the multidisciplinary nature of the book, scientific contributions are from computer science, physics, psychology, statistics, mathematics, electrical engineering, and communication science. The contributors to this volume are leading authorities in their respective fields. We are grateful to them for accepting our invitation and making (through their participation) the book a worthwhile effort. We would like to thank the Springer project coordinator for books production Mr. **Ayyasamy Gowrishankar**, the Springer executive editor Dr. **Thomas Ditzinger**, and the editor assistant Mr. **Holger Schaepe**, for their outstanding support and availability. In addition we would like to express our deep appreciation and gratitude to the editors-in-chief of the Springer series Smart Innovation, Systems and Technologies, Profs. **Jain Lakhmi C. and Howlett Robert James**, for supporting our initiative and giving credit to our efforts.

Caserta, Vietri sul Mare, Italy   Anna Esposito
Mataro, Spain   Marcos Faudez-Zanuy
Reggio Calabria, Italy   Francesco Carlo Morabito
Torino, Italy   Eros Pasero

Organization

SIREN Executive Committee

Anna Esposito, Università degli Studi della Campania "Luigi Vanvitelli" and IIASS (IT)
Marcos Faundez-Zanuy, Pompeu Fabra University (UPF) (ES)
Francesco Carlo Morabito, Università Mediterranea di Reggio Calabria (IT)
Francesco AN Palmieri, Seconda Università di Napoli (IT)
Eros Pasero, Politecnico di Torino (IT)
Stefano Rovetta, Università di Genova (IT)
Stefano Squartini, Università Politecnica delle Marche (IT)
Aurelio Uncini, Università di Roma "La Sapienza" (IT)
Salvatore Vitabile, Università di Palermo (IT)

SIREN International Advisory Committee

Metin Akay, Arizona State University, USA
Pierre Baldi, University of California in Irvine, USA
Piero P. Bonissone, Computing and Decision Sciences
Leon O. Chua, University of California at Berkeley, USA
Jaime Gil-Lafuente, University of Barcelona, Spain
Giacomo Indiveri, Institute of Neuroinformatics, Zurich, Switzerland
Nikola Kasabov, Auckland University of Technology, New Zealand
Vera Kurkova, Academy of Sciences, Czech Republic
Shoji Makino, NTT Communication Science Laboratories, USA
Dominic Palmer-Brown, London Metropolitan University, UK
Witold Pedrycz, University of Alberta, Canada
Harold H. Szu, Army Night Vision Electronic Sensing Directory, UA
Jose Principe, University of Florida at Gainesville, USA

Alessandro Villa, Università Joseph Fourier, Grenoble 1, France
Fredric M. Ham, Florida Institute of Technology, USA
Cesare Alippi, Politecnico di Milano, Italy
Marios M. Polycarpou, Politecnico of Milano, Italy

Program Committee

Alippi Cesare, Politecnico di Milano
Altilio Rosa, Università di Roma "La Sapienza"
Angiulli Giovanni, Università Mediterranea di Reggio Calabria
Apicella Ilenia, Università di Salerno
Beugeling Willem, Tata Steel
Bevilacqua Vitoantonio, Politecnico di Bari
Bramanti Alessia, Istituto di Scienze Applicate e Sistemi Intelligenti ISASI-CNR"Eduardo Caianiello" Messina
Camastra Francesco, Università Napoli Parthenope
Campolo Maurizio, Università degli Studi Mediterranea Reggio Calabria
Capuano Vincenzo, Seconda Università di Napoli
Cauteruccio Francesco, Università degli Studi della Calabria
Ciaramella Angelo, Università Napoli Parthenope
Ciccarelli Valentina, Università di Roma "La Sapienza"
Cirrincione Giansalvo, UPJV
Colla Valentina, Scuola Superiore S. Anna
Comajuncosas Andreu, Tecnocampus
Commimiello Danilo, Università di Roma "La Sapienza"
Committeri Giorgia, Università di Chieti
Cordasco Gennaro, Seconda Università di Napoli
de Candia Antonio, Università di Napoli Federico II
De Carlo Domenico, Università Mediterranea di Reggio Calabria
De Felice Domenico, Università Napoli Parthenope
Dell'Orco Silvia, Università degli Studi della Basilicata
Dettori Stefano, Scuola Superiore S. Anna
Droghini Diego, Università Politecnica delle Marche
Esposito Anna, Università degli Studi della Campania "Luigi Vanvitelli" and IIASS
Esposito Antonietta Maria, sezione di Napoli Osservatorio Vesuviano
Esposito Francesco, Università di Napoli Parthenope
Esposito Marilena, International Institute for Advanced Scientific Studies (IIASS)
Faundez-Zanuy Marcos, Tecnocampus Universitat Pompeu Fabra
Fiasché Maurizio, Politecnico di Milano
Fragoulis Nikos, IRIDA Labs S.A.
Ieracitano Cosimo, Università degli Studi Mediterranea Reggio Calabria
Invitto Sara, Università del Salento

La Foresta Fabio, Università degli Studi Mediterranea Reggio Calabria
Lo Console Claudio, Scuola Superiore Sant' Anna
Maldonato Mauro, Università degli Studi della Basilicata
Mammone Nadia, IRCCS Centro Neurolesi Bonino-Pulejo, Messina
Manghisi Vito, Politecnico di Bari
Matarazzo Olimpia, Seconda Università di Napoli
Militello Carmelo, Consiglio Nazionale delle Ricerche (IBFM-CNR), Cefalù (PA)
Morabito Francesco Carlo, Università Mediterranea di Reggio Calabria
Narejo Sanam, Politecnico di Torino
Nastasi Gianluca, Scuola Superiore S. Anna
Panella Massimo, Università di Roma "La Sapienza"
Parisi Raffaele, Università di Roma "La Sapienza"
Piraino Giulia, Istituto Santa Chiara, Lecce
Portero-Tresserra Marta, Universitat Pompeu Fabra
Principi Emanuele, Università Politecnica delle Marche
Qi Wen, Politecnico di Milano
Randazzo Vincenzo, Politecnico di Torino
Reyneri Leonardo, Politecnico di Torino
Rundo Leonardo, Università degli Studi di Milano-Bicocca
Russo Giorgio, Consiglio Nazionale delle Ricerche (IBFM-CNR), Cefalù (PA)
Salerno Giuseppe, General Electric Oil and Gas
Sappey-Marinier Dominique, Université de Lyon
Scardapane Simone, Università di Roma "La Sapienza"
Scarpiniti Michele, Università di Roma "La Sapienza"
Scibelli Filomena, Università di Napoli "Federico II", and Seconda Università di Napoli
Senese Vincenzo Paolo, Seconda Università di Napoli
Sgrò Annalisa, Università Mediterranea di Reggio Calabria
Sisca Francesco Giovanni, Politecnico di Milano
Squartini Stefano, Università Politecnica delle Marche
Staiano Antonino, Università Napoli Parthenope
Stamile Claudio, Université de Lyon
Taisch Marco, Politecnico di Milano
Terracina Giorgio, Università della Calabria
Theoharatos Christos, Computer Vision Systems, IRIDA Labs S.A.
Troncone Alda, Seconda Università di Napoli
Uncini Aurelio, Università di Roma "La Sapienza"
Ursino Domenico, Università Mediterranea di Reggio Calabria
Vannucci Marco, Scuola Superiore Sant'Anna
Vecchiotti Paolo, Università Politecnica delle Marche
Vesperini Fabio, Università Politecnica delle Marche
Vitabile Salvatore, Università degli Studi di Palermo
Zucco Gesualdo, Università di Padova

Sponsoring Institutions

International Institute for Advanced Scientific Studies (IIASS) of Vietri S/M (Italy)
Department of Psychology, Università degli Studi della Campania "Luigi Vanvitelli" (Italy)
Provincia di Salerno (Italy)
Comune di Vietri sul Mare, Salerno (Italy)
International Neural Network Society (INNS)
Università Mediterranea di Reggio Calabria (Italy)

Contents

Part I
Introduction

Chapter 1
Redefining Information Processing Through Neural Computing Models

Anna Esposito, Marcos Faundez-Zanuy, Francesco Carlo Morabito and Eros Pasero

Abstract Artificial Neural Networks (ANN) are currently exploited in many scientific domains. They had shown to act as doable, practical, and fault tolerant computational methodologies. They are equipped with solid theoretical background and proved to be effective in many demanding tasks such as approximating complex functions, optimizing search procedures, detecting changes in behaviors, recognizing familiar patterns, identifying data structures. ANNs computational limitations, essentially related to the presence of strong nonlinearities in the data and their poor generalization ability when provided of fully connected architectures, have been hammered by more sophisticated models, such as Modular Neural Networks (MNNs), and more complex learning procedures, such as deep learning. Given the multidisciplinary nature of their use and the interdisciplinary characterization of the problems they approach, ranging from medicine to psychology, industrial and social robotics, computer vision, and signal processing (among many others) ANNs may provide the bases for a redefinition of the concept of information processing. These reflections are supported by theoretical models and applications presented in the chapters of this book.

Keywords Neural network models · Learning · Artificial intelligent methods

A. Esposito
Dipartimento di Psicologia, Università della Campania "Luigi Vanvitelli", Caserta, Italy
e-mail: iiass.annaesp@tin.it

A. Esposito
International Institute for Advanced Scientific Studies (IIASS), Vietri sul Mare, Italy

M. Faundez-Zanuy (✉)
Tecnocampus, Pompeu Fabra University, Mataró, Barcelona, Spain
e-mail: faundez@tecnocampus.cat

F.C. Morabito
Università degli Studi "Mediterranea" di Reggio Calabria, Calabria, Italy
e-mail: morabito@unirc.it

E. Pasero
Dipartimento Elettronica e Telecomunicazioni, Politecnico di Torino, Turin, Italy
e-mail: eros.pasero@polito.it

© Springer International Publishing AG 2018
A. Esposito et al. (eds.), *Multidisciplinary Approaches to Neural Computing*,
Smart Innovation, Systems and Technologies 69,
DOI 10.1007/978-3-319-56904-8_1

1.1 Introduction

There are a number of theories on information processing that argue on different meanings attributed to the concept of information and on different uses of it. Information processing is a research branch of communication theory that attempts to quantify how well signals encode meaningful information and how well [receiving] systems process [it] [5]. The conventional approach (introduced by Shannon [7-8] consider *"information"* as a probabilistic measure of the amount of signals emitted by an information source (the information entropy) and transmitted over a communication channel. This theoretical definition does not account for signal's meanings (emitted signals may be only noise), and rely on the receiver's subjective ability to decode the received information. In order to overcome these limitations, alternative information theories have been proposed, such as the generalized information theory [6] which include theories of imprecise probabilities [2] and fuzzification of uncertainty theories [10], and adaptive online learning algorithms [9] which include independent component analysis [3] and unsupervised factorial learning [4].

In this context ANNs are considered information processing systems consisting of a network of elementary computational units distributed on different layers and linked together, paralleling to some extent the structure of the human brain cortex. The computation that takes place over the several hidden layers of the network allows the implementation of complex learning algorithms (by examples, or associations), and provide generalized representations and solutions even in the case of nonstationary and nonlinear input-output relationships. Limitations due to the fact that large and fully connected networks may never converge to a solutions because they may develop nonsensical interactions among nodes are surmounted by new biologically inspired models such as modular neural networks (MNN) which exploit the concept of ANN modules "that have identifiable inputs and outputs from other modules" [1]. ANNs, being equipped with specific communication abilities, specific activation functions, and adaptation characteristics, as well as, offering the freedom to complicate their structure for the generation of subnetworks, connections, and weights to satisfy specific constraints, are computing machines mimicking the human way to achieve information processing.

1.2 Content of This Book

The themes approached in this book are multidisciplinary in nature, even though, they are all closely connected in their final aims to identify features from dynamic realistic signal exchanges and invariant machine representations in order to automatically identify, detect, analyze, and process real data. The content of book is organized in sections, each dedicated to a specific topic, and including peer reviewed chapters reporting applications and/or research results in neural computing from different scientific disciplines. The book is a follow-up of the International Workshop on Neural Networks (WIRN 2016) held in Vietri sul Mare, Italy, from the 18th to the 20th of May 2016. The workshop, being at its 28th edition is nowadays a historical and traditional scientific event bringing together scientists from different countries, and scientific disciplines. Each chapter is a peer reviewed extended version of the original work presented at the workshop and not published elsewhere.

Section I introduces the *concept of information processing* under the neural computing umbrella through a short chapter proposed by Esposito and colleagues.

Section II is dedicated to artificial neural networks algorithms adopting either supervised or unsupervised learning approaches. The section includes 5 short chapters discussing respectively on *Unsupervised Learning Techniques for Hybrid Seismic Events Discrimination* (proposed by Esposito AM and colleagues), *a Fuzzy C-Means Clustering Algorithm for Multispectral MR Image Segmentation* (proposed by Rundo and colleagues), *a Clustering Algorithm (QuickBundles) for Tractography Representations* (proposed by Cauteruccio and colleagues), *a Micro-Genetic Algorithm for Accurate Computation of Drude-Lorentz Coefficients* (proposed by Sgrò and colleagues), and the *Adam's Algorithm Applied to a Blind Source Separation of Signals* (proposed by Scarpiniti and colleagues).

Section III is devoted to the effectiveness of neural networks in identifying and discriminating desired patterns in the data. This section includes 10 short chapters where ANNs are successfully applied to daily life problems. In particular, NNs are exploited for Hand Pose Recognition (by De Felice and Camastra), detection of Simple Juggling Movements (by Camastra and colleagues), modeling of Cortical Phase Transitions (by Apicella and colleagues), Fall Detection (by Droghini and colleagues), Transformation of Non-Linear and Time-Variant Signals (by Alippi and colleagues), Data Mining for Distributed Medical Scenarios (by Scardapane and colleagues), Rule Base Reduction Using Conflicting and Reinforcement Measures (proposed by Anzilli and Giove), Internet Traffic Prediction (proposed by Narejo and Pasero), Dimensionality Reduction of Nonstationary Data (proposed Cirrincione and Randazzo), and Voice Activity Detection in a Multiroom Environment (proposed by Vecchiotti and colleagues).

Section IV is a special session devoted to Computational Intelligent Approaches to Industrial Processes. The section includes 7 short chapters dedicated respectively to: NN-Based Classification in Industrial Context (proposed by Cateni and Colla), Neural Networks Systems for Unbalanced Industrial Datasets (proposed byVannucci and Colla), Quantum-Inspired Evolutionary Multiobjective Optimization (proposed by Fiasché and colleagues), Neural Networks for Steam Turbine Monitoring (proposed by Dettori and colleagues), Predictive Neural Networks for Fuel Consumption (proposed by Aliev and colleagues), SOM-Based Analysis to Relate Non-Uniformities in Magnetic Measurements to Hot Strip Mill Process Conditions (proposed by Nastasi and colleagues), and Vision-Based Mapping and Micro-Localization of Industrial Components in the Fields of Laser Technology (proposed by Theoharatos and colleagues).

Section V is a special session devoted to "Social and Biometric Data for Applications in Human-Machine Interactions: Models and Algorithms". The topic is rooted in the current trends of Information Communication Technology (ICT) for social information processing in order to develop ICT interfaces facilitating learning, education, aging and well-being. The section includes 13 short chapters describing perceptual, social, and behavioral data. In this context the 13 chapters are dedicated respectively to: A Neural Network Analysis of ERP in Intimate Partner Violence (Invitto and colleagues), EEG Signals' Wavelet Coherence in Mild Cognitive Impaired and Alzheimer's Patients (Ieracitano and colleagues), EEG Signal Classication by a SVM Based Algorithm (Saccà and colleagues), Automatic Removal of Artifacts from the EEGs of Alzheimer's Patients (proposed by Mammone), An OERP Study on Smell and Meaning (Invitto and colleagues), Detection of Depressive States from Speech (Mendiratta and colleagues), Effects of Gender and Luminance Backgrounds on the Recognition of Neutral Facial Expressions (by Capuano and colleagues), Semantic Maps of Twitter Conversations (Ciaramella and colleagues), Cursive Handwriting Learning in Schools (Comajuncosas and colleagues), Implicit and Explicit Attitudes on Voting Behaviour (Di Conza and colleagues), Gambler's Fallacy (Matarazzo and colleagues), Intuitive Decisions Deliver Better Results than a Deliberate Approach (Maldonato and colleagues), and How AI is Changing Human Evolution (by Maldonato and Valerio).

1.3   Conclusion

A number of different research communities within the psychological and computational sciences have tried to characterize information processing in terms of human neural behaviors, formal and informal social signals, communication modes, hearing and vision processes, as well as, task optimization rules for detecting environmental events, improving industrial processes, helping in decision makings. There has been substantial progress in these different communities and surprising convergence in the field of Neural Computing. This growing interest makes the current intellectual climate an ideal one for the unveiling of this book devoted to collect current progresses in dynamic signal exchanges, information processing and social information processing by artificial neural networks. Key aspects considered are the integration of algorithms and procedures for the recognition of dynamic (faces, speech, gaits, EEGs, seismic waves) signals, in anticipation of the implementation of intelligent avatars, interactive dialog systems, and reliable complex autonomous systems for facilitation learning, decision making, aging, and improve the common well-being.

References

1. Caelli, T., Guan, L., Wilson Wen, W.: Modularity in neural computing. Proc. IEEE **87**(9), 1497–1518 (1999)
2. Cattaneo, M.: A continuous updating rule for imprecise probabilities. In: Laurent, A., et al. (eds.) Information Processing and Management of Uncertainty in Knowledge Based Systems, pp. 426–435. Springer (2014)
3. Comon, P.: Independent component analysis, a new concept? Sig. Process. **36**, 287–314 (1994)
4. Deco, G., Brauer, W.: Nonlinear higher-order statistical decorrelation by volume-conserving neural architectures. Neural Netw. **8**(4), 525–535 (1996)
5. Esposito, A.: The amount of information on emotional states conveyed by the verbal and nonverbal channels: some perceptual data. In: Stylianou, Y., et al. (eds.) LNCS 4391, 249–268. Springer-Verlag, Berlin (2007)
6. Klir, G.J.: Generalized information theory: aims, results, and open problems. Reliab. Eng. Syst. Saf. **85**(1–3), 21–38 (2004)
7. Shannon, C. E.: A mathematical theory of communication. First published: July 1948 (1948). doi:10.1002/j.1538-7305.1948.tb01338.x
8. Shannon, C., Weaver, W.: The mathematical theory of communication. University of Illinois Press, Urbana, IL (1963)
9. Yang, H.H., Amari, S.: Adaptive on-line learning algorithms for blind separation—maximum entropy and minimum mutual information. Neural Comput. **9**, 1457–1482 (1997)
10. Yang, M., Chen, T., Wu, K.: Generalized belief function, plausibility function, and Dempster's combination rule to fuzzy sets. Int. J. Intel. Syst. **18**(8), 925–937 (2003)

Part II
Algorithms

Chapter 2
A Neural Approach for Hybrid Events Discrimination at Stromboli Volcano

Antonietta M. Esposito, Flora Giudicepietro, Silvia Scarpetta
and Sumegha Khilnani

Abstract Stromboli volcano is considered one of the most active volcanoes in the world. During its effusive phases, it is possible to record a particular typology of events named "hybrid events", that rarely are observed in the daily volcano activity. These ones are often associated to fault failure in the volcanic edifice due to magma movement and/or pressurization. Their identification, analysis and location can improve the volcano eruptive process comprehension. However, it is not easy to distinguish them from the other usually recorded events, i.e. explosion-quakes, through a visual seismogram analysis. Thus, we present an automatic supervised procedure, based on a Multi-layer Perceptron (MLP) neural network, to identify and discriminate them from the explosions-quakes. The data are encoded by using LPC coefficients and then adding to this coding waveform features. The 99% of accuracy was reached when waveform features are coded together with LPC coefficients as input to the network, emphasizing the importance of temporal features for discriminating hybrid events from explosion-quakes. The results allow us to assert that the proposed neural strategy can be included in a more complex automatic system for the monitoring of Stromboli volcano and of other volcanoes in the world.

A.M. Esposito (✉) · F. Giudicepietro
Istituto Nazionale di Geofisica e Vulcanologia,
Sezione di Napoli Osservatorio Vesuviano, Naples, Italy
e-mail: antonietta.esposito@ingv.it

F. Giudicepietro
e-mail: flora.giudicepietro@ingv.it

S. Scarpetta
Dipartimento di Fisica "E.R.Caianiello", INFN gruppo coll. di Salerno,
Università di Salerno Fisciano (SA), sez. Di Napoli, Italy
e-mail: sscarpetta@unisa.it

S. Khilnani
Indian Institute of Technology Delhi, Delhi, India
e-mail: sumeghakhilnani@gmail.com

S. Khilnani
Istituto Internazionale per gli Alti Studi Scientifici (IIASS), Vietri sul Mare (SA), Italy

© Springer International Publishing AG 2018   11
A. Esposito et al. (eds.), *Multidisciplinary Approaches to Neural Computing*,
Smart Innovation, Systems and Technologies 69,
DOI 10.1007/978-3-319-56904-8_2

Keywords Hybrid events · Explosion-quakes · Seismic waves · Volcano
monitoring · Neural learning · MLP network

2.1 Introduction

The volcanic activity of Stromboli, one of the most active volcanoes in the world, is
described by small to medium size explosions occurring at the eruptive vents on the
top of the volcano [11]. Usually, the explosion frequency rate ranges between 6 and
18 events per hour [4, 5]. This persistent eruptive activity is sometimes interrupted
by effusive phases which lead to the formation of lava flows on the Sciara del
Fuoco, the NW flank of the volcano characterized by lateral collapse structures
[16]. In recent decades, two major effusive phases occurred in 2002–2003 [3] and
2007 [24], after which the scientific community and civil protection authorities
made a big effort to improve the monitoring [6] and knowledge of Stromboli
volcano dynamics [1, 18]. Also the landslides, which caused the tsunami of 2002
[21, 26–28] and were precursors of the effusive phase on February, 2007 [24], were
investigated, in particular their detection and discrimination has been exploited by
using neural networks [8–10, 17]. Some minor episodes of lava overflow from the
summit craters occurred in later years [14].

Also hybrid events, typically recorded during Stromboli effusive phases, can be
considered significant signals as they provide important information on the volcano
status. Their sources are usually very shallow [12, 22]. Esposito et al. [12] suggested a relationship between the formation of a fracture system at the summit of
the volcano (6–8 March, 2007) and the source of the hybrid events, recorded as
swarms in that period and located in the same position. Hybrid event waveforms are
hardly distinguishable from those of explosion-quakes which are characteristic of
the Stromboli wave-field. Thus, we propose an automatic system, based on a
Multi-Layer Perceptron network, to discriminate hybrid signals from the
explosion-quakes. In the following we introduce the seismic signals recorded at
Stromboli volcano, explain the parametrization chosen for the data encoding,
illustrate the adopted neural strategy, describe the MLP results and finally present
our conclusions.

2.2 Seismic Signals at Stromboli Volcano

The semi-persistent explosive activity of the Stromboli volcano produces the typical
explosion-quake signals (Fig. 2.1). The monitoring seismic network also records
regional and teleseismic earthquakes, which occur at a great distance from the
island, and local seismic transients due to volcano-tectonic events or landslides.
However, they are not investigated at this time.

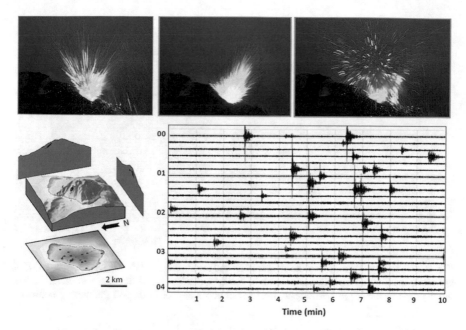

Fig. 2.1 Examples of explosion-quakes at Stromboli volcano: the first three panels are photos of explosion-quakes (courtesy of Rosario Peluso). Below, on the *right*, a 4-hour long recording window of powerful explosion-quakes is depicted; while on the *left* the explosion-quake location on the map of Stromboli island is visualized (small *red points*)

Hybrid events are also recorded, generally during volcano effusive phases and rarely during its usual explosive activity. They can be considered a particular typology of seismic events observed at several volcanoes. The left panel of Fig. 2.2 shows the location, with the geometry of the seismic network, of some hybrid events recorded during the swarm occurred on March 6–8, 2007, while on the right their representation on the seismogram is depicted.

The aim of this work is to discriminate hybrid events from explosion-quakes, since they are not easily distinguishable only through a visual analysis of their seismograms. Figure 2.3 shows, on the top panel, the seismograms of an hybrid event (on the left) and of an explosion-quake (on the right) and their corresponding spectrograms in the frequency domain, on the bottom panel. As observed in the figure, the hybrid event (on the left) presents an initial part with a high frequency content and a second part with a narrow frequency band, while the explosion-quake signal (on the right) exhibits no distinct seismic phases and has a frequency range of 1–6 Hz.

Observing the hypocenters of the events of 6–8 March 2007 swarm (Fig. 2.2), we can see that they are concentrated near the volcano surface, at an elevation raging between 600–800 m a.s.l., indicating a shallow source of these signals. In the same period, the formation of a fracture system at the summit of the volcano was observed (Fig. 2.4).

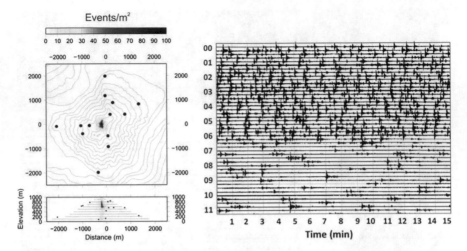

Fig. 2.2 A temporal window of the swarm of hybrid events recorded during March 6–8, 2007, on the *right*, and the relative location of some of them, on the *left* (after Longobardi et al. 2012). The *black points* in the *left panel* indicate the 13 seismic stations of Stromboli monitoring network

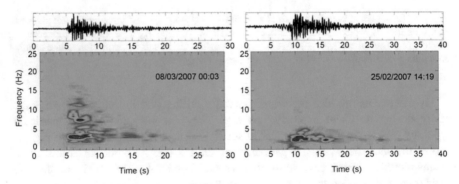

Fig. 2.3 The waveforms and the associated spectrograms of an hybrid event (on the *left*) and of an explosion-quakes (on the *right*)

2.3 Data Parametrization

The exploited dataset is of 884 events partitioned into two classes, i.e. 455 hybrid events and 429 explosion-quakes. For both classes each record has a duration of 18 s i.e. a vector of 900 samples with a sampling frequency of 50 Hz.

In order to obtain a significant and discriminating data encoding, the following plots of the signals have been performed and analyzed: the Amplitude v/s Time plot (Fig. 2.5), the LPC [20, 23] Spectrum (Fig. 2.6) and the Spectrograms (Fig. 2.7) both for an explosion-quake (on the left) that for an hybrid event (on the right) respectively.

Fig. 2.4 The fracture system at the crater terrace of Stromboli during the 2007 effusive phase highlighted by the *red dashed line* (photo by Tullio Ricci)

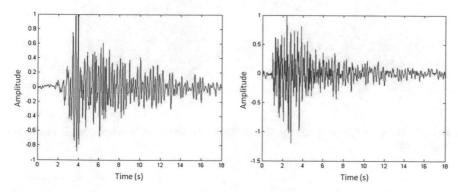

Fig. 2.5 The amplitude v/s time plot of an explosion-quake (on the *left*) and of a hybrid event (on the *right*) respectively

From the above plots we can infer the following observations: first, the LPC spectra of the explosion-quake (Fig. 2.6) shows some peaks and then a sudden and consistent decrease; on the contrary, that of the hybrid event presents clear peaks with no sudden decline. Second, looking at the spectrograms (Fig. 2.7), peaks of amplitude are observed at low frequency for the explosion-quake, whereas at the beginning of the hybrid event we can observe high frequency peaks. The same can be noticed from the seismograms (Fig. 2.5). These remarks led us to use a data representation that considers not only the LPC [20, 23] coefficients, but also the waveform information obtained as:

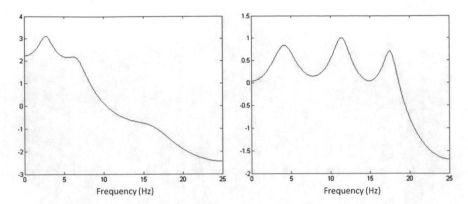

Fig. 2.6 The LPC spectrum of an explosion-quake (on the *left*) and of a hybrid event (on the *right*) respectively

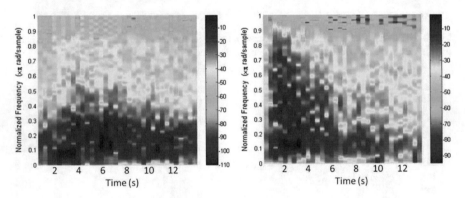

Fig. 2.7 The spectrograms of an explosion-quake (on the *left*) and of a hybrid event (on the *right*) respectively

$$W_i = (S_{imax} - S_{imin}) * N / \Sigma (S_{imax} - S_{imin}) \tag{1}$$

where:

S_{imax} = is the maximum value in the i-th window,
S_{imin} = is the minimum value in the i-th window,
N = is the number of windows

We use a window length of 90 samples (i.e. 1.8 s). Figure 2.8 visualizes the waveform coefficients plotted in the mid-point of time windows for an explosion-quake signal.

At the end of the preprocessing stage, each signal will be encoded by a vector of LPC and waveform coefficients. The number of coefficients was not fixed in order to test different representation and select the best one.

Fig. 2.8 The waveform
coefficients of an
explosion-quake signal

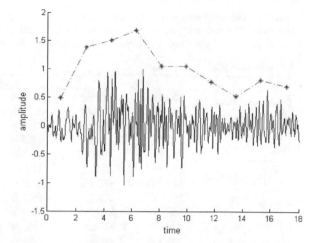

Table 2.1 The dataset distribution for the MLP training, validation and testing

	Training (50%)	Validation (20%)	Testing (30%)
Explosion-quakes (429)	214	85	130
Hybrid events (455)	227	92	136
	441	177	266

2.4 MLP Technique

For the discrimination task we adopted a Multi-Layer Perceptron (MLP) [2] network. Neural networks have proved to be effective in several applications and fields [7, 8, 13, 15, 17, 25, 29, 30] thanks to their learning ability form the experience and to be data-driven methods. Moreover, they are non-linear models and can represent complex data relationships.

In our experiments, the dataset for each class of signals was partitioned into training, validation and testing sets as shown in Table 2.1. In particular, the training set was composed of 441 input vectors (227 hybrid events and 214 explosion-quakes). The same training set was used for all the experimental conditions detailed in Table 2.2. The validation process is done in order to find the optimal number of iterations to be used in the training process.

Then, the net parameters adjustment was realized taking into account previous works [13, 17] and using a trial and error procedure. Regarding to the MLP architecture, we used a variable number of hidden nodes, a nonlinear hyperbolic-tangent function for the hidden nodes and a logistic sigmoidal activation function for the output unit. The weight optimization was carried out through two algorithms, i.e. the Quasi-Newton and the Conjugate Gradient [2]. Finally, during the training, instead of the conventional mean square error (MSE) function,

Table 2.2 The MLP performances obtained varying the number of the hidden nodes (X) and the input vector dimension expressed as the number of LPC coefficients (Y) with or without a fixed number (i.e. 10) of waveform coefficients. The best performances are in cyan color

Conjugate Gradient

Y/X	2	4	6	8	10
6	88.90%	88.55%	88.55%	88.75%	87.75%
8	80.65%	82.70%	85.75%	86.45%	60.90%
10	89.65%	87.40%	88.90%	90.20%	84.95%
12	91.75%	88.35%	88.90%	88.70%	88.00%
14	90.05%	90.95%	90.60%	90.05%	91.00%
16	88.75%	90.00%	91.75%	90.60%	91.35%
8+10(wave)	97.00%	98.50%	97.55%	97.55%	98.70%
10+10(wave)	98.10%	98.30%	98.85%	98.85%	97.35%
12+10(wave)	96.95%	97.35%	98.90%	97.95%	98.10%
14+10(wave)	97.95%	97.90%	99.05%	97.20%	99.25%
16+10(wave)	98.30%	98.85%	99.05%	98.30%	97.00%
18+10(wave)	98.65%	97.75%	98.15%	97.55%	98.45%
20+10(wave)	98.10%	98.30%	98.90%	98.15%	97.95%

Quasi New

Y/X	2	4	6	8	10
6	90.00%	87.75%	86.50%	88.90%	87.60%
8	72.60%	80.45%	87.00%	85.35%	70.70%
10	87.05%	87.20%	89.30%	89.30%	84.40%
12	93.05%	87.95%	88.90%	90.20%	88.00%
14	89.50%	90.75%	90.80%	89.65%	91.00%
16	89.10%	90.80%	91.90%	91.00%	91.35%
8+10(wave)	96.85%	98.65%	96.60%	97.35%	98.70%
10+10(wave)	98.50%	98.30%	98.30%	98.50%	97.90%
12+10(wave)	96.95%	97.90%	98.90%	98.60%	97.35%
14+10(wave)	97.75%	98.85%	98.90%	97.20%	99.25%
16+10(wave)	98.30%	99.20%	99.05%	98.45%	97.35%
18+10(wave)	98.15%	96.95%	97.95%	97.40%	98.10%
20+10(wave)	97.55%	98.30%	97.90%	98.15%	97.20%

we minimize the Cross-Entropy Error Function [2]. Combining logistic output units and this error function, the network response indicates the probability that a certain input belong or not to one of two classes, providing a probabilistic interpretation.

2.5   Results

In the following we illustrate the results of the MLP discrimination on the testing set by using the Conjugate Gradient and the Quasi-Newton learning algorithm respectively (Table 2.2). The performances are obtained varying the number of the hidden nodes and the input vector dimension. In particular, we first used only the LPC coefficients, then we added a fixed number of waveform coefficients (i.e. 10). The best performances are indicated in cyan color. In the Table 2.2, X indicates the number of the hidden nodes, while Y is the number of the LPC coefficients.

2.6 Conclusions

A neural strategy is proposed in order to detect and discriminate hybrid events recorded at Stromboli volcano from the typical explosion-quakes. Hybrid signals are a particular typology of events often associated to fault failure in the volcanic edifice due to magma movement and/or pressurization [19]. Their analysis can improve the eruptive process comprehension. However, it is difficult to distinguish them from the explosion-quakes by using only a visual inspection of their seismograms. So, to accomplish this aim, first we encoded data by using their discriminating features. In particular, we applied the LPC technique [20, 23] for extracting the spectral content of both signals, and a waveform representation to obtain their temporal features. Then, we selected a Multi-Layer Perceptron [2] network to realize the discrimination task.

As visualized in Table 2.2, the best performances was obtained with an input dimension between 22 (i.e. 12 LPC coefficients + 10 waveform coefficients) and 26 (i.e. 16 LPC coefficients + 10 waveform coefficients) and with a number of hidden nodes between 4 and 6. Moreover, a sudden increase in accuracy, from 88–94% to 97–99%, was reached when the waveform features are coded together with LPC coefficients as input to the network, emphasizing the importance of temporal features for discriminating hybrid events from explosion-quakes.

The achieved results demonstrate that the proposed method, based on the MLP network, well discriminate the two classes of signals. So, it could be included in an advanced automated system for the monitoring of Stromboli and of other volcanoes in the world.

References

1. Auger, E., D'Auria., L, Martini, M., Chouet, B., Dawson, P.: Real-time monitoring and massive inversion of source parameters of very long period seismic signals: An application to Stromboli volcano, Italy. Geophys. Res. Lett. **33**(4) (2006). doi:10.1029/2005GL024703
2. Bishop, C.: Neural Networks for Pattern Recognition, p. 500. Oxford University Press, New York (1995)
3. Calvari, S., Spampinato, L., Lodato, L., Harris, A.J.L., Patrick, M.R., Dehn, J., Burton, M.R., Andronico, D.: Chronology and complex volcanic processes during the 2002–2003 flank eruption at Stromboli volcano (Italy) reconstructed from direct observations and surveys with a handheld thermal camera. J. Geophys. Res. **110**(B02), 201 (2005). doi:10.1029/2004JB003129

4. Calvari, S., Büttner, R., Cristaldi, A., Dellino, P., Giudicepietro, F., Orazi, M., Peluso, R., Spampinato, L., Zimanowski, B., Boschi, E.: The 7 September 2008 vulcanian explosion at Stromboli volcano: multiparametric characterization of the event and quantification of the ejecta. J. Geophys. Res. **117**(B05), 201 (2012). doi:10.1029/2011JB009048

5. Chouet, B., Dawson, P., Ohminato, T., Martini, M., Saccorotti, G., Giudicepietro, F., De Luca, G., Milana, G., Scarpa, R.: Source mechanisms of explosions at Stromboli Volcano, Italy, determined from moment-tensor inversion of very-long period data. J. Geophys. Res. **108**(B1) (2003). doi:10.1029/2002JB001919

6. De Cesare, W., Peluso, O.R.M., Scarpato, G., Caputo, A., D'Auria, L., Giudicepietro, F., Martini, M., Buonocunto, C., Capello, M., Esposito, A.M.: The broadband seismic network of Stromboli volcano. Italy. Seism Res Lett **80**(3), 435–439 (2009). doi:10.1785/gssrl.80.3.435

7. Del Pezzo, E., Esposito, A., Giudicepietro, F., Marinaro, M., Martini, M., Scarpetta, S.: Discrimination of earthquakes and underwater explosions using neural networks. Bull. Seism. Soc. Am. **93**(1), 215–223 (2003)

8. Esposito, A.M., Giudicepietro, F., Scarpetta, S., D'Auria, L., Marinaro, M., Martini, M.: Automatic discrimination among landslide, explosion-quake and microtremor seismic signals at Stromboli volcano using neural networks. Bull Seismol. Soc. Am. (BSSA), **96**(4), 1230–1240 (2006a). doi:10.1785/0120050097

9. Esposito, A.M., Scarpetta, S., Giudicepietro, F., Masiello, S., Pugliese, L., Esposito, A.: Nonlinear exploratory data analysis applied to seismic signals. In: Apolloni, B., et al. (eds.) WIRN/NAIS 2005. LNCS, vol. 3931, 70–77. Springer, Berlin, Heidelberg (2006b)

10. Esposito, A., Esposito, A.M., Giudicepietro, F., Marinaro, M., Scarpetta, S.: Models for identifying structures in the data: a performance comparison. Knowl. Based Intell. Info. Eng. Syst. **4694**, 275–283 (2007). doi:10.1007/978-3-540-74829-834

11. Esposito, A.M., Giudicepietro, F., D'Auria, L., Scarpetta, S., Martini, M.G., Moltelli, M., Marinaro, M.: Unsupervised neural analysis of very-long-period events at Stromboli volcano using the self-organizing maps. BSSA **98**(5), 2449–2459 (2008). doi:10.1785/0120070110

12. Esposito, A.M., D'Auria, L., Giudicepietro, F., Longobardi, M., Martini, M.: Clustering of hybrid events at Stromboli volcano (Italy). In: Apolloni, B., et al. (eds.) Frontiers in Artificial Intelligence and Applications, Vol. 234 (2011), Neural Nets WIRN 2011—Proceedings of the 21th Italian Workshop on Neural Nets, IOS Press, December 2011, ISBN: 978-1-60750-971-4. doi:10.3233/978-1-60750-972-1-56

13. Esposito, A.M., D'Auria, L., Giudicepietro, F., Peluso, R., Martini, M.: Automatic recognition of landslide seismic signals based on neural network analysis of seismic signals: an application to the monitoring of Stromboli volcano (Southern Italy). Pure Appl. Geophys. pageoph © Springer Basel (2012). doi:10.1007/s00024-012-0614-1

14. Esposito, A.M., D'Auria, L., Giudicepietro, F., Peluso, R., Martini, M.: Automatic recognition of landslides based on neural network analysis of seismic signals: an application to the monitoring of Stromboli volcano (Southern Italy). Pure. Appl. Geophys. **170**, 1821–1832 (2013a). doi:10.1007/s00024-012-0614-1

15. Esposito, A.M., D'Auria, L., Giudicepietro, F., Caputo, T., Martini, M.: Neural analysis of seismic data: applications to the monitoring of Mt. Vesuvius, Special Issue "Mt. Vesuvius monitoring: the state of the art and perspectives". Ann. Geophys. **56**(4), S0446 (2013b) doi:10.4401/ag-6452

16. Falsaperla, S., Neri, M., Pecora, E., Spampinato, S.: Multidisciplinary study of flank instability phenomena at Stromboli volcano, Italy. Geophys. Res. Lett. **33**(9) (2006). doi:10.1029/2006GL025940

17. Giacco, F., Esposito, A.M., Scarpetta, S., Giudicepietro, F., Marinaro, M.: Support vector machines and MLP for automatic classification of seismic signals at Stromboli volcano. In: Apolloni, B., et al. (eds.) Neural Nets WIRN 2009 Proceedings of the 19th Italian Workshop on Neural Nets, Vietri sul Mare, Salerno, Italy, May 28–30. IOS Press (2009)

18. Giudicepietro, F., D'Auria, L., Martini, M., Orazi, M., Peluso, R., De Cesare, W., Scarpato, G.: Changes in the vlp seismic source during the 2007 Stromboli eruption. J. Volcanol. Geotherm. Res. **182**, 162–171 (2009). doi:10.1016/j.jvolgeores.2008.11.008

19. Harrington, R.M., Brodsky, E.E. Volcanic hybrid earthquakes that are brittle-failure events. Geoph. Res. Lett. **34**, L06308 (2007). doi:10.1029/2006GL028714
20. Junqua, J.C., Haton, J.P.: Robustness in Automatic Speech Recognition: Fundamentals and Applications. Springer (2012)
21. La Rocca, M., Galluzzo, D., Saccorotti, G., Tinti, S., Cimini, G., Del Pezzo, E.: Seismic signals associated with landslides and with a tsunami at Stromboli volcano, Italy. BSSA **94**: 1850–1867 (2004)
22. Longobardi, M., D'Auria, L., Esposito, A.M.: Relative location of hybrid events at Stromboli volcano, Italy. Acta Vulcanologica, 23/24 (2012)
23. Makhoul, T.: Linear Prediction: a Tutorial Review. Proceeding of IEEE, 561–580 (1975)
24. Martini, M., Giudicepietro, F., DAuria, L., Esposito, A.M., Caputo, T., Curciotti, R., De Cesare, W., Orazi, M., Scarpato, G., Caputo, A., Peluso, R., Ricciolino, P., Linde, A., Sacks, S.: Seismological monitoring of the February 2007 effusive eruption of the Stromboli volcano. Ann. Geophys. **50**(6) (2007). doi:10.4401/ag-3056
25. Scarpetta, S., Giudicepietro, F., Ezin, E.C., Petrosino, S., Del Pezzo, E., Martini, M., Marinaro, M.: Automatic classification of seismic signals at Mt. Vesuvius Volcano, Italy using neural networks. Bull. Seism. Soc. Am. **95**, 185–196 (2005)
26. Tibaldi, A.: Multiple sector collapses at Stromboli volcano, Italy: how they work. Bull Volcanol. **63**(2/3), 112–125 (2001). doi:10.1007/s004450100129
27. Tinti, S., Manucci, A., Pagnoni, G., Armigliato, A., Zaniboni, F.: The 30 December 2002 landslide-induced tsunamis in Stromboli: sequence of the events reconstructed from the eyewitness accounts. Nat. Hazards Earth Syst. Sci. **5**, 763–775 (2005). doi:10.5194/nhess-5-763-2005
28. Tinti, S., Pagnoni, G., Zaniboni, F.: The landslides and tsunamis of the 30th of December 2002 in Stromboli analysed through numerical simulations. Bull Volcanol. **68**, 462–479 (2006). doi:10.1007/s00445-005-0022-9
29. Widrow, B., Rumelhart, D.E., Lehr, M.A.: Neural networks Applications in industry, business and science. Commun. ACM **37**(3), 93–105 (1994)
30. Zhang, G., Patuwo, B.E., Hu, M.Y.: Forecasting with artificial neural networks: the state of the art. Intern. J. Forecast. **14**, 35–62 (1998), Elsevier Science B.V. PII S0169-2070(97) 00044-7

Chapter 3
Fully Automatic Multispectral MR Image Segmentation of Prostate Gland Based on the Fuzzy C-Means Clustering Algorithm

Leonardo Rundo, Carmelo Militello, Giorgio Russo, Davide D'Urso, Lucia Maria Valastro, Antonio Garufi, Giancarlo Mauri, Salvatore Vitabile and Maria Carla Gilardi

Abstract Prostate imaging is a very critical issue in the clinical practice, especially for diagnosis, therapy, and staging of prostate cancer. Magnetic Resonance Imaging (MRI) can provide both morphologic and complementary functional information of tumor region. Manual detection and segmentation of prostate gland and carcinoma on multispectral MRI data is not easily practicable in the clinical routine because of the long times required by experienced radiologists to analyze several types of imaging data. In this paper, a fully automatic image segmentation method, exploiting an unsupervised *Fuzzy C-Means (FCM)* clustering technique for multispectral T1-weighted and T2-weighted MRI data processing, is proposed. This approach enables prostate segmentation and automatic gland volume calculation. Segmentation trials have been performed on a dataset composed of 7 patients affected by prostate cancer, using both area-based and distance-based metrics for its

L. Rundo (✉) · G. Mauri
Dipartimento di Informatica, Sistemistica e Comunicazione (DISCo),
Università degli Studi di Milano-Bicocca, Milan, Italy
e-mail: leonardo.rundo@disco.unimib.it

L. Rundo · C. Militello · G. Russo · D. D'Urso · M.C. Gilardi
Istituto di Bioimmagini e Fisiologia Molecolare - Consiglio Nazionale delle Ricerche
(IBFM-CNR), Cefalù (PA), Italy

G. Russo · L.M. Valastro · A. Garufi
Azienda Ospedaliera per l'Emergenza Cannizzaro, Catania, Italy

G. Russo · L.M. Valastro
Laboratori Nazionali del Sud (LNS), Istituto Nazionale di Fisica Nucleare (INFN),
Catania, Italy

D. D'Urso
Università degli Studi di Catania, Catania, Italy

S. Vitabile
Dipartimento di Biopatologia e Biotecnologie Mediche (DIBIMED),
Università degli Studi di Palermo, Palermo, Italy

© Springer International Publishing AG 2018   23
A. Esposito et al. (eds.), *Multidisciplinary Approaches to Neural Computing*,
Smart Innovation, Systems and Technologies 69,
DOI 10.1007/978-3-319-56904-8_3

evaluation. The achieved experimental results are encouraging, showing good segmentation accuracy.

Keyword Fully automatic segmentation · Multispectral MR imaging · Prostate gland · Prostate cancer · Unsupervised Fuzzy C-Means clustering

3.1 Introduction

Prostate imaging is a critical issue in diagnosis, therapy, and staging of prostate cancer (PCa). The use of different imaging modalities, such as Transrectal Ultrasound (TRUS), Computed Tomography (CT) or Magnetic Resonance Imaging (MRI), depends on the clinical context. More recently, MRI is emerging more and more. For instance, PCa treatment by radiotherapy requires an accurate localization of the prostate. In addition, neighboring tissues and organs at risk, i.e., rectum and bladder, must be carefully preserved from radiation, whereas the tumor should receive a prescribed dose [1]. This applies also for prostate tumor ablation using High Intensity Focused Ultrasound (HIFU) [2]. CT was primarily used for treatment planning, but increasingly MRI is used because of its superior soft-tissue contrast and multispectral capabilities [3]. However, CT images are currently used in prostate brachytherapy to locate the radioactive seeds.

Manual detection and segmentation of both prostate gland and prostate carcinoma on multispectral MRI is a time-consuming task, which has to be performed by experienced radiologists. In fact, in addition to conventional structural MRI protocols T1-weighted (T1w) and T2-weighted (T2w) MR images, complementary and powerful functional information of the tumor can be extracted from: Dynamic Contrast Enhanced (DCE), Diffusion Weighted Imaging (DWI), and Magnetic Resonance Spectroscopic Imaging (MRSI) [4–6]. A standardized interpretation of multiparametric MRI is very difficult and significant inter-observer variability has been reported in the literature [7]. Therefore, automated and computer-aided segmentation methods are needed to ensure result repeatability.

Despite the technological developments in MRI scanners and coils, prostate images are prone to artifacts related to magnetic susceptibility. Although the transition from 1.5T to 3T magnetic field strength systems theoretically results in a doubled Signal to Noise Ratio (SNR), it also involves different T1 and T2 relaxation times and greater magnetic field inhomogeneity of inside organ and tissue of interest [8, 9]. On the other hand, 3T MRI scanners allow to use pelvic coils instead of endorectal coils, obtaining good quality images and avoiding invasiveness as well as prostate gland compression and deformation.

Accurate prostate segmentation is a mandatory step in clinical activities. First of all, prostate volume, which can be directly calculated from the prostate Region of Interest (ROI) segmentation, aids in diagnosis of benign prostate hyperplasia and prostate bulging. Prostate boundaries are then used in several treatments of prostate diseases: radiation therapy, brachytherapy, HIFU ablation, cryotherapy, and trans-urethral microwave therapy [4]. Lastly, the computed prostate ROI is very useful for focusing the subsequent processing for prostate cancer segmentation and characterization [6].

In this paper, a fully automatic segmentation method, based on an unsupervised *Fuzzy C-Means (FCM)* clustering technique, for multispectral MRI morphologic data processing is presented. The proposed approach enables prostate segmentation and automatic gland volume calculation. The clustering procedure is performed on co-registered T1w and T2w MR image series. To the best of our knowledge, this is the first work introducing T1w MR image information to enhance prostate gland segmentation. Preliminary segmentation tests have been performed on an MRI dataset composed of 7 patients affected by PCa, using both area-based and distance-based metrics to evaluate the effectiveness of the proposed approach.

This manuscript is organized as follows: Sect. 3.2 introduces an overview of the state of the art about MRI prostate segmentation methods; Sect. 3.3 describes the MRI data as well as the proposed automatic segmentation approach; Sect. 3.4 reports the experimental results obtained in the segmentation tests; finally, some conclusive remarks and possible future developments are provided in Sect. 3.5.

3.2 Related Works

In this section, the most representative literature works on prostate segmentation in MRI are briefly outlined.

In [1] an automatic method, which relies on the non-rigid registration of a set of pre-labeled atlas images, for delineating the prostate in 3D MR scans was presented. After co-registering each atlas image to the target patient image, the resulting deformed atlas label images are fused to obtain a single segmentation for the patient image. The method was evaluated on 50 clinical scans, previously segmented by three experts. Dice similarity index (*DSI*) and the shortest Euclidean distance between the manual and automatic ROIs were computed. In most cases, the accuracy of the automatic segmentation method was close to the level of inter-observer variability.

The authors of [10] proposed a fully automatic system for delineating the prostate on MR images, by employing an Active Appearance Model (AAM) constructed from 50 manually segmented examples provided by the MICCAI 2012 PROMISE12 team [11]. Accurate correspondences for the model are defined using a Minimum Description Length (MDL) groupwise image registration method, by means of a multi-start optimization scheme to increase the generalization capabilities of the system. *DSI*, mean absolute distance and the 95th% Hausdorff distance (*HD*) were used to evaluate the performance. The model was cross-validated, using the standard leave-one-out technique, on the training data obtaining a good degree of accuracy and successfully segmented 30 test data.

A Pattern Recognition approach to detect prostate regions was described in [12]. This method integrates three types of features (anatomical, intensity and texture), by registering the T2w image and the quantitative Apparent Diffusion Coefficient (ADC) map simultaneously, which can differentiate between the central gland and peripheral zone. It was compared with a multiparametric multi-atlas based method on 48 MRI series. *DSI* values showed good results in prostate zonal segmentation.

In [13] a unified shape-based framework for prostate extraction on MR images was proposed. The authors represent the shapes of the training set as point clouds. Accordingly, more global aspects of registration can be considered by using a particle filtering scheme. After shape registration, a cost functional is defined to combine both the local image statistics and the learnt shape prior (i.e., constraint for the segmentation task). Satisfying experimental results were obtained on several challenging clinical datasets, considering *DSI* and the 95th% *HD*.

Martin et al. in [14] introduced a semi-automatic MRI prostate segmentation approach based on the registration of an anatomical atlas (generated from a dataset of 18 MRI series) against a patient image. The authors employed a hybrid registration scheme, by combining an intensity-based registration and the Robust Point-Matching (RPM) algorithm, for both atlas construction and atlas registration. Statistical shape information is taken into account to increase and regularize the achieved segmentation performance, by using Active Shape Models (ASMs). The method was validated on the same dataset employed for atlas construction, through the leave-one-out method. Better segmentation accuracy, in terms of volume-based and distance-based metrics, was obtained in the apical and central zones with respect to the base of the prostate. The same authors proposed a fully automatic algorithm, based on a probabilistic atlas [15]. This anatomical atlas is computed according to the transformation fields, which map a set of manually segmented images to a given reference system. The segmentation is then performed in two phases. First, the analyzed image is co-registered against the probabilistic atlas and a probabilistic segmentation with a spatial constraint is carried out. Afterwards, a deformable surface refines the prostate boundaries by combining the probabilistic segmentation with both an image feature model and a statistical shape model. This method including a spatial constraint was evaluated on 36 MRI series and achieved more robust results with respect to a deformable surface employing an image appearance model alone.

3.3 Patients and Methods

In the current section, firstly MRI data concerning the subjects are reported, and then a novel fully automatic multispectral MRI segmentation approach is presented. We exploit traditional methods in an original way, proposing an advanced clustering application for prostate gland ROI segmentation.

(a)   (b)

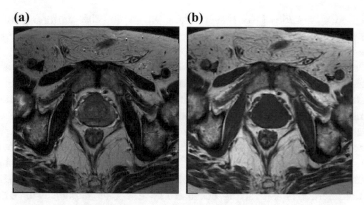

Fig. 3.1 Instance of input MR axial slices: **a** T2-weighted MR image; **b** corresponding T1-weighted MR image

Table 3.1 Acquisition parameters of the MRI prostate dataset

MRI protocol	TR (ms)	TE (ms)	Matrix size (pixels)	Pixel spacing (mm)	Slice thickness (mm)	Interslice gap (mm)	Slice number
T1w	515.3	10	256 × 256	0.703	3	4	18
T2w	3035.6	80	288 × 288	0.625	3	4	18

3.3.1   Patient Dataset Composition

The initial dataset is composed of 7 patients affected by PCa, who underwent radiation therapy. The analyzed MR images were T1w and T2w Fast Spin Echo (FSE) sequences scanned with 3.0 T MRI equipment (Philips Medical Systems, Eindhoven, the Netherlands) using a SENSE XL Torso coil (16 elements phased-array pelvic coil). Two examples of T1w and T2w MR images are shown in Fig. 3.1. T2w and T1w MRI acquisition parameters are reported in Table 3.1.

3.3.2   The Proposed Prostate Segmentation Method

In this section, a fully automatic segmentation approach of the whole prostate gland from axial MRI slices is presented.

T2w FSE imaging is currently the standard protocol for depicting the anatomy of the prostate. These sequences are sensitive to susceptibility artifacts, for instance due to the presence of air in the rectum, which may affect the correct detection of tissue boundaries [1]. On the other hand, because the prostate has uniform intermediate signal intensity at T1w imaging, the zonal anatomy cannot be clearly identified on T1w series [16]. Nevertheless, we exploit this uniform gray appearance, by combining T2w and T1w prostate imaging for clustering result enhancement.

The overall prostate segmentation method is outlined in Fig. 3.2, and each processing phase is described in the following paragraphs.

MR Image Co-registration. Although T1w and T2w MRI series are included in the same study, they are not acquired contextually because of the different employed extrinsic parameters that determine T1 and T2 relaxation times. Thus, an *image co-registration* step on multispectral MR images is required. Moreover, T1w and T2w images have different Field of Views (FOVs), in terms of pixel spacing and matrix size (see Table 3.1 in Sect. 3.1). However, in our case a 2D image registration method is satisfactory because T1w and T2w sequences have the same number of slices as well as slice thickness and interslice gap values.

Image co-registration is accomplished by using an iterative process to align a moving image (T1w) along with a reference image (T2w). In this way, the two images will be represented in the same reference system, so enabling quantitative analysis and processing on fused imaging data. We chose T2w MRI as reference image to consider its own reference system during the possible subsequent processing phases for differentiating prostate anatomy and for PCa detection and characterization.

In intensity-based algorithms, the accuracy of the registration is evaluated iteratively according to an image similarity metrics. We used mutual information that is an information theoretic measure of the correlation degree between two random variables [17]. High mutual information means a large reduction in the uncertainty (entropy) between the two probability distributions, signaling that the images are likely better aligned [18]. An evolutionary algorithm with one-to-one optimization strategy is utilized to find a set of transformation that yields the sub-optimal registration result. In this strategy, both the number of parents and the population size (i.e., number of offspring) are set to one [19]. Affine transformations are applied to the moving image to be aligned, consisting of a series of translation, rotation, scale, and shear.

Pre-Processing. Some pre-processing operations are applied on T2w and co-registered T1w input MR series concerning the same study. First of all, *stick filtering* is applied to remove the speckle noise that affects MR images, due to the MRI reconstruction procedure based on the inverse discrete Fourier transform of the raw data [20]. By considering a neighborhood around each pixel, the intensity value of the current pixel is replaced by the maximum of all the stick projections, which are calculated by the convolution of the input image with a long narrow average filter at various orientations [20]. Thereby, smoothing is performed on homogeneous regions, while preserving resolvable features and enhancing weak edges and linear structures, without producing too many false edges [21]. Stick filtering parameters are a good balance: the sticks have to be so short that they can approximate the edges in images, but long enough that the speckle along the sticks is uncorrelated. Thus, the projections along the sticks can smooth speckle noise, without destroying the edges in the image. Experimentally, considering the

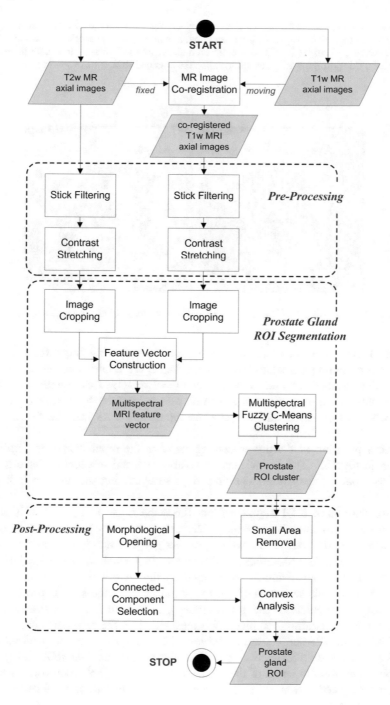

◀Fig. 3.2   Flow diagram of the proposed prostate gland segmentation method. The pipeline can be subdivided in three main stages: (i) pre-processing, required to remove speckle noise and enhance *FCM* clustering process; (ii) *FCM* clustering on multispectral MR images, to extract the prostate gland ROI; (iii) post-processing, useful to refine the obtained ROI segmentation results

(a)   **(b)**

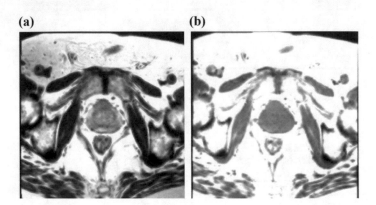

Fig. 3.3   Pre-Processing phase: **a** T2w image and **b** T1w image after pre-processing steps

obtained filtering results in order to be the most suitable input for the *Fuzzy C-Means* clustering algorithm and enhance the achieved classification, the selected stick filter parameters are: stick thickness = 1 pixel; stick length = 5 pixels. Afterwards, a *contrast stretching* operation is performed by means of a linear intensity transformation that converts the input intensities into the full dynamic range.

The application of the pre-processing steps on the input images in Fig. 3.1 is shown in Fig. 3.3. The image noise is visibly reduced in order to have a more suitable input for the subsequent clustering, so extracting the prostate ROI more easily.

Prostate Gland ROI Segmentation. Since the organ to be imaged is always positioned approximately near the isocenter of the principal magnetic field to minimize MRI distortions, the whole prostate gland is represented in the imaged FOV center [22]. After dividing the entire image in 9 equal-sized and fixed tiles, we considered the cropped image represented by the central one. Therefore, only a cropped image, whose size is 1/9 of the initial input image size, is processed for ROI image segmentation. Using this strategy, user input is not required and processing time is certainly optimized. We only relied on the prostate position in the FOV center, possible problems may be solved by using an interactive ROI selection tool, such as a draggable rectangle. Considering the prostate ROIs imaged on morphologic MRI, in terms of the segmented area in each slice, we observed that the prostate gland can be suitably contained in a patch with size 1/9 of the initial input image size.

In Pattern Recognition, features are defined as measurable quantities that could be used to distinguish two or more regions. In fact, more than one feature could be used to classify different regions and an array of these features is known as a feature vector. Therefore, the feature vector is constructed by concatenating corresponding pixels on T2w and T1w MR images, after the image co-registration step.

The goal of unsupervised clustering methods is to determine an intrinsic partitioning in a set of unlabeled data, which are associated with a feature vector. Distance metrics are used to group data into clusters of similar types and the number of clusters is assumed to be known [4].

Fuzzy C-Means (FCM) cluster analysis is an iterative algorithm that classifies a set of data (i.e., a digital image) into partitions (regions) [23]. The aim is to divide a set $X = \{\mathbf{x}_1, \mathbf{x}_2, \ldots, \mathbf{x}_N\}$ of N objects (statistical samples represented by n-dimensional vectors belonging to $\mathbb{R}^n$) into C clusters, identified by C centroids $V = \{\mathbf{v}_1, \mathbf{v}_2, \ldots, \mathbf{v}_C\}$ [24]. A partition P is defined as a set family $P = \{Y_1, Y_2, \ldots, Y_C\}$.

The crisp version (*K-Means*) assumes that: (i) the clusters must be a proper subset X: $\emptyset \subset Y_i \subset X, \forall i$; (ii) the set union of the clusters must also reconstruct the whole dataset: $\bigcup_{i=1}^{C} Y_i = X$. Moreover, the various clusters are mutually exclusive: $Y_i \cap Y_j = \emptyset, \forall i \neq j$ (i.e., each feature vector belongs to only one group). In the *FCM* formulation, a fuzzy partition P is defined as a fuzzy set family $P = \{Y_1, Y_2, \ldots, Y_C\}$ such that each object can have a partial membership to multiple clusters. Let $\mathbf{U} = [u_{ik}] \in \mathbb{R}^{C \times N}$ be the matrix that defines a fuzzy C-partition of the set X through C membership functions $u_i: X \to [0, 1]$ whose values $u_{ik}: = u_i(x_k) \in [0, 1]$ are as the membership degrees of each element $\mathbf{x}_k$ to the ith fuzzy set (cluster) Y_i and have to satisfy the constraints in (3.1).

$$\begin{cases} 0 \leq u_{ik} \leq 1 \\ \sum_{i=1}^{C} u_{ik} = 1, \; \forall k \in \{1, 2, \ldots, N\} \\ 0 < \sum_{k=1}^{N} u_{ik} < N, \; \forall i \in \{1, 2, \ldots, C\} \end{cases} \quad (3.1)$$

The sets of all the fuzzy C-partitions of the input X are defined by $M_{fuzzy}^{(C)} = \{\mathbf{U} \in \mathbb{R}^{C \times N}: u_{ik} \in [0, 1]\}$. Computationally, *FCM* is an optimization problem where the objective function in (3.2) must be iteratively minimized using the least-squares method. The weighting exponent m controls the fuzziness of the classification process. After the identification of the cluster containing prostate ROI, during the defuzzification step, the pixels that have the maximum membership with the ROI cluster are selected.

$$J_m(\mathbf{U}, V; X): = \sum_{i=1}^{C} \sum_{k=1}^{N} (u_{ik})^m \|\mathbf{x}_k - \mathbf{v}_i\|^2 \quad (3.2)$$

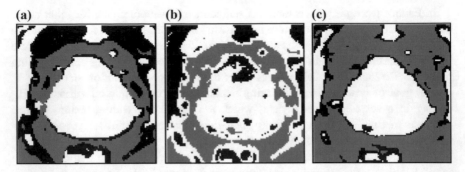

Fig. 3.4 Prostate gland segmentation using *FCM* clustering with $C = 3$ on the input images in Fig. 3.1a, b: **a** result on the multispectral T1w and T2w co-registered MR images; **b** result on T2w MR image alone; **c** result on T1w MR image alone. It is possible to appreciate the multispectral MR imaging contribution with respect to the single series clustering outputs

The clustering procedure is performed on the feature vector composed of corresponding T2w and T1w pixel values. The cluster number is always set to $C = 3$, by considering both structural and morphologic properties of the processed prostate MR images. As a matter of fact, three different classes can essentially be seen by visual inspection according to gray levels. This choice was also justified and endorsed by experimental trials. Figure 3.4 shows the ROI segmentation results achieved using the *FCM* clustering algorithm with three classes on three different input MRI data. It is appreciable how the *FCM* clustering output on multispectral MRI data (Fig. 3.4a) considerably enhances the achieved results. Especially, the introduction of T1w imaging, in the feature vector construction step, yields a more uniform overall clustering output by decisively separating the ROI from the other surrounding tissues and organs (i.e., bladder, rectum, levator ani muscle).

Post-Processing. A sequence of morphological operations is applied to the obtained ROI cluster, in order to resolve possible ambiguities resulting from the clustering process and to improve the quality of the segmented prostate gland ROI. According to Fig. 3.2, the post-processing steps are the following:

- a *small area removal* operation is employed to delete any unwanted connected-components, whose area is less than 500 pixels, which are included into the ROI cluster. These small regions can have similar gray levels to the prostate ROI and were classified into the same cluster by the *FCM* clustering algorithm. The areas less than 500 pixels certainly do not represent the prostate ROI, since the prostate gland in MRI slices has always a greater area, regardless of the used protocol. Thus, these small areas can be removed from the prostate ROI cluster to avoid a wrong connected-component selection in the next processing steps;
- a *morphological opening*, using a 5×5 pixel square structuring element, is a good compromise between precision in the actual detected contours and capability for unlinking poorly connected regions to the prostate ROI;

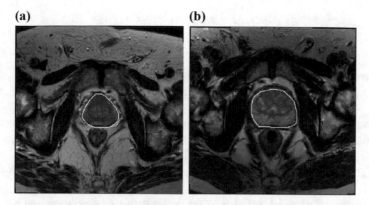

Fig. 3.5 Two examples of prostate gland ROI segmentation results on axial MR slices: the *white* contours are superimposed on the corresponding T2w MR images

- connected-components are determined using a *flood-fill algorithm* in order to select the nearest connected-component to the cropped image center, since prostate is always imaged at the center of the FOV;
- prostate gland appearance is always convex. Considering this, a *convex hull algorithm* is suitable to enclose the segmented ROI with the smallest convex polygon that contains it, so considering possible adjacent regions excluded by the *FCM* clustering output. Finally, a *morphological opening* operation with a circular structuring element, is performed for smoothing ROI boundaries. The used circular structuring element with 3 pixel radius allows for smoother and more realistic ROI boundaries without deteriorating significantly the output yielded by the *FCM* clustering. Accordingly, the value of the radius does not affect the segmentation results and it is not dependent on image resolution.

Two examples of prostate gland ROI obtained by the proposed processing pipeline are shown in Fig. 3.5.

3.4 Segmentation Results

Several measures (area-based and distance-based metrics) were calculated to evaluate the accuracy of the proposed segmentation method [25, 26]. Supervised evaluation is used to quantify the goodness of the segmentation outputs, by comparing the automatically segmented ROIs with the target volume manually contoured by an experienced radiologist (ground truth). As evidenced in Fig. 3.5a and b, the proposed segmentation approach obtains accurate results even with inhomogeneous input data.

3.4.1 Evaluation Metrics and Achieved Experimental Results

Area-based metrics quantify the similarity between the segmented regions through the proposed method (R_A) and the "gold standard" (R_T). Thus, the regions containing "true positives" $(R_{TP} = R_A \cap R_T)$, "false positives" $(R_{FP} = R_A - R_{TP})$, and "false negatives" $(R_{FN} = R_T - R_{TP})$ are defined. In our experimental trials, according to the formulations in [25], we used the most important evaluation measures: Dice similarity index $DSI = 2|R_{TP}|/(|R_A| + |R_T|) \times 100$, Jaccard index $JI = |R_A \cap R_T|/|R_A \cup R_T| \times 100$, Sensitivity $SE = |R_{TP}|/|R_T| \times 100$, and Specificity $SP = (1 - |R_{FP}|/|R_A|) \times 100$.

In addition, the spatial distance between the automatically generated boundaries (defined by the vertices $A = \{\mathbf{a}_i : i = 1, 2, \ldots, K\}$) and the manually traced boundaries (determined by $T = \{\mathbf{t}_j : j = 1, 2, \ldots, N\}$) was also evaluated. Accordingly, the distance between each element of the automatic prostate gland A contour and the set T must be defined: $d(\mathbf{a}_i, T) = \min_{j \in \{1, 2, \ldots, N\}} \|\mathbf{a}_i - \mathbf{t}_j\|$. The following measures were calculated: Mean Absolute Difference $MAD = (1/K) \sum_{i=1}^{K} d(\mathbf{a}_i, T)$, Maximum Difference $MAXD = \max_{i \in \{1, 2, \ldots, K\}} \{d(\mathbf{a}_i, T)\}$, and Hausdorff distance $HD = \max\{h(T, A), h(A, T)\}$ (where $h(T, A) = \max_{\mathbf{t} \in T} \{\min_{\mathbf{a} \in A} \{d(\mathbf{t}, \mathbf{a})\}\}$ and $d(\mathbf{t}, \mathbf{a}) = \sqrt{(t_x - a_x)^2 + (t_y - a_y)^2}$).

The values of both area-based and distance-based metrics achieved in the experimental ROI segmentation trials are reported in Table 3.2. DSI and JI values show overall good segmentation accuracy. Sensitivity and specificity average values prove the correct detection of the "true" ROI and the capability of not detecting "false" parts outside the actual prostate, respectively. Distance-based metrics are consistent with the area-based ones and corroborate the experimental findings. Satisfying segmentation results are also observed when dealing with prostate glands characterized by irregular cancer regions or inhomogeneous peripheral zone.

3.5 Conclusions and Future Works

In this paper, a fully automatic method for prostate segmentation based on the *FCM* clustering algorithm on multispectral MRI morphologic data is proposed. This approach can be used to support prostate gland delineation, such as in radiation therapy. The main key novelty is that the clustering procedure is performed on co-registered T1w and T2w MR image series. The *FCM* clustering results on multispectral MRI anatomical data considerably enhance the achieved prostate ROI segmentations. Our method uses an unsupervised Machine Learning technique, without requiring a training phase. On the contrary, the other literature works are based on atlases [1, 12, 14, 15], AAMs [10], or statistical shape models [13–15],

Table 3.2 Values of area-based and distance-based metrics obtained on each segmented prostate gland in the dataset. The last row reports the average values ± standard deviations

Patient	DSI$_\%$	JI$_\%$	SE$_\%$	SP$_\%$	MAD	MAXD	HD
#1	79.24	69.83	87.91	75.50	5.248	11.489	3.941
#2	95.77	91.91	94.60	97.13	1.442	4.742	3.778
#3	93.25	88.09	96.33	91.44	2.543	8.185	4.113
#4	92.80	86.85	98.28	88.40	2.397	8.832	4.211
#5	83.51	72.15	76.23	94.05	5.111	16.267	4.643
#6	90.48	83.39	95.07	87.88	3.190	10.058	4.394
#7	85.71	76.59	86.58	89.20	4.775	14.224	4.022
AVG ± SD	88.68 ± 6.00	81.26 ± 8.48	90.71 ± 7.72	89.09 ± 6.85	3.529 ± 1.51	10.542 ± 3.86	4.157 ± 0.29

which require manual labeling of a significant image sample set performed by expert physicians. The implemented approach was tested on a dataset made up of 7 patients, considering T2w and T1w MRI series. Area-based and distance-based metrics were calculated to evaluate the performance. The achieved preliminary results are very encouraging, as confirmed by the previous section.

Future works will be aimed to a more selective segmentation technique, distinguishing between central gland and peripheral zone of the prostate anatomy [16]. The advanced segmentation method can be definitely employed in a two-step prostate cancer delineation approach, in order to focus on pathological regions in the central gland and in the peripheral zone.

References

1. Klein, S., van der Heide, U.A., Lips, I.M., van Vulpen, M., Staring, M., Pluim, J.P.: Automatic segmentation of the prostate in 3D MR images by atlas matching using localized mutual information. Med. Phys. **35**(4), 1407–1417 (2008). doi:10.1118/1.2842076
2. Rouvière, O., Lyonnet, D., Raudrant, A., Colin-Pangaud, C., Chapelon, J.Y., Bouvier, R., Dubernard, J.M., Gelet, A.: MRI appearance of prostate following transrectal HIFU ablation of localized cancer. Eur. Urol. **40**, 265–274 (2001). doi:10.1159/000049786
3. Villeirs, G.M., De Meerleer, G.O.: Magnetic resonance imaging (MRI) anatomy of the prostate and application of MRI in radiotherapy planning. Eur. J. Radiol. **63**(3), 361–368 (2007). doi:10.1016/j.ejrad.2007.06.030
4. Ghose, S., Oliver, A., Martí, R., Lladó, X., Vilanova, J.C., Freixenet, J., Mitra, J., Sidibé, D., Meriaudeau, F.: A survey of prostate segmentation methodologies in ultrasound, magnetic resonance and computed tomography images. Comput. Meth. Prog. Bio. **108**(1), 262–287 (2012). doi:10.1016/j.cmpb.2012.04.006
5. Chilali, O., Ouzzane, A., Diaf, M., Betrouni, N.: A survey of prostate modeling for image analysis. Comput. Biol. Med. **53**, 190–202 (2014). doi:10.1016/j.compbiomed.2014.07.019
6. Lemaître, G., Martí, R., Freixenet, J., Vilanova, J.C., Walker, P.M., Meriaudeau, F.: Computer-aided detection and diagnosis for prostate cancer based on mono and multi-parametric MRI: A review. Comput. Biol. Med. **60**, 8–31 (2015). doi:10.1016/j.compbiomed.2015.02.009
7. Rosenkrantz, A.B., Lim, R.P., Haghighi, M., Somberg, M.B., Babb, J.S., Taneja, S.S.: Comparison of interreader reproducibility of the prostate imaging reporting and data system and likert scales for evaluation of multiparametric prostate MRI. Am. J. Roentgenol. **201**(4), W612–W618 (2013). doi:10.2214/AJR.12.10173
8. Caivano, R., Cirillo, P., Balestra, A., Lotumolo, A., Fortunato, G., Macarini, L., Zandolino, A., Vita, G., Cammarota, A.: Prostate cancer in magnetic resonance imaging: diagnostic utilities of spectroscopic sequences. J. Med. Imag. Radiat. On. **56**(6), 606–616 (2012). doi:10.1111/j.1754-9485.2012.02449.x
9. Rouvière, O., Hartman, R.P., Lyonnet, D.: Prostate MR imaging at high-field strength: Evolution or revolution? Eur. Radiol. **16**(2), 276–284 (2006). doi:10.1007/s00330-005-2893-8
10. Vincent, G., Guillard, G., Bowes, M.: Fully automatic segmentation of the prostate using active appearance models. In: Medical Image Computing and Computer Assisted Intervention (MICCAI) Grand Challenge: Prostate MR Image Segmentation 2012, Nice, France, 7 p. (2012)
11. Litjens, G., Toth, R., van de Ven, W., Hoeks, C., Kerkstra, S., van Ginneken, B., et al.: Evaluation of prostate segmentation algorithms for MRI: The PROMISE12 challenge. Med. Image Anal. **18**(2), 359–373 (2014). doi:10.1016/j.media.2013.12.002

12. Litjens, G., Debats, O., van de Ven, W., Karssemeijer, N., Huisman, H.: A pattern recognition approach to zonal segmentation of the prostate on MRI. In: Medical Image Computing and Computer-Assisted Intervention (MICCAI), p. 413–420. Springer, Berlin Heidelberg. (2012). doi:10.1007/978-3-642-33418-4_51

13. Gao, Y., Sandhu, R., Fichtinger, G., Tannenbaum, A.R.: A coupled global registration and segmentation framework with application to magnetic resonance prostate imagery. IEEE T. Med. Imaging 29(10), 1781–1794 (2010). doi:10.1109/TMI.2010.2052065

14. Martin, S., Daanen, V., Troccaz, J.: Atlas-based prostate segmentation using an hybrid registration. Int. J. Comput. Assist. Radiol. Surg. 3(6), 485–492 (2008). doi:10.1007/s11548-008-0247-0

15. Martin, S., Troccaz, J., Daanen, V.: Automated segmentation of the prostate in 3D MR images using a probabilistic atlas and a spatially constrained deformable model. Med. Phys. 37(4), 1579–1590 (2010). doi:10.1118/1.3315367

16. Choi, Y.J., Kim, J.K., Kim, N., Kim, K.W., Choi, E.K., Cho, K.S.: Functional MR imaging of prostate cancer. Radiographics 27(1), 63–75 (2007). doi:10.1148/rg.271065078

17. Pluim, J.P.W., Maintz, J.B.A., Viergever, M.A.: Mutual-information-based registration of medical images: a survey. IEEE T. Med. Imaging 22(8), 986–1004 (2003). doi:10.1109/TMI.2003.815867

18. Mattes, D., Haynor, D.R., Vesselle, H., Lewellen, T., Eubank, W.: Non-rigid multimodality image registration. In: Medical Imaging 2001: Image Processing, 1609, Proceedings of SPIE 4322, pp. 1609–1620 (2001). doi:10.1117/12.431046

19. Styner, M., Brechbuhler, C., Szckely, G., Gerig, G.: Parametric estimate of intensity inhomogeneities applied to MRI. IEEE T. Med. Imaging 19(3), 153–165 (2000). doi:10.1109/42.845174

20. Czerwinski, R.N., Jones, D.L., O'Brien, W.D.: Line and boundary detection in speckle images. IEEE T. Image Process. 7(12), 1700–1714 (1998). doi:10.1109/83.730381

21. Xiao, C.Y., Zhang, S., Cheng, S., Chen, Y.Z.: A novel method for speckle reduction and edge enhancement in ultrasonic images. In: Electronic Imaging and Multimedia Technology IV, 469. Proceedings of SPIE 5637, 28 February, 2005. doi:10.1117/12.575389

22. Lagendijk, J.J.W., Raaymakers, B.W., Van den Berg, C.A.T., Moerland, M.A., Philippens, M.E., van Vulpen, M.: MR guidance in radiotherapy. Phys. Med. Biol. 59, R349–R369 (2014). doi:10.1088/0031-9155/59/21/R349

23. Bezdek, J.C., Ehrlich, R., Full, W.: FCM: The fuzzy C-means clustering algorithm. Comput. Geosci. 10(2), 191–203 (1984). doi:10.1016/0098-3004(84)90020-7

24. Militello, C., Vitabile, S., Rundo, L., Russo, G., Midiri, M., Gilardi, M.C.: A fully automatic 2D segmentation method for uterine fibroid in MRgFUS treatment evaluation. Comput. Biol. Med. 62, 277–292 (2015). doi:10.1016/j.compbiomed.2015.04.030

25. Fenster, A., Chiu, B.: Evaluation of segmentation algorithms for medical imaging. In: 27th Annual International Conference of the Engineering in Medicine and Biology Society, IEEE-EMBS 2005, pp. 7186–7189 (2005). doi:10.1109/IEMBS.2005.1616166

26. Rundo, L., Militello, C., Vitabile, S., Russo, G., Pisciotta, P., Marletta, F., Ippolito, M., D'Arrigo, C., Midiri, M., Gilardi, M.C.: Semi-automatic brain lesion segmentation in gamma knife treatments using an unsupervised fuzzy c-means clustering technique. In: Bassis, S., Esposito, A., Morabito, F.C., Pasero, E. (eds.) Advances in Neural Networks: Computational Intelligence for ICT, Smart Innovation, Systems and Technologies, vol. 54, pp. 15–26, Springer International Publishing (2016). doi:10.1007/978-3-319-33747-0_2

Chapter 4
Integrating QuickBundles into a Model-Guided Approach for Extracting "Anatomically-Coherent" and "Symmetry-Aware" White Matter Fiber-Bundles

Francesco Cauteruccio, Claudio Stamile, Giorgio Terracina,
Domenico Ursino and Dominique Sappey-Marinier

Abstract This paper presents a novel approach aiming at improving the White Matter (WM) fiber-bundle extraction approach described in (Stamile C et al Brain Informatics and Health: 8th International Conference, BIH, 2015). This provides anatomically coherent fiber-bundles, but it is unable to distinguish symmetric fiber-bundles. The new approach we are proposing here overcomes this limitation by integrating QuickBundles (QB) into it. As a matter of fact, QB has features complementary to those of the approach of (Stamile C et al Brain Informatics and Health: 8th International Conference, BIH, 2015), because it is capable of distinguishing symmetric fiber-bundles but, often, it does not return anatomically coherent fiber-bundles. We also present some experiments showing that the Precision, the Recall and the F-Measure of this new approach improve by 9.76, 3.08 and 8.96%, compared to the corresponding ones of the approach of (Stamile C et al Brain Informatics and Health: 8th International Conference, BIH, 2015), which, in their turn, were shown to be better than the ones of QB.

F. Cauteruccio (✉) · G. Terracina
DEMACS, University of Calabria, Rende, Italy
e-mail: cauteruccio@mat.unical.it

G. Terracina
e-mail: terracina@mat.unical.it

C. Stamile · D. Sappey-Marinier
CREATIS; CNRS UMR5220; INSERM U1044; Université de Lyon,
INSA-Lyon, Université Lyon 1, Lyon, France
e-mail: stamile@creatis.insa-lyon.fr

D. Sappey-Marinier
e-mail: dominique.sappey-marinier@univ-lyon1.fr

D. Ursino
DICEAM, University Mediterranea of Reggio Calabria, Reggio Calabria, Italy
e-mail: ursino@unirc.it

© Springer International Publishing AG 2018
A. Esposito et al. (eds.), *Multidisciplinary Approaches to Neural Computing*,
Smart Innovation, Systems and Technologies 69,
DOI 10.1007/978-3-319-56904-8_4

Keywords Brain analysis · Magnetic resonance imaging · White matter fibers · Model-based fiber characterization · QuickBundles

4.1 Introduction

In the investigation of brain, the capability of extracting and visualizing White Matter (WM) fibers plays a key role. In fact, the analysis of WM structures can largely benefit from the so called WM fiber-bundles. These consist of subsets of fibers that belong to the White Matter region of interest [2]. The extraction activity is often carried out manually by experts. These are required to define the regions of interest by specifying inclusion and exclusion criteria [5]. However, in presence of a large number of subjects to examine, this kind of approach cannot be adopted, and semi-automatic or automatic approaches appear compulsory.

For this reason, researchers have striven to define (semi-)automatic approaches for the isolation and the extraction of WM fiber-bundles [4, 7, 8]. These approaches can be categorized in two groups. The former consists of atlas-based algorithms, which operate by exploiting an a priori knowledge regarding the position of specific WM regions [7, 8]. The latter comprises algorithms that do not require this information [4].

The approaches based on a priori knowledge exploit pre-labeled atlases of WM fiber-bundles belonging to the image of a subject into consideration. They are fast and simple. Nevertheless, they present some problems. In fact, they can extract only fiber-bundles conform to provided atlases. Furthermore, the quality of the fibers returned by them is strictly connected to the atlas registration algorithm. We evidence that these approaches may be integrated with enhancing techniques (think, for instance, of clustering [8]), supervised by experts through the setting of some input parameters.

The approaches not based on a priori knowledge formalize and, then, exploit some specific measures of similarity and proximity in $\mathbb{R}^3$. These measures have been conceived to help: (i) the assignment to the same set of the fibers having the same structure; (ii) the assignment to different sets of the fibers having different structures. The most known approach belonging to this group is QuickBundles (QB) [4]. QB is based on a very simple idea. In fact, at each iteration, it either assigns a given fiber to a pre-existing cluster or generates a new cluster. Initially, it assigns the first fiber to a first cluster containing only it. Then, it assigns a fiber to a cluster according to a certain threshold θ. If the distance between the current fiber and the centroid of at least one cluster is less than θ, QB assigns the fiber to the cluster with the minimum distance. Otherwise, QB creates a new cluster and assigns the current fiber to it. QB repeats this process until to each fiber into examination has been assigned to exactly one cluster. The distance between two fibers is measured by means of a new metric called Minimum Average Direct Flip (MDF). Differently from most clustering algorithms, QB has no re-assignment or updating step. As a consequence, in QB, after a fiber has been assigned to a certain cluster, it cannot be re-assigned to a different one.

Thanks to its simplicity, QB returns good results in terms of fiber-bundle extraction and execution time.

The totally unsupervised approach characterizing QB could cause the extraction of anatomically incoherent areas. This because the fiber generation approach of QB does not exploit prior information provided by neuroanatomists. Actually, this last information could be extremely precious for leading to more satisfying results. As a consequence, the results returned by QB could be biased in real cases, where the knowledge of anatomical information might prove extremely precious for analysis. Beside its quickness, one of the main positive features of QB is its capability of distinguishing symmetric fiber-bundles.

It would be extremely interesting to design an approach that avoids the extraction of anatomically incoherent regions and, at the same time, maintains the capability of extracting symmetric fiber-bundles. In [6], we proposed a model-guided string-based approach to extracting WM fiber-bundles. It provides anatomically coherent fiber-bundles but it is incapable of distinguishing symmetric fiber-bundles.

Here, we are proposing a novel approach that integrates the one presented in [6] and QB with the goal of extracting "anatomically coherent" and "symmetry-aware" WM fiber-bundles.

Our approach receives a set $F = \{f_1, f_2, \ldots, f_n\}$ of WM fibers to cluster and a set $M = \{m_1, m_2, \ldots, m_k\}$ of models. It performs the following steps: (i) it applies the approach proposed in [6] and constructs a set T (resp., V) of the strings corresponding to F (resp., M); (ii) it constructs a matrix D whose generic element $D[i, j]$ denotes the dissimilarity degree between the string corresponding to f_i and the one associated with m_j; (iii) it exploits D to assign each fiber of F to at most one model of M; in this way, it produces a set $B = \{b_1, b_2, \ldots, b_m\}$ of WM fiber-bundles; at this stage, symmetrical structures cannot be distinguished; (iv) it applies QB for facing this last problem.

We performed several tests aiming at comparing the performances of both our approach and the one proposed in [6] (which, in its turn, showed better performances than QB [6]). As we will show below, we obtained very encouraging results.

The rest of this paper is as follows. In Sect. 4.2, we describe our approach. In Sect. 4.3, we illustrate the tests performed to evaluate it. Finally, in Sect. 4.4, we draw our conclusion and have a look at the future.

4.2   Overview of the Approach for Extracting "Anatomically-Coherent" and "Symmetry-Aware" WM Fiber-Bundle

Our approach needs that one or more models of the WM fiber-bundles of interest are provided as input. Each model can be considered as an approximation of the shape the extracted fiber-bundle should conform to. This approximation can be obtained either with a spline curve directly drawn by the user through a visual interface, or by exploiting pre-defined models corresponding to pre-elaborated or already known sets of fiber-bundles.

In our approach, a key issue is that fibers (and models), originally represented as ordered lists of voxels in 3D, must be expressed as strings. In order to explain how this crucial task is carried out, let us first concentrate on one single voxel v_r. In particular, consider that v_r is characterized by an orientation in the 3D space. If we assign each axis of the 3D space to a color (e.g., Red, Blue, and Green), decompose v_r on its three axes, and assign a color intensity to each projection on the corresponding axis, then a composite RGB color c_{v_r} can be obtained from the three basic intensities. c_{v_r} will represent v_r in all the subsequent steps. Now, let Σ be a set of symbols (with $|\Sigma| = s$). It is easy to show that every c_{v_r} can be mapped by discretization onto one of the symbols in Σ. As a consequence, since a fiber is a sequence of voxels, and since each voxel can be represented by a symbol in Σ, each fiber can be represented as a string on Σ. As pointed out in [6], there is no loss in generality if we assume that all fibers are described by the same amount of voxels.

We are now able to describe our approach. Let $M = \{m_1, m_2, \ldots, m_k\}$ be a set of models of interest for the user, and let $F = \{f_1, f_2, \ldots, f_n\}$ be a set of fibers. Assume that both M and F are expressed as strings, by using the transformation method outlined above. The following actions are carried out:

1. A matrix Δ, having n rows (one for each fiber) and k columns (one for each model) is computed. $\Delta[i, j]$ stores the level of dissimilarity existing between f_i and m_j. In order to compute this information, the SBED metric, defined in [6], can be applied.
2. Models in M are exploited to group fibers in F through the following computations:

    2.1 For $1 \leq i \leq n$, compute $\mu(i) = min_{1 \leq j \leq k}\{D[i, j]\}$ and $j_\mu(i) = arg\ min_{1 \leq j \leq k}\{D[i, j]\}$.
    2.2 For $1 \leq i \leq n$, if $\mu(i) < Th$ assign fiber f_i to a model $m_{j_\mu(i)}$; if $\mu(i) \geq Th$ leave f_i unassigned.
    2.3 Build the set of fiber-bundles $B = \{b_1, b_2, \ldots, b_k\}$, such that each b_i stores the fibers that have been assigned to m_i.

3. Now, since SBED considers symmetrical fibers as similar [6], it is necessary to refine the set B. We exploit QB to carry out this task. In particular:

    3.1 Given a bundle $b_i \in B$, apply QB to it. QB associates a set $CS = \{C_1, C_2, \ldots, C_q\}$ of clusters with b_i. If $q = 1$, no symmetrical structures were present in b_i and no other actions must be carried out.
    3.2 Otherwise:
        3.2.1 Compute the set of distances $DS = \{d_1, d_2, \ldots, d_q\}$ as the average SBED obtained by comparing fibers in C_m with b_i ($1 \leq m \leq q$). Only the fibers corresponding to the cluster C_μ showing the lowest average distance are eventually associated with b_i.

3.2.2 Let $\overline{CS} = CS - \{C_\mu\}$, and let $\overline{M} = M - \{b_i\}$. Given a cluster $C_\beta \in \overline{CS}$, derive the set $\overline{DS}_\beta = \{d_1, d_2, \ldots, d_{k-1}\}$ such that d_l is the average SBED between fibers in C_β and the model b_l in $\overline{M}$. Assign the fibers in C_β to the model d_v showing the lowest distance in $\overline{DS}_\beta$.

4.3 Experiments

In order to give an experimental proof about the capability of our method to improve the algorithms already proposed in literature, we performed two different kinds of experiment, namely: (i) algorithm parameter tuning, (ii) comparison of the performances of the proposed algorithm with our previous approach described in [6]. For these tasks we used simulated diffusion phantoms generated with Phantomas [1]. The comparison of the performances was performed by computing Precision, Recall, F-Measure and Overall [3].

4.3.1 Tuning of Algorithm Parameters

In this section we focus our attention mainly on the study of the parameter Th used by our algorithm. In order to have a golden standard to extract quantitative performance measurements, we used the simulated diffusion phantom described in Fig. 4.1a. Since the phantom contains 4 major populations of fibers, we manually delineate those regions in order to generate our ground truth.

Identification of the best value, and the best range of values, for the parameter Th was performed by running multiple tests using different values of Th. For each value

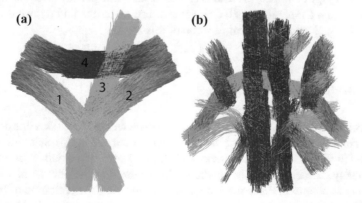

Fig. 4.1 Graphical representation of the simulated diffusion phantoms used for our experiments

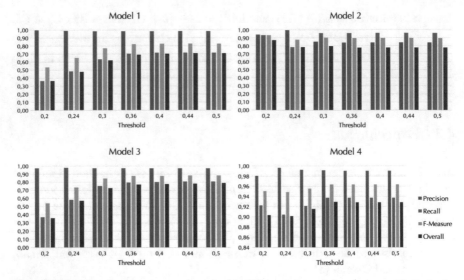

Fig. 4.2 Changes of performance measures with different values of the parameter *Th*. Each plot shows performance values computed for each model of the phantom

of *Th*, the performance measures described in Sect. 4.3 were computed. We tested our algorithm with values $Th \in [0.20, 0.24, 0.30, 0.36, 0.40, 0.44, 0.50]$.

Performances obtained using different values of *Th* are visually reported in Fig. 4.2. From the plots visible in this figure, it is possible to see how, in the range $0.20 \leq Th \leq 0.36$, our method suffers from a large variability in the performances. In particular, in this range, variations in Precision or Recall are visible when the value of *Th* is modified. As side effect, the values of F-Measure and Overall, which depend on the ones of Precision and Recall, change.

In the range $0.36 \leq Th \leq 0.50$, our approach reaches its best performances. Moreover, in this range of values, the results are stable since the performances do not show variations when the parameter *Th* changes. According to this finding, we decided to set $Th = 0.44$, since it is the central value of this range.

4.3.2   Method Comparison

As a second experiment, we compared our approach with our previous one described in [6]. For this comparison, we used the simulated diffusion phantom visually described in Fig. 4.1b. Like the previous phantom, in order to obtain a ground truth, we manually extracted 17 fiber-bundles, which represent the populations of fibers visible in the phantom. Since the proposed method needs the definition of the parameter *Th*, we exploited the results obtained in the previous section. In particular, in all the tests, we used $Th = 0.44$ (Table 4.1).

Table 4.1 Table describing the performances reached by the proposed approach and the one of [6], when applied on the phantom of Fig. 4.1a

Model	Our approach				Approach of [6]			
	Precision	Recall	F-Measure	Overall	Precision	Recall	F-Measure	Overal
1	0.99	0.72	0.83	0.71	0.75	0.94	0.83	0.62
2	0.84	0.96	0.90	0.78	0.75	0.94	0.83	0.62
3	0.97	0.81	0.88	0.78	0.81	0.42	0.54	0.30
4	0.99	0.94	0.96	0.93	0.96	0.97	0.96	0.92
Avg values	0.95	0.85	0.89	0.80	0.82	0.82	0.79	0.62

Table 4.2 Comparison of the performance metrics obtained with our approach and the one described in [6]

Model	Our approach				Approach of [6]			
	Precision	Recall	F-Measure	Overall	Precision	Recall	F-Measure	Overal
1	1.00	0.46	0.63	0.46	0.98	0.43	0.60	0.42
2	1.00	0.53	0.69	0.53	0.97	0.54	0.69	0.52
3	0.55	0.76	0.64	0.14	0.21	0.78	0.33	−2.15
4	0.65	0.62	0.63	0.29	0.41	0.63	0.50	−0.28
5	0.85	0.64	0.73	0.53	0.97	0.59	0.73	0.57
6	0.77	0.71	0.74	0.50	0.42	0.75	0.54	−0.29
7	0.94	0.68	0.79	0.64	0.94	0.66	0.78	0.62
8	0.99	0.56	0.72	0.55	0.95	0.53	0.68	0.50
9	0.83	0.92	0.87	0.73	0.73	0.92	0.81	0.58
10	0.78	0.68	0.73	0.49	0.58	0.67	0.62	0.18
11	1.00	0.54	0.70	0.54	1.00	0.46	0.63	0.46
12	1.00	0.55	0.71	0.55	0.95	0.52	0.67	0.49
13	1.00	0.63	0.77	0.63	0.94	0.63	0.75	0.59
14	1.00	0.49	0.66	0.49	0.96	0.52	0.67	0.50
15	0.99	0.77	0.87	0.76	0.95	0.75	0.84	0.71
16	1.00	0.62	0.77	0.62	0.99	0.61	0.75	0.60
17	0.94	0.70	0.80	0.66	0.92	0.70	0.80	0.64
Avg values	0.90	0.65	0.73	0.54	0.82	0.63	0.67	0.28

The performances obtained for fiber-bundle extraction using the two methods are reported in Table 4.2. From this table it is possible to see how the approach proposed here outperforms the one of [6]. Indeed, the average values of Precision, Recall, F-Measure and Overall reached by it are higher than the corresponding one of [6]. Moreover, it is important to underline how our method outperforms also QB. Indeed, the approach described in [6] had already shown better average performance values than QB.

4.4   Conclusion

In this paper, we have proposed a new approach to extracting anatomically-coherent and symmetry-aware WM fiber-bundles. We have shown that our approach integrates the WM fiber-bundle extraction approach presented in [6] (which provides anatomically coherent fiber-bundles but is unable to distinguish symmetric ones) with QB (which, by contrast, can distinguish symmetric fiber-bundles but often returns anatomically incoherent ones). Our approach maintains the strong points of both the approach of [6] and QB and, at the same time, avoids the weaknesses of both of them. We have also presented several experiments showing that the performances of our approach are better than the ones of the approach of [6], which, in their turn, were proved to be better than the ones of QB.

The approach presented in this paper could have several future developments. The most immediate one consists of extending our tests from phantoms to real cases. Then, we plan to investigate a possible improvement of QB in such a way as to make it capable of correcting the assignment of a fiber to a cluster when this action, in a second time, appears incorrect. Finally, we plan to optimize the usage of SBED constraints in such a way as to favor the extraction of specific WM fiber-bundles.

Acknowledgements This research was supported by EU MC ITN TRANSACT 2012 #316679. This work was partially supported by Aubay Italia S.p.A.

References

1. Caruyer, E., Daducci, A., Descoteaux, M., Houde, J.-C., Thiran, J.: Phantomas: a flexible software library to simulate diffusion mr phantom. In: ISMRM (2014)
2. Catani, M., Thiebaut de Schotten, M.: A diffusion tensor imaging tractography atlas for virtual in vivo dissections. Cortex **44**(8), 1105–1132 (2008)
3. Do, H., Melnik, S., Rahm, E.: Comparison of Schema Matching Evaluations. In: Proceedings of Web, Web-Services, and Database Systems, pp. 221–237. LNCS (2002)
4. Garyfallidis, E., Brett, M., Correia, M.M., Williams, G.B., Nimmo-Smith, I.: Quickbundles, a method for tractography simplification. Front. Neurosci. **6**, 175 (2012)
5. Mårtensson, J., Nilsson, M., Ståhlberg, F., Sundgren, P.C., Nilsson, C., van Westen, D., Larsson, E.-M., Lätt, J.: Spatial analysis of diffusion tensor tractography statistics along the inferior fronto-occipital fasciculus with application in progressive supranuclear palsy. MAGMA **26**(6), 527–537 (2013)
6. Stamile, C., Cauteruccio, F., Terracina, G., Ursino, D., Kocevar, G., Sappey-Marinier, D.: A model-guided string-based approach to white matter fiber-bundles extraction. In: Brain Informatics and Health: 8th International Conference, BIH, pp. 135–144. LNCS (2015)
7. Yeatman, J.D., Dougherty, R.F., Myall, N.J., Wandell, B.A., Feldman, H.M.: Tract profiles of white matter properties: automating fiber-tract quantification. PLoS One **7**(11), e49790 (2012)
8. Zhang, S., Correia, S., Laidlaw, D.H.: Identifying white-matter fiber bundles in dti data using an automated proximity-based fiber-clustering method. IEEE Trans. Vis. Comput. Graph. **14**(5), 1044–1053 (2008)

Chapter 5
Accurate Computation of Drude-Lorentz Model Coefficients of Single Negative Magnetic Metamaterials Using a Micro-Genetic Algorithm Approach

Annalisa Sgrò, Domenico De Carlo, Giovanni Angiulli, Francesco Carlo Morabito and Mario Versaci

Abstract Metamaterials are artificial materials having uncommon physical properties. For a fast and careful design of these structures, the development of simple and faithful models able to reproduce their electromagnetic behavior is a key factor. Very recently a quick method for the extraction of Drude-Lorentz models for electromagnetic metamaterials has been presented [1]. In this work we improve that approach, introducing a novel procedure exploiting a micro-genetic algorithm (μGA). Numerical results obtained for the case of a split ring resonator structure cleary show a better reconstruction behaviour for equivalent magnetic permittivity μ_{eff} than those provided by [1].

Keywords Single negative magnetic metamaterials · Drude-lorentz model · Genetic algorithms

A. Sgrò · D. De Carlo
TEC Spin-In - DICEAM, University Mediterranea, via Graziella - Loc. Feo di Vito,
89121 Reggio Calabria, Italy
e-mail: annalisa.sgro@unirc.it

D. De Carlo
e-mail: domenico.decarlo@unirc.it

G. Angiulli (✉)
DIIES, University Mediterranea, via Graziella - Loc. Feo di Vito, 89121
Reggio Calabria, Italy
e-mail: giovanni.angiulli@unirc.it

F.C. Morabito · M. Versaci
DICEAM, University Mediterranea, via Graziella - Loc. Feo di Vito, 89121
Reggio Calabria, Italy
e-mail: morabito@unirc.it

M. Versaci
e-mail: mario.versaci@unirc.it

© Springer International Publishing AG 2018
A. Esposito et al. (eds.), *Multidisciplinary Approaches to Neural Computing*,
Smart Innovation, Systems and Technologies 69,
DOI 10.1007/978-3-319-56904-8_5

5.1 Introduction

During the last decade, there has been a growing interest in the scientific community on Metamaterials (MMs), this due to their revolutionary physical properties [2]. Broadly speaking, electromagnetic MMs can be defined as artificial materials showing exotic electromagnetic properties (unfounded in nature) when excited by an outer electromagnetic field.

They are designed by periodically embedding suitable metallic inclusions (usually identical to each other) into a dielectric medium, named as *host medium* [2, 3]. Accordingly, a MM can be viewed as an ensemble of elementary unities named as *meta-atoms* [3]. Since the wavelength of the outer excitation field is usually far more larger than the dimension of meta-atoms, MMs can be considered to all intents and purposes as homogeneous dispersive media characterized by an effective electrical permittivity $\epsilon_{eff}(\omega) = \epsilon'_{eff}(\omega) + i\epsilon''_{eff}(\omega)$ and a magnetic permeability $\mu_{eff}(\omega) = \mu'_{eff}(\omega) + i\mu''_{eff}(\omega)$ [2].

It is common practice to classify MMs on the basis on the sign of the real part of their dielectric permittivity $\epsilon'_{eff}(\omega)$ and magnetic permeability $\mu'_{eff}(\omega)$, having thus single negative electric MMs ($\epsilon'_{eff}(\omega) < 0$), single negative magnetic MMs ($\mu'_{eff}(\omega) < 0$), (named simply as single negative MMs (SN-MMs)), and double negative magnetic metamaterials (DN-MMs) ($\epsilon'_{eff}(\omega) < 0$ and $\mu'_{eff}(\omega) < 0$) [3]. The design of a MM, i.e. the implementation of a well specified behavior of $\epsilon_{eff}(\omega)$ and/or $\mu_{eff}(\omega)$ as a function of the angular frequency ω, is engineered working on the geometric dimensions of its meta-atom and on the host medium characteristics [1, 3]. More in detail, an iterative procedure involving a great number of full wave simulations is carried out on the meta-atom, in order to evaluate its scattering matrix [1–4]. Effective parameters $\epsilon_{eff}(\omega)$ and/or $\mu_{eff}(\omega)$ are then extracted from this data and the meta-atom geometry is tuned until the desired electromagnetic behavior for $\epsilon_{eff}(\omega)$ and/or $\mu_{eff}(\omega)$ is achieved. During the last years, the design process of a MM has become increasingly time consuming due to the growing complexity of the desidered electromagnetic behaviors and by the arising interest in manufacturing inhomogeneous and anisotropic artificial structures [5].

In order to lighten the computational burden involved in the MM design process, an approach commonly adopted is to set up an analytical equivalent circuit model for the meta-atom [6]. However, this method can only be applied to a narrow classes of structures and is often unable to model accurately the behavior of the effective electromagnetic parameters of the MM at hand. Alternately, a more accurate approach based on the extraction of the Drude-Lorentz model for the meta-atoms has recently been developed in [1].

In this work, we employ for the first time an evolutionary optimization procedure based on a micro-genetic algorithm (μGA) approach in order to improve notably the accuracy of the models provided by the method proposed in [1]. During the years,

μGAs have been fruitfully employed to solve a wide wariety of engineering problems ranging from microwave energy harvesting [7] to electricity pricing signal forecasting [8]. To demonstrate the effectiveness of the μGA approach, numerical results relevant to the Drude-Lorentz model for a SN magnetic metamaterial, i.e. a split ring resonator structure [4], are reported and discussed.

5.2 The Drude-Lorentz Model of a SN Magnetic Metamaterial in a Nutshell

The electromagnetic response of a SN magnetic MM can be controlled by the knowledge of the functional relationship existing between its effective magnetic permittivity $\mu_{eff}(\omega)$ and the geometric parameters of the meta-atoms with which the SN magnetic MM is set up [4]. This functional relationship is usually built up numerically from the computation of the scattering matrix of the meta-atom by using the Nicholson Ross Weird (NRW) method [1, 3, 4]. However, in this way, a lot of repeated time consuming simulations are necessary to tailor the response of the SN magnetic MM at hand.

To overcome this problem, a method based on the extraction of an analytical model, named as *Drude-Lorentz* model, from the knowledge of the numerical data for $\mu_{eff}(\omega)$, has been introduced in [1]. The starting point for this result arises from the observation that the majority of SN magnetic MMs are realized by using resonant meta-atoms which shape for $\mu_{eff}(\omega)$ can be analytically described as follow

$$\mu_{eff}(\omega) = \hat{\mu}_0 \left(1 - \frac{F_u \omega^2}{\omega^2 + i\gamma_u \omega - \omega_{0u}^2} \right), \tag{5.1}$$

where $\hat{\mu}_0$ is the background permeability, F_u is the magnetic resonant intensity, ω_{0u} is the magnetic resonance frequency, and γ_u is the magnetic damping factor. The above coefficients are unknowns that have to be evaluated for the meta-atom at hand.

In [1] a simple procedure to extract the tuple $\{\hat{\mu}_0, F_u, \omega_{0u}, \gamma_u\}$, has been introduced. At the initial stage of this procedure, the coefficient $\hat{\mu}_0$ is setted equal to unity. The term ω_{0u} is evaluated determining the angular frequency value for which $\mu''_{eff}(\omega)$ reaches its maximum value, whereas F_u is evaluated employing the following approximate relation [1]

$$F_u = 1 - \left(\frac{\omega_{0u}}{\omega_{pu}} \right)^2, \tag{5.2}$$

where ω_{pu} is the angular frequency for which $\mu'_{eff}(\omega)$ reaches its minimum value. The coefficient γ_u is determined putting into Eq. (5.1), the values of ω_{0u} and F_u above calculated. Afterwards, the values of $\hat{\mu}_0$ and γ_u are refined by solving (considering alternatively these parameters as an unknown) the following overdetermined linear system

$$F_u \omega_i^2 = \left(1 - \frac{\mu_{eff,i}}{\hat{\mu}_0} \right) (\omega_{0u}^2 + i\gamma_u \omega_i - \omega_i^2) \qquad i \in (1, \ldots, M), \tag{5.3}$$

until stable values for these coefficients are achieved.

5.3 Genetic and Micro-Genetic Algorithms

Genetic algorithms (GAs) are heuristic self adaptive global optimizing probability search algorithms which mimic some of the natural evolution processes [9]. Thanks to their characteristics they have demonstrated a great ability to treat and solve a broad range of complex problems [9]. Unlike classical mathematical optimization algorithms, GAs do not use any initial guess for the solution or derivatives for the objective function to optimize.

Basically, for solving a given optimization problem, a GA operates creating a population of individuals or *chromosomes*. This population evolves over multiple generations, under specified rules (*selection, crossover, mutation*) which give more chances to reproduce to those chromosomes which are the better candidates to solve the optimization problem at hand, in order to find the individual that that minimize the *fitness* i.e., minimizes the considered objective function [9]. The workflow of a GA is depicted in Fig. 5.1. Usually, when a GA is exploited to cope with a high dimensionality optimization problem, a large population has to be employed. This involves a great number of objective function evaluations, which could result to be too time consuming to reach convergence within a given margin of error. An alternative to overcome this issue is the use of Micro genetic algorithm (μGA) [10].

A μGA [10] exploits a tiny population for solving an optimization problem, which evolves as in a standard GA, converging in few generations, due its small dimensionality, to a best chromosome. Keeping this best fit individual, a μGA generates a new random population restarting the overall evolutionary procedure, avoiding in this way the possibility of convergence to a local minimum in the search space and evolving to the better solution in a small computational time [10].

Fig. 5.1 The flow chart of a GA

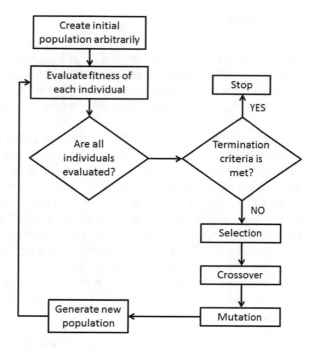

5.4 Improving the Drude-Lorentz Model for a SN Magnetic Metamaterial by Using a μGA

The fitting of the meta-atom response provided by the Drude-Lorentz model defined by Eq. (5.1) can show poor accuracy in some cases [1]. In order to improve its fitting property, the Drude-Lorentz coefficients $\hat{\mu}_0$, F_u, ω_{0u}, and γ_u computed by means of the procedure described in Sect. 5.2, can be quickly refined by using a suitable μGA algorithm approach. The optimal coefficients $\hat{\mu}_0'$, F_u', ω_{0u}', γ_u' able to provide a more accurate fitting for the effective permeability μ_{eff} of a SN magnetic meta-atom, can be evaluated searching for the minimum the following objective function

$$||\mu_{eff}^{num}(\omega) - \mu_{eff}(\hat{\mu}_0, F_u, \omega_{0u}, \gamma_u)||_2, \qquad (5.4)$$

where $\mu_{eff}^{num}(\omega)$ is the magnetic response obtained numerically by means of full-wave simulations for the considered meta-atom, $\mu_{eff}(\hat{\mu}_0, F_u, \omega_{0u}, \gamma_u)$ is the magnetic response described by the Drude-Lorentz model (5.1), now considered as a four-variable function of its coefficients $\hat{\mu}_0'$, F_u', ω_{0u}', γ_u' (ranging on a suitable subset of $\mathbb{R}^4$ built by means of the values of $\hat{\mu}_0$, F_u, ω_{0u}, γ_u computed by using the procedure described in Sect. 5.2) and $|| \cdot ||_2$ is the standard euclidean norm.

5.5   Numerical Results

The suitability and the effectiveness of the μGA approach to guarantee the minimization of (5.4) with a small computational time, when compared to the standard GA approach, have been investigated considering the Split Ring Resonator meta-atom depicted in Fig. 5.2 (all dimensions and the operation frequency range are reported in [4]).

The behavior of $\mu_{eff}^{num}(\omega)$ for this structure has been reconstructed by using the Nicholson-Ross-Weird method on its S parameters data computed by means of a finite element full-wave code. All genetic optimization experiments were executed by using the MATLAB R2015a Global Optimization Toolbox running on a AMD(R) ATHLON(R) 3.00 GHz machine equipped with 4 GB of RAM. The genetic parameters employed in our numerical experiments with GA and μGA algorithms are reported in Table 5.1. In Table 5.2 are reported the results for the best values obtained for the minimum of (5.4) (as a function of the genetic parameters) and the related CPU time. They point out the suitability of μGA (see row #23 in Table 5.2) to minimize accurately the objective function (5.4) with a small CPU time than compared with conventional GA. Figures 5.3 and 5.4 show the comparison among the effective

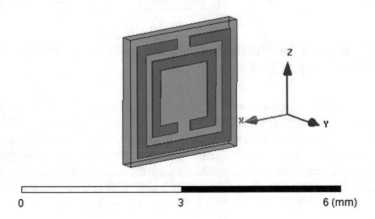

0   3   6 (mm)

Fig. 5.2   A Split Ring Resonator (SRR) meta-atom

Table 5.1   Range for the GA and μGA parameters adopted in this study

Population size	$10 \div 300$
Elite count	$1 \div 25$
Mutation function	Gaussian
Crossover function	Scattered
	Single point
	Two point
	Arithmetic

Table 5.2 Minimum norm value for (5.4) an related CPU time for different GA and μGA parameters

#	Population Size	Elite Count	Mutation Function	Crossover Function	Fitness value	CPU Time [sec]
1	300	24	Gaussian	Arithmetic	11.031109	2467
2	300	6	Gaussian	Arithmetic	11.031112	1073
3	300	3	Gaussian	Arithmetic	11.031128	337
4	300	3	Gaussian	Two point	11.031145	280
5	300	6	Gaussian	Scattered	11.031159	795
6	100	8	Gaussian	Arithmetic	11.031175	1026
7	300	6	Gaussian	Single point	11.031227	874
8	300	15	Gaussian	Scattered	11.031249	1496
9	300	3	Gaussian	Single point	11.031254	225
10	300	24	Gaussian	Scattered	11.031256	2198
11	300	15	Gaussian	Two point	11.031292	1746
12	300	15	Gaussian	Arithmetic	11.031321	1860
13	100	8	Gaussian	Single point	11.031335	957
14	100	2	Gaussian	Single point	11.031358	376
15	300	24	Gaussian	Single point	11.033136	2274
16	300	15	Gaussian	Single point	11.031457	1594
17	300	24	Gaussian	Two point	11.031457	2352
18	300	6	Gaussian	Two point	11.031457	952
19	60	2	Gaussian	Scattered	11.031555	179
20	60	2	Gaussian	Arithmetic	11.031563	258
21	60	2	Gaussian	Single point	11.031577	321
22	60	5	Gaussian	Arithmetic	11.031574	608
23	30	1	Gaussian	Arithmetic	11.031157	95

magnetic permeability calculated (*i*) by means of Nicholson-Ross Weird algorithm, (*ii*) by means of Drude-Lorentz model [1], and (*iii*) by means of our μGA approach. The value assumed by (5.4) by using the standard Drude-Lorentz is reduced of about 2.26 times by using the Drude-Lorentz model optimized by means of a μGA. These results clearly show the superiority of our approach. The final optimized coefficients for the Drude-Lorentz model are reported in Table 5.3.

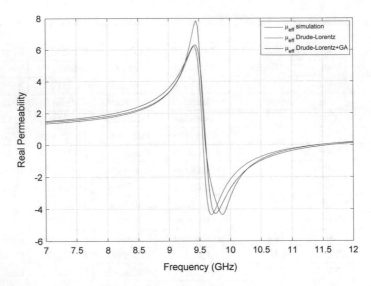

Fig. 5.3 Real part of μ_{eff} versus frequency for the SRR meta-atom. *Blue line*: Nicholson-Ross Weird method, *purple line*: Drude-Lorentz [1], *black line*: our method

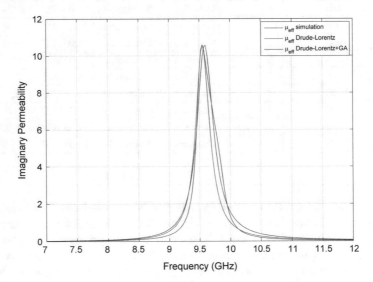

Fig. 5.4 Imaginary part of μ_{eff} versus frequency for the SRR meta-atom. *Blue line*: Nicholson-Ross Weird method, *purple line*: Drude-Lorentz [1], *black line*: our method

Table 5.3 μGA optimized coefficients (computed using the genetic parameters reported by row #23 in Table 5.2)

$\hat{\mu}_0$	F_u	ω_{0u} [rad/s]	γ_u [rad/s]
1.1063	0.3228	$2\pi \cdot 9.59 \cdot 10^9$	$2\pi \cdot 3.232 \cdot 10^8$

5.6 Conclusions

In this paper, we have introduced for the first time an evolutionay computation procedure based on a μGA for the accurate computation of the Drude-Lorentz model coefficients of SN magnetic metamaterials.

The effectiveness of the proposed approach has been tested for the case of a Split Ring Resonator meta-atom structure. Numerical results show as the Drude-Lorentz model thus obtained is able to model more accurately than the standard Drude-Lorentz model [1] the behavior of the numerically reconstructed Split Ring Resonator magnetic permittivity $\mu_{eff}^{num}(\omega)$.

Future researches will be devoted to exploit the proposed technique to improve the predictive capabilities of the generalized Drude-Lorentz models for MMs [1].

References

1. Cui, T.J., Smith, D.R., Liu, R.: Metamaterials. Theory, Design, and Applications. Springer Science & Business Media (2009)
2. Shivola, A.: Metamaterials in electromagnetics. Metamaterials **1**(1), 2–11 (2007)
3. Capolino, F. (ed.): Theory and Phenomena of Metamaterials. CRC Press (2009)
4. Smith, D.R., Vier, D.C., Koschny, T., Soukoulis, C.M.: Electromagnetic parameter retrieval from inhomogeneous metamateials. Phys. Rev. E **3**(71), 1–11 (2005)
5. Zhang, X., Wu, Y.: Effective medium theory for anisotropic metamaterials. Sci. Rep. (5) (2015)
6. Bilotti, F., Toscano, A., Vegni, L., Aydin, K., Alici, K.B., Ozbay, E.: Equivalent circuit models for the design od metamaterial based on artificial magnetic inclusions. IEEE Trans. Microw. Theory Techn. **12**(55), 1865–2873 (2007)
7. Mori, T., Murakami, R., Sato, Y., Campelo, F., Igarashi, H.: Shape optimization of wideband antennas for microwave energy harvesters using FDTD. IEEE Trans. Mag. **3**(51), 1–4 (2014)
8. Alamaniotis, M., Bargiotas, D., Bourbakis, D., Tsoukalas, L.H.: Genetic optimal regression of relevance vector machines for electricity pricing signal forecasting in smart grids. IEEE Trans. Smart Grids **6**(6), 2997–3005 (2015)
9. Goldberg, D.E.: Genetic Algorithms. Pearson Education (2006)
10. Koeppen, M., Schaefer, G., Abraham, A.: Intelligent Computation Optimization in Engineering: Techniques & Applications. Springer Science & Bussiness Media (2011)

Chapter 6
Effective Blind Source Separation Based on the Adam Algorithm

Michele Scarpiniti, Simone Scardapane, Danilo Comminiello, Raffaele Parisi and Aurelio Uncini

Abstract In this paper, we derive a modified InfoMax algorithm for the solution of Blind Signal Separation (BSS) problems by using advanced stochastic methods. The proposed approach is based on a novel stochastic optimization approach known as the Adaptive Moment Estimation (Adam) algorithm. The proposed BSS solution can benefit from the excellent properties of the Adam approach. In order to derive the new learning rule, the Adam algorithm is introduced in the derivation of the cost function maximization in the standard InfoMax algorithm. The natural gradient adaptation is also considered. Finally, some experimental results show the effectiveness of the proposed approach.

Keywords Blind source separation · Stochastic optimization · Adam algorithm · Infomax algorithm · Natural gradient

6.1 Introduction

Blind Source Separation (BSS) is a well-known and well-studied field in the adaptive signal processing and machine learning [5–7, 10, 16, 20]. The problem is to recover original and unknown sources from a set of mixtures recorded in an unknown

M. Scarpiniti (✉) · S. Scardapane · D. Comminiello · R. Parisi · A. Uncini
Department of Information Engineering, Electronics and Telecommunications (DIET),
"Sapienza" University of Rome, Via Eudossiana 18, 00184 Rome, Italy
e-mail: michele.scarpiniti@uniroma1.it

S. Scardapane
e-mail: simone.scardapane@uniroma1.it

D. Comminiello
e-mail: danilo.comminiello@uniroma1.it

R. Parisi
e-mail: raffaele.parisi@uniroma1.it

A. Uncini
e-mail: aurelio.uncini@uniroma1.it

© Springer International Publishing AG 2018
A. Esposito et al. (eds.), *Multidisciplinary Approaches to Neural Computing*,
Smart Innovation, Systems and Technologies 69,
DOI 10.1007/978-3-319-56904-8_6

environment. The term *blind* refers to the fact that both the sources and the mixing environment are unknown.

Several well-performing approaches exist when the mixing environment is instantaneous [3, 7], while some problems still arise in convolutive environments [2, 4, 17]. Different approaches were proposed to solve BSS in linear and instantaneous environment. Some of these approaches perform separation by using high order statistics (HOS) while others exploit information theoretic (IT) measures [6]. One of the well-know algorithms in this latter class is the InfoMax one proposed by Bell and Sejnowski in [3]. The InfoMax algorithm is based on the maximization of the joint entropy of the output of a single layer neural network and it is very efficient and easy to implement since the gradient of the joint entropy can be evaluated simply in a closed form. Moreover, in order to avoid numerical instability, a natural gradient modification to InfoMax algorithm has also been proposed [1, 6].

Unfortunately, all these solutions perform slowly when the number of the original sources to be separated is high and/or bad scaled. The separation becomes impossible if the number of sources is equal or greater than ten. In addition, the convergence speed problem worsen in the case of additive sensor noise to mixtures or when the mixing matrix is close to be ill-conditioned. However, specially when working with speech and audio signals, fast convergence speed is an important task to be performed. Many authors have tried to overcome this problem: some solutions consist in incorporating a momentum term in the learning rule [13], in a self-adjusting variable step-size [18] or in a scaled natural gradient algorithm [8].

Recently, a novel algorithm for gradient based optimization of stochastic cost functions has been proposed by Kingma and Ba in [12]. This algorithm is based on the adaptive estimates of the first and second order moments of the gradient, and for this reason has been called the Adaptive Moment Estimation (Adam) algorithm. The authors have demonstrated in [12] that Adam is easy to implement, computationally efficient, invariant to diagonal rescaling of the gradients and well suited for problems with large data and parameters.

The Adam algorithm combines the advantages of other state-of-the-art optimization algorithms, like AdaGrad [9] and RMSProp [19], outperforming the limitations of these algorithms. In addition, Adam can be related to the natural gradient (NG) adaptation [1], employing a preconditioning that adapts to the geometry of data.

In this paper we propose a modified InfoMax algorithm based on the Adam algorithm [12] for the solution of BSS problems. We derive the proposed modified algorithm by using the Adam algorithm instead of the standard stochastic gradient ascent rule. It is shown that the novel algorithm has a faster convergence speed with respect to the standard InfoMax algorithm and usually also reaches a better separation. Some experimental results, evaluated in terms of the Amari Performance index (PI) [6], show the effectiveness of the proposed idea.

The rest of the paper is organized as follows. In Sect. 6.2 we briefly introduce the problem of BSS. Then, we give some details on the Adam algorithm in Sect. 6.3. The main novelty of this paper, the extension of InfoMax algorithm with Adam is provided in Sect. 6.4. Finally, we validate our approach in Sect. 6.5. We conclude with some final remarks in Sect. 6.6.

6.2 The Blind Source Separation Problem

Let us consider a set of N unknown and statistically independent sources denoted as $\mathbf{s}[n] = \left[s_1[n], \ldots, s_N[n]\right]^T$, such that the components $s_i[n]$ are zero-mean and mutually independent. Signals received by an array of M sensors are denoted by $\mathbf{x}[n] = \left[x_1[n], \ldots, x_M[n]\right]^T$ and are called mixtures. For simplicity, here we consider the case of $N = M$.

In the case of a linear and instantaneous mixing environment, the mixture can be described in a matrix form as

$$\mathbf{x}[n] = \mathbf{A}\mathbf{s}[n] + \mathbf{v}[n], \tag{6.1}$$

where the matrix $\mathbf{A} = \left(a_{ij}\right)$ collects the mixing coefficients a_{ij}, $i,j = 1, 2, \ldots, N$ and $\mathbf{v}[n] = \left[v_1[n], \ldots, v_N[n]\right]^T$ is a noise vector, with correlation matrix $\mathbf{R_v} = \sigma_v^2 \mathbf{I}$ and noise variance σ_v^2.

The separated signals $\mathbf{u}[n]$ are obtained by a separating matrix $\mathbf{W} = \left(w_{ij}\right)$ as described by the following equation

$$\mathbf{u}[n] = \mathbf{W}\mathbf{x}[n]. \tag{6.2}$$

The transformation in (6.2) is such that $\mathbf{u}[n] = \left[u_1[n], \ldots, u_N[n]\right]^T$ has components $u_k[n]$, $k = 1, 2, \ldots, N$, that are as independent as possible.

Moreover, due to the well-known permutation and scaling ambiguity of the BSS problem, the output signals $\mathbf{u}[n]$ can be expressed as

$$\mathbf{u}[n] = \mathbf{P}\mathbf{D}\mathbf{s}[n], \tag{6.3}$$

where $\mathbf{P}$ is an $N \times N$ permutation matrix and $\mathbf{D}$ is an $N \times N$ diagonal scaling matrix. The weights w_{ij} can be adapted by maximizing or minimizing some suitable cost function [5, 6]. A particularly good approach is to maximize the joint entropy of a single layer neural network [3], as shown in Fig. 6.1, leading to the Bell and Sejnowski InfoMax algorithm. In this network each output $y_i[n]$ is a nonlinear transformation of each signal $u_i[n]$:

$$y_i[n] = h_i\left(u_i[n]\right). \tag{6.4}$$

Each function $h_i(\cdot)$ is known as activation function (AF).

With reference to Fig. 6.1, using the equation relating the probability density function (pdf) of a random variable $p_\mathbf{x}(\mathbf{x})$ and nonlinear transformation of it $p_\mathbf{y}(\mathbf{y})$ [14], the joint entropy $H(\mathbf{y})$ of the network output $\mathbf{y}[n]$ can be evaluated as

$$H(\mathbf{y}) \equiv -E\left\{\ln p_\mathbf{y}(\mathbf{y})\right\} = H(\mathbf{x}) + \ln \det \mathbf{W} + \sum_{i=1}^{N} \ln \left|h_i'\right|, \tag{6.5}$$

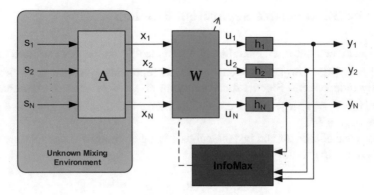

Fig. 6.1 Unknown mixing environment and the InfoMax network

where h_i' is the first derivative of the i-th AF with respect its input $u_i[n]$.

Evaluating the gradient of (6.5) with respect to the separating parameters $\mathbf{W}$, after some not too complicated manipulations, leads to

$$\nabla_\mathbf{W} H(\mathbf{y}) = \mathbf{W}^{-T} + \boldsymbol{\Psi}\mathbf{x}^T[n], \tag{6.6}$$

where $(\cdot)^{-T}$ denotes the transpose of the inverse and $\boldsymbol{\Psi} = \left[\Psi_1, \dots, \Psi_N\right]^T$ is a vector collecting the terms $\Psi_k = h_k''/h_k'$, defined as the ratio of the second and the first derivatives of the AFs.

In order to avoid the numerical problems of the matrix inversion in (6.6) and the possibility to remain blocked in a local minimum, Amari has introduced the Natural Gradient (NG) adaptation [1] that overcomes such problems. The NG adaptation rule, can be obtained simply by right multiplying the stochastic gradient for the term $\mathbf{W}^T\mathbf{W}$. Hence, after multiplying (6.6) for this term, the NG InfoMax is simply

$$\nabla_\mathbf{W} H(\mathbf{y}) = \left(\mathbf{I} + \boldsymbol{\Psi}\mathbf{u}^T[n]\right)\mathbf{W}. \tag{6.7}$$

Regarding the selection of the AFs shape, there are several alternatives. However, especially in the case of audio and speech signals, a good nonlinearity is represented by the $\tanh(\cdot)$ function. With this choice, the vector $\boldsymbol{\Psi}$ in (6.6) and (6.7) is simply evaluated as $-2\mathbf{y}[n]$.

In summary, using the $\tanh(\cdot)$ AF, the InfoMax and natural gradient InfoMax algorithms are described by the following learning rules:

$$\mathbf{W}_t = \mathbf{W}_{t-1} + \mu \left(\mathbf{W}_{t-1}^{-T} - \mathbf{y}[n]\mathbf{x}^T[n]\right), \tag{6.8}$$

$$\mathbf{W}_t = \mathbf{W}_{t-1} + \mu \left(\mathbf{I} - \mathbf{y}[n]\mathbf{u}^T[n]\right)\mathbf{W}_{t-1}, \tag{6.9}$$

where μ is the learning rate or step-size.

6.3 The Adam Algorithm

Le us denote with $f(\theta)$ a noisy cost function to be minimized (or maximized) with respect to the parameters θ. The problem is considered stochastic for the random nature of data samples or for inherent function noise. In the following, the noisy gradient vector at time t of the cost function $f(\theta)$ with respect to the parameters θ, will be denoted with $\mathbf{g}_t = \nabla_\theta f_t(\theta)$.

The Adam algorithm performs the gradient descent (or ascent) optimization by evaluating the moving averages of the noisy gradient $\mathbf{m}_t$ and the square gradient $\mathbf{v}_t$ [12]. These moment vectors are updated by using two scalar coefficients β_1 and β_2 that control the exponential decay rates:

$$\mathbf{m}_t = \beta_1 \mathbf{m}_{t-1} + \left(1 - \beta_1\right) \mathbf{g}_t, \tag{6.10}$$

$$\mathbf{v}_t = \beta_2 \mathbf{v}_{t-1} + \left(1 - \beta_2\right) \mathbf{g}_t \odot \mathbf{g}_t, \tag{6.11}$$

where $\beta_1, \beta_2 \in [0, 1)$ and $\odot$ denotes the element-wise multiplication, while $\mathbf{m}_0$ and $\mathbf{v}_0$ are initialized as zero vectors. These vectors represent the mean and the uncentered variance of the gradient vector $\mathbf{g}_t$. Since the estimates of $\mathbf{m}_t$ and $\mathbf{v}_t$ are biased towards zero, due to their initialization, a bias correction is computed on these moments

$$\hat{\mathbf{m}}_t = \frac{\mathbf{m}_t}{1 - \beta_1^t}, \tag{6.12}$$

$$\hat{\mathbf{v}}_t = \frac{\mathbf{v}_t}{1 - \beta_2^t}. \tag{6.13}$$

The vector $\hat{\mathbf{v}}_t$ represents an approximation of the diagonal of the Fisher information matrix [15]. Hence Adam can be related to the natural gradient algorithm [1].

Finally, the parameter vector θ_t at time t, is updated by the following rule

$$\theta_t = \theta_{t-1} - \eta \frac{\hat{\mathbf{m}}_t}{\sqrt{\hat{\mathbf{v}}_t} + \varepsilon}, \tag{6.14}$$

where η is the step size and ε is a small positive constant used to avoid the division for zero. In the gradient ascent, the minus sign in (6.14) is substituted with the plus sign.

6.4 Modified InfoMax Algorithm

In this section, we introduce the modified Bell and Sejnowski InfoMax algorithm, based on the Adam optimization method. Since Adam algorithm uses a vector of parameters, we perform a vectorization of the gradient (6.6) or (6.7)

$$\mathbf{w} = \text{vec}\left(\nabla_{\mathbf{W}} H\left(\mathbf{y}\right)\right) \in \mathbb{R}^{N^2 \times 1}, \tag{6.15}$$

where $\text{vec}\,(\mathbf{A})$ is the vectorization operator, that forms a vector by stacking the columns of the matrix $\mathbf{A}$ below one another. The gradient vector is evaluated on a number of blocks N_B extracted from the signals.

At this point, the mean and variance vectors are evaluated from the knowledge of the gradient $\mathbf{w}_t$ at time t by using Eqs. (6.10)–(6.13). Then, using (6.14), the gradient vector (6.15) is updated for the maximization of the joint entropy by

$$\mathbf{w}_t = \mathbf{w}_{t-1} + \eta \frac{\hat{\mathbf{m}}_t}{\sqrt{\hat{\mathbf{v}}_t} + \varepsilon}. \tag{6.16}$$

Finally, the vector $\mathbf{w}_t$ is reshaped in matrix form, by

$$\mathbf{W}_t = \text{mat}\left(\mathbf{w}_t\right) \in \mathbb{R}^{N \times N}, \tag{6.17}$$

where $\text{mat}\,(\mathbf{a})$ reconstructs the $N \times N$ matrix by unstacking the columns from the vector $\mathbf{a}$. The whole algorithm is in case repeated for a certain number of epochs N_{ep}.

The pseudo-code of the modified InfoMax algorithm with Adam, called here Adam InfoMax, is described in Algorithm 1.

Algorithm 1: Pseudo-code for the Adam InfoMax algorithm

Data: Mixture signals $\mathbf{x}[n]$, η, β_1, β_2, ε, N_B, N_{ep}.

1 **Initialization:** $\mathbf{W}_0 = \mathbf{I}$, $\mathbf{m}_0 = \mathbf{0}$, $\mathbf{v}_0 = \mathbf{0}$, $\mathbf{w}_0 = \mathbf{0}$

2 $P = N_B N_{ep}$

3 **for** $t = 1{:}P$ **do**

4   Extract the t-th block $\mathbf{x}_t$ from $\mathbf{x}[n]$

5      $\mathbf{u}_t = \mathbf{W}_{t-1}\mathbf{x}_t$

6      $\mathbf{y}_t = \tanh\left(\mathbf{u}_t\right)$

7   Update gradient $\nabla_{\mathbf{W}} H\left(\mathbf{y}_t\right)$ in (6.6) or (6.7)

8      $\mathbf{w}_t = \text{vec}\left(\nabla_{\mathbf{W}} H\left(\mathbf{y}_t\right)\right)$

9      $\mathbf{m}_t = \beta_1 \mathbf{m}_{t-1} + \left(1 - \beta_1\right)\mathbf{w}_t$

10     $\mathbf{v}_t = \beta_2 \mathbf{v}_{t-1} + \left(1 - \beta_2\right)\mathbf{w}_t \odot \mathbf{w}_t$

11     $\hat{\mathbf{m}}_t = \frac{\mathbf{m}_t}{1 - \beta_1^t}$

12     $\hat{\mathbf{v}}_t = \frac{\mathbf{v}_t}{1 - \beta_2^t}$

13     $\mathbf{w}_t = \mathbf{w}_{t-1} + \eta \frac{\hat{\mathbf{m}}_t}{\sqrt{\hat{\mathbf{v}}_t} + \varepsilon}$

14     $\mathbf{W}_t = \text{mat}\left(\mathbf{w}_t\right)$

15 **end**

Result: Separated signals: $\mathbf{u}[n] = \mathbf{W}_P \mathbf{x}[n]$

6.5 Experimental Results

In this section, we propose some experimental results to demonstrate the effectiveness of the proposed idea. We perform separation of mixtures of both synthetic and real-world data. The results are evaluated in terms of the Amari Performance Index (PI) [6], defined as

$$PI = \frac{1}{N(N-1)} \sum_{i=1}^{N} \left[\left(\sum_{k=1}^{N} \frac{|q_{ik}|}{\max_j |q_{ij}|} - 1 \right) + \left(\sum_{k=1}^{N} \frac{|q_{ki}|}{\max_j |q_{ji}|} - 1 \right) \right], \quad (6.18)$$

where q_{ij} are the elements of the matrix $\mathbf{Q} = \mathbf{WA}$. This index is close to zero if the matrix $\mathbf{Q}$ is close to the product of a permutation matrix and a diagonal scaling matrix.

The performances of the proposed algorithm were also compared with the standard InfoMax algorithm [3] and the Momentum InfoMax described in [13]. In this last algorithm, the α parameter is set to 0.5 in all experiments.

In a first experiment, we perform the separation of five mixtures obtained as linear combination of the following bad-scaled independent sources

$$s_1[n] = 10^{-6} \cdot \sin(350n) \sin(60n),$$
$$s_2[n] = 10^{-5} \cdot \text{tri}(70n),$$
$$s_3[n] = 10^{-4} \cdot \sin(800n) \sin(80n),$$
$$s_4[n] = 10^{-5} \cdot \cos(400n + 4\cos(60n)),$$
$$s_5[n] = \xi[n],$$

where $\text{tri}(\cdot)$ denotes a triangular waveform and $\xi[n]$ is a uniform noise in the range $[-1, 1]$. Each signal is composed of $L = 30,000$ samples. The mixing matrix is a 5×5 Hilbert matrix, which is extremely ill-conditioned. All simulations have been performed by MATLAB 2015a, on an Intel Core i7 3.10 GHz processor at 64 bit with 8 GB of RAM. Parameters of the algorithms have been found heuristically.

We perform separation by the Adam modification of the standard InfoMax algorithm, with gradient in (6.6). We use a block length of $B = 30$ samples (hence $N_B = \lfloor L/B \rfloor = 1,000$), while the other parameters are set as: $N_{ep} = 200$, $\beta_1 = 0.5$, $\beta_2 = 0.75$, $\varepsilon = 10^{-8}$, $\eta = 0.001$ and the learning rate of the standard InfoMax and the Momentum InfoMax to $\mu = 5 \times 10^{-5}$. Performance in terms of the PI in (6.18) is reported in Fig. 6.2, that clearly shows the effectiveness of the proposed idea.

A second and third experiments are performed on speech audio signals sampled at 8 kHz. Each signal is composed of $L = 30,000$ samples. In the second experiment a male and a female speech are mixed with a 2×2 random matrix with entries uniformly distributed in the interval $[-1, 1]$ while in the third one, two male and two

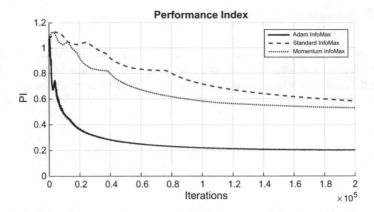

Fig. 6.2 Performance Index (PI) of the first proposed experiment

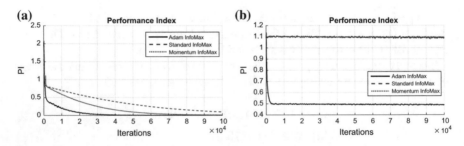

Fig. 6.3 Performance Index (PI) of the: second (**a**) and third (**b**) proposed experiment

female speeches are mixed with a 4×4 ill-conditioned Hilbert matrix. In addition, an additive white noise with 30 dB of SNR is added to the mixtures in both cases.

We perform separation by the Adam modification of the NG InfoMax algorithm, with gradient in (6.7). We use a block length of $B = 30$ samples, while the other parameters are set as: $N_{ep} = 100$, $\beta_1 = 0.9$, $\beta_2 = 0.999$, $\varepsilon = 10^{-8}$, $\eta = 0.001$ and the learning rate of the standard InfoMax and the Momentum InfoMax to $\mu = 0.001$. Performances in terms of the PI, for the second and third experiments, are reported in Fig. 6.3a, b, respectively, that clearly show also in these cases the effectiveness of the proposed idea. In particular, Fig. 6.3b confirms that the separation obtained by using the Adam InfoMax algorithm in the third experiment is quite satisfactory, while the standard and the Momentum InfoMax give worse solutions.

Finally, a last experiment is performed on real data. We used an EEG signal recorded according the 10–20 system, consisting of 19 signals with artifacts. ICA is a common approach to deal with the problem of artifact removal from EEG [11]. We use a block length of $B = 30$ samples, while the other parameters are set as: $N_{ep} = 270$, $\beta_1 = 0.9$, $\beta_2 = 0.999$, $\varepsilon = 10^{-8}$, $\eta = 0.01$ and the learning rate of the standard InfoMax and the Momentum InfoMax to $\mu = 10^{-6}$. Since we used real data and the mixing matrix **A** is not available, the PI cannot be evaluated. Hence,

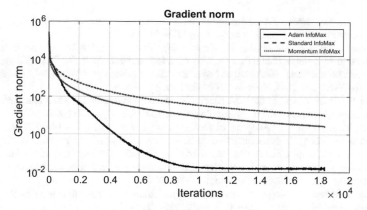

Fig. 6.4 Norm of the gradient of the cost function in the fourth proposed experiment

we decided to evaluate the performance by the norm of the gradient of the cost function. As it can be seen from Fig. 6.4, also in this case the Adam InfoMax algorithm achieves better results in a smaller number of iterations with respect to the compared algorithms.

6.6 Conclusions

In this paper a modified InfoMax algorithm for the blind separation of independent sources, in a linear and instantaneous environment, has been introduced. The proposed approach is based on a novel and advanced stochastic optimization method known as Adam and it can benefit from the excellent properties of the Adam approach. In particular, it is easy to implement, computationally efficient, and it is well suited when the number of sources is high and bad-scaled, the mixing matrix is close to be ill-conditioned and some additive noise is considered. Some experimental results, evaluated in terms of the Amari Performance Index and compared with other state-of-the-art approaches, have shown the effectiveness of the proposed approach.

References

1. Amari, S.: Natural gradient works efficiently in learning. Neural Comput. **10**(3), 251–276 (1998)
2. Araki, S., Mukai, R., Makino, S., Nishikawa, T., Saruwatari, H.: The fundamental limitation of frequency domain blind source separation for convolutive mixtures of speech. IEEE Trans. Speech Audio Process. **11**(2), 109–116 (2003)

3. Bell, A.J., Sejnowski, T.J.: An information-maximisation approach to blind separation and blind deconvolution. Neural Comput. **7**(6), 1129–1159 (1995)
4. Boulmezaoud, T.Z., El Rhabi, M., Fenniri, H., Moreau, E.: On convolutive blind source separation in a noisy context and a total variation regularization. In: Proceedings of IEEE Eleventh International Workshop on Signal Processing Advances in Wireless Communications (SPAWC2010), pp. 1–5. Marrakech (20–23 June 2010)
5. Choi, S., Cichocki, A., Park, H.M., Lee, S.Y.: Blind source separation and independent component analysis: a review. Neural Inf. Process. Lett. Rev. **6**(1), 1–57 (2005)
6. Cichocki, A., Amari, S.: Adaptive Blind Signal and Image Processing. Wiley (2002)
7. Comon, P., Jutten, C. (eds.): Handbook of Blind Source Separation. Springer (2010)
8. Douglas, S.C., Gupta, M.: Scaled natural gradient algorithm for instantaneous and convolutive blind source separation. In: IEEE International Conference on Acoustics, Speech and Signal Processing (ICASSP2007), vol. 2, pp. 637–640 (2007)
9. Duchi, J., Hazan, E., Singer, Y.: Adaptive subgradient methods for online learning and stochastic optimization. J. Mach. Learn. Res. **12**(7), 2121–2159 (2011)
10. Haykin, S. (ed.): Unsupervised Adaptive Filtering, vol. 2: Blind Source Separation. Wiley (2000)
11. Inuso, G., La Foresta, F., Mammone, N., Morabito, F.C.: Wavelet-ICA methodology for efficient artifact removal from electroencephalographic recordings. In: Proceedings of International Joint Conference on Neural Networks (IJCNN2007)
12. Kingma, D.P., Ba, J.L.: Adam: a method for stochastic optimization. In: International Conference on Learning Representations (ICLR2015), pp. 1–13 (2015). arXiv:1412.6980
13. Liu, J.Q., Feng, D.Z., Zhang, W.W.: Adaptive improved natural gradient algorithm for blind source separation. Neural Comput. **21**(3), 872–889 (2009)
14. Papoulis, A.: Probability, Random Variables and Stochastic Processes. McGraw-Hill (1991)
15. Pascanu, R., Bengio, Y.: Revisiting natural gradient for deep networks. In: International Conference on Learning Representations (April 2014)
16. Scarpiniti, M., Vigliano, D., Parisi, R., Uncini, A.: Generalized splitting functions for blind separation of complex signals. Neurocomputing **71**(10–12), 2245–2270 (2008)
17. Smaragdis, P.: Blind separation of convolved mixtures in the frequency domain. Neurocomputing **22**(21–34) (1998)
18. Thomas, P., Allen, G., August, N.: Step-size control in blind source separation. In: International Workshop on Independent Component Analysis and Blind Source Separation, pp. 509–514 (2000)
19. Tieleman, T., Hinton, G.: Lecture 6.5—RMSProp. Technical report, COURSERA: Neural Networks for Machine Learning (2012)
20. Vigliano, D., Scarpiniti, M., Parisi, R., Uncini, A.: Flexible nonlinear blind signal separation in the complex domain. Int. J. Neural Syst. **18**(2), 105–122 (2008)

Part III
ANN Applications

Chapter 7
Depth-Based Hand Pose Recognizer Using Learning Vector Quantization

Domenico De Felice and Francesco Camastra

Abstract The paper describes a depth-based hand pose recognizer by means of a *Learning Vector Quantization* (*LVQ*) classifier. The hand pose recognizer is composed of three modules. The first module segments the scene isolating the hand. The second one carries out the feature extraction, representing the hand by a set of 8 features. The third module, the classifier, is a LVQ. The recognizer, tested on a dataset of 6500 hand poses, carried out by people of different sex and physical aspect, has shown an accuracy larger than 99% recognition rate. The hand pose recognizer accuracy is among highest presented in literature for hand pose recognition.

7.1 Introduction

Humans use gesture as a means for transmitting information. According to Kendon [1], *the information amount conveyed by gesture increases when the information quantity sent by the human voice decreases*. Gestures can be divided in two big families: dynamic gestures and static gestures or *hand poses*. This work aims to develop a hand pose recognizer based on a *Kinect* sensor [2] and a *Learning Vector Quantization* (*LVQ*) classifier. The recognizer is formed by three modules. The first one, whose input is the depth image acquired by Kinect, isolates the hand pose in the scene. The second module is a feature extractor and it represents the hand pose by a eight-dimensional feature vector. The third module, the classifier, is LVQ.

The paper is organized as follows: the hand pose recognizer is presented in Sect. 7.2.1; Sect. 7.3 reports some experimental results; in Sect. 7.4 some conclusions are drawn.

D. De Felice · F. Camastra (✉)
Department of Science and Technology, University of Naples Parthenope,
Centro Direzionale Isola C4, 80143 Naples, Italy
e-mail: camastra@ieee.org

D. De Felice
e-mail: domenico.defelice@alice.it

© Springer International Publishing AG 2018   69
A. Esposito et al. (eds.), *Multidisciplinary Approaches to Neural Computing*,
Smart Innovation, Systems and Technologies 69,
DOI 10.1007/978-3-319-56904-8_7

7.2 The Hand Pose Recognizer

The hand pose recognizer is composed of three components, the segmentation module, the feature extractor and the classifier.

7.2.1 Segmentation Module

The first component of the hand pose recognizer is the *segmentation* module. It receives, as input, the depth image acquired by the Kinect and carries out the segmentation procedure isolating the hand image in the scene. The segmentation procedure has three different phases. During the first phase, the image of person body is identified in the scene, using a thresholding algorithm, and the background of the scene is removed. In the second phase the hand image in the person body is isolated using a K-means-based [3] clustering approach and finally, the depth image is transformed in a binary image containing only the hand image (see Fig. 7.1b). In the last phase, in order to remove the morphological noise, the hand image is gone through a *morphological opening* followed by a *morphological closure* [4]. Both morphological operations use as structuring element the circle of radius three and area of 37 pixels, showed in Fig. 7.2.

7.2.2 Feature Extractor

The second component of the hand pose recognizer is the feature extractor. It receives, as input, the binary image containing the hand and performs the feature

(a)   **(b)**

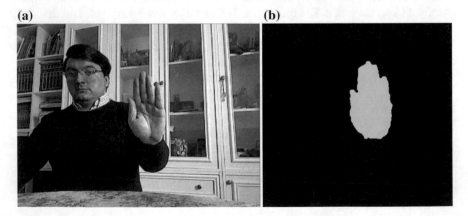

Fig. 7.1 The original image (**a**). The image after the segmentation process (**b**). The person in **a** is the first author paper, who gives his consent to show his image in the paper

Fig. 7.2 The structuring element, in *grey*, used in the morphological opening and closure. The *center* of the element is denoted by a cross

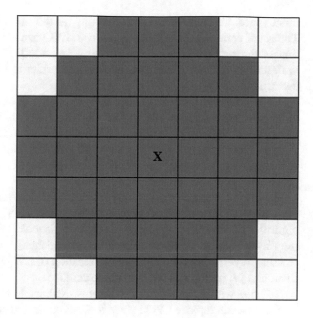

extraction procedure, representing, in this way, the hand image by means of eight numerical features. The feature extraction procedure has the following stages. The first stage computes the first seven features, represented by the respective *Hu invariant moments* [5] of the binary hand image. In the second stage the *convex hull*[1] of the hand image is computed using Slansky's algorithm [6]. In the third stage, the eighth feature is extracted computing the number of *convexity defects*[2] [7] in the convex hull by *OpenCV* library [8]. Finally, in the last and four stage, all features are normalized. It is worth to observe that the extracted features, by construction, are invariant by rotation and translation in the plane of Kinect camera.

7.2.3 The Classifier

The classifier is the third and last component of the hand pose recognizer. *Learning Vector Quantization (LVQ)* [9] has been used as classifier since it requires moderate computational resources and, therefore it results very suitable for a real-time implementation of the hand pose recognizer. Moreover, LVQ efficacy, as classifier, is testified by its usage in many and disparate domains such as the classification of nonstationary power signals [10], cursive characters [11], and arrythmias [12] and the gesture recognition [13]. LVQ is a supervised vector quantizer that has the aim of representing the input data by means of a much smaller number of vectors in the

---

[1]The convex hull of an image is the minimum polygon enclosing the image itself.

[2]A convexity defect is a point of the image contour where the contour is not convex anymore.

input space, i.e., *codevectors*, minimizing, at the same time, the misclassification. The set of codevectors is called *codebook*. LVQ resides in the consecutive application of three diverse learning algorithms, i.e., LVQ1, LVQ2 and LVQ3. We limit to describe them, in general, remanding the reader to [9, 14] for a comprehensive description.

LVQ training is formed by two steps. The former consists in applying LVQ1, the latter in using LVQ2, or alternatively LVQ3. LVQ1 is the first learning algorithm to be used and adopts for classification the winner-takes-all rule. It compares the class of the input vector with the class of the closest codevector. If the classes are the same, the codevector is approached to the input vector, otherwise it is pushed away. No rule is applied to the rest of codebook.

However, the continual application of LVQ1 rule pushes the codebook vectors away from the *Bayes' rule* optimal decision surfaces [15], therefore a further learning algorithm, LVQ2, must be applied to the codebook. LVQ2 approximates roughly the Bayes' rule by pairwise arrangement of codebook vectors belonging to adjacent classes. Nevertheless, the repeated application of LVQ2 rule may generate instable dynamics [9] in the codebook, in some cases.

To cope with these stability problems, Kohonen proposed a further algorithm, LVQ3, that can replace LVQ2 since it is not subject to stability problems.

In practical applications, it is desiderable that the recognition of a hand pose, is followed by the performing of a given action, e.g., the door opening or the the lamp lightning in a domotic house. In this applicative scenario, the classifier must classify a hand pose, only when the misclassification probability is negligible. When the misclassification probability is significant, the classifier must *reject* the hand pose, namely it does not classify. In the classifier a rejection scheme is implemented as follows.

Let d_E be the Euclidean distance between the input and the closest codevector, the following rule is applied:

$$\text{If } d_E \leq \theta \text{ then classify else reject} \tag{7.1}$$

where θ is a parameter, that has to be fixed properly, for managing the trade-off between error and rejection.

7.3 Experimental Results

In the experimental validation of the hand pose recognizer it has been considered the same 13 hand poses, used in [16]. The handposes, invariant by translation and rotation and translation, are shown in Fig. 7.3. A handpose database was constructed by performing each pose for 1000 times by people of different sex and constitution. The database was divided randomly in training and test set. Each set had the same cardinality, i.e., 6500.

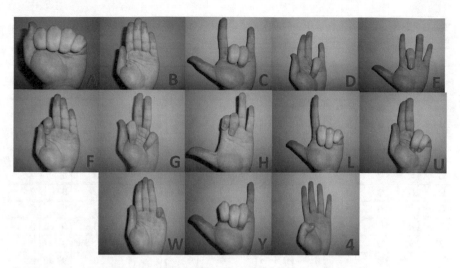

Fig. 7.3 Hand poses represented in the database

Table 7.1 Accuracy of LVQ classifiers on the test set. In knn, the number of neighbors k was fixed, by crossvalidation to 3

Algorithm	Accuracy (%)	Error (%)
knn	91.94	8.06
LVQ1	99.63	0.37
LVQ1+LVQ2	99.68	0.32
LVQ1+LVQ3	**99.68**	0.32

In the validation several LVQs were trained by using diverse set of training parameters and diverse number of codevectors. The number of classes was set equal to the number of the different handposes, i.e., 13, present in the dataset. The best LVQ net was chosen by *cross-validation* [17]. The number of classes used in the experiments was 13, namely the number of the different hand poses in our database. LVQ trainings were carried out by using *LVQ-pak* [18] software library. Table 7.1 reports the classifier performances on the test set, measured in terms of accuracy. The best result in terms of accuracy was 99.68% and it is among highest presented in literature for hand pose recognition. Table 7.2 presents the confusion matrix for LVQ1+LVQ3 classifier.

Then, the rejection rule (see Sect. 7.2.3) was applied to the best classifier, namely LVQ1+LVQ3. Table 7.3 shows the results achieved for different values of the rejection threshold θ. It can observe that the hand pose recognizer, in presence of a negligible error, i.e., lower than 0.2%, can still provide an accuracy bigger than 90%. The hand pose recognizer, implemented in C++, needs 62 CPU ms to recognize a single hand pose on a PC with i5-2410M Processor 2.30 GHz, 6 GB RAM and 64 bit Windows 10 Microsoft operating system.

Table 7.2 The confusion matrix of LVQ1+LVQ3 classifier, without rejection, on the test set. The values are expressed in terms of percentage rates

	A	B	C	D	E	F	G	H	L	U	W	Y	4
A	100	0	0	0	0	0	0	0	0	0	0	0	0
B	0.40	99.60	0	0	0	0	0	0	0	0	0	0	0
C	0	0	100	0	0	0	0	0	0	0	0	0	0
D	0	0	0	99.60	0	0.40	0	0	0	0	0	0	0
E	0	0	0	0	100	0	0	0	0	0	0	0	0
F	0	0	0	0.40	0	99.60	0	0	0	0	0	0	0
G	0	0	0	0.20	0	0.60	99.20	0	0	0	0	0	0
H	0	0	0	0	0	0	0	100	0	0	0	0	0
L	0	0	0	0	0	0	0	0	100	0	0	0	0
U	0	0	0	0	0	0	0	0	1.20	98.80	0	0	0
W	0	0	0	0.6	0	0	0	0	0	0	99.40	0	0
Y	0.20	0	0	0	0	0	0	0	0	0	0	99.80	0
4	0	0	0	0	0	0	0	0	0.20	0	0	0	99.80

Table 7.3 Accuracy, Error and Reject rates on the test set of LVQ1+LVQ3 for different rejection threshold θ

θ	Accuracy (%)	Error (%)	Reject (%)
1	99.68	0.32	0.00
0.60	99.68	0.31	0.01
0.40	98.18	0.26	1.56
0.20	90.89	0.17	8.94
0.15	73.36	0.06	26.58
0.112	47.00	0	53.00

7.4 Conclusions

The paper has presented a hand pose recognition system based on a Kinect sensor and a Learning Vector Quantization. The hand pose recognizer is composed of three components. The first component isolates the hand image in the scene. The second component carries out the feature extraction representing the hand pose image by 8 features, invariant by translation and rotation in the sensor camera plane. The third component is Learning Vector Quantization that performs the hand pose classification. The hand pose recognizer, tested on a dataset of 6500 hand poses, performed by people of different sex and physical aspect, has shown an accuracy bigger than 99.5%: This accuracy is among highest presented in literature for hand pose recognition. The system needs an average time of 62 CPU ms for recognizing a single hand pose. In the next future, we plan to investigate its usage in ambient assist living as a tool for helping disabled and aged people.

Acknowledgements Domenico De Felice developed part of the work, during his M.Sc. thesis in Computer Science at University of Naples Parthenope, with the supervision of Francesco Camastra. The research was funded by *Sostegno alla ricerca individuale per il triennio 2015–17* project of University of Naples Parthenope.

References

1. Kendon, A.: How gestures can become like words. In: Crosscultural Perspectives in Nonverbal Communication, Toronto, Hogrefe, pp. 131–141 (1988)
2. Zhang, Z.: Microsoft kinect sensor and its effect. IEEE Trans. Multimedia **19**(2), 4–10 (2012)
3. Lloyd, S.: An algorithm for vector quantizer design. IEEE Trans. Commun. **28**(1), 84–95 (1982)
4. Gonzales, R., Woods, R.: Digital Image Processing, 3rd edn. Pearson/Prentice-Hall, Upper Saddle River (2008)
5. Hu, M.K.: Visual pattern recognition by moment invariants. IRE Trans. Inf. Theory **8**(2), 179–187 (1962)
6. Slansky, J.: Finding the convex hull of a simple polygon. Pattern Recogn. Lett. **1**(2), 79–83 (1982)
7. Homma, K., Takenaka, E.: An image processing method for feature extraction of space-occupying lesions. J. Nucl. Med. **26**, 1472–1477 (1985)
8. Bradski, G., Kaehler, A.: Learning OpenCV: Computer Vision with the OpenCV Library. O'Reilly, Cambridge (USA) (2008)
9. Kohonen, T.: Learning vector quantization. In: The Handbook of Brain Theory and Neural Networks, pp. 537–540. MIT Press (1995)
10. Biswal, B., Biswal, M., Hasan, S., Dash, P.: Nonstationary power signal time series data classification using LVQ classifier. Appl. Soft Comput. **18**, 158–166 (2014)
11. Camastra, F., Vinciarelli, A.: Cursive character recognition by learning vector quantization. Pattern Recogn. Lett. **22**(6–7), 625–629 (2001)
12. Melin, P., Amezcua, J., Valdez, F., Castillo, O.: A new neural network model based on the LVQ algorithm for multi-class classification of arrhythmias. Inf. Sci. **279**, 483–497 (2014)
13. Lamberti, L., Camastra, F.: Real-time hand gesture recognition using a color glove. In: Proceedings of the 16th International Conference on Image Analysis and Processing, ICIAP 2011, pp. 365–373. Springer (2011)
14. Lamberti, L., Camastra, F.: Handy: a real-time three color glove-based gesture recognizer with learning vector quantization. Expert Syst. Appl. **39**(12), 10489–10494 (2012)
15. Duda, R., Hart, P., Stork, D.: Pattern Classification. Wiley, New York (2001)
16. Camastra, F., De Felice, D.: LVQ-based hand gesture recognition using a data glove. Smart Innov. Syst. Technol. **19**, 159–168 (2013)
17. Hastie, T., Tibshirani, R., Friedman, R.: The Elements of Statistical Learning, 2nd edn. Springer (2009)
18. Kohonen, T., Hynninen, J., Kangas, J., Laaksonen, J., Torkkola, K.: LVQ-PAK: the learning vector quantization program package. Technical Report A30, Helsinki University of Technology, Laboratory of Computer and Information Science (1996)

Chapter 8
Correlation Dimension-Based Recognition of Simple Juggling Movements

Francesco Camastra, Francesco Esposito and Antonino Staiano

Abstract The last decade of technological development has given raise to a myriad of new sensing devices able to measure in many ways the movements of human arms. Consequently, the number of applications in human health, robotics, virtual reality and gaming, involving the automatic recognition of the arm movements, has notably increased. The aim of this paper is to recognise the arm movements performed by jugglers during their exercises with three and four balls, on the basis of few information on the arm orientation given by Euler Angles, measured with a cheap sensor. The recognition is obtained through a linear Support Vector Machine after a feature extraction phase in which the reconstruction of the system dynamics is performed, thus estimating three *Correlation Dimensions*, corresponding to Euler Angles. The effectiveness of the proposed system is assessed through several experimentations.

8.1 Introduction

Several social deep impact application domains, e.g., human health, robot design, video games and virtual reality, just to name a few [15], involve the study of human movements. Nowadays, thanks to technology, there is a plethora of sensing devices for measuring and analyzing the human movements [2, 10]. Thus, an ever increasing number of applications, in particular in the e-health domain, have been developing, exploiting several kind of sensors (e.g., body sensor or wireless sensor networks). For instance, they are devoted to the recognition of elderly activities in ambient assisted living by using etherogeneous machine learning techniques [3, 6, 7, 20], involving

F. Camastra · F. Esposito · A. Staiano (✉)
Department of Science and Technology, University of Naples "Parthenope",
Isola C4, Centro Direzionale, 80143 Napoli, NA, Italy
e-mail: staiano@ieee.org

F. Camastra
e-mail: camastra@ieee.org

F. Esposito
e-mail: francescoesposito7@gmail.com

© Springer International Publishing AG 2018
A. Esposito et al. (eds.), *Multidisciplinary Approaches to Neural Computing*,
Smart Innovation, Systems and Technologies 69,
DOI 10.1007/978-3-319-56904-8_8

generally simple and cheap sensors [11]. In this context, our paper is aimed at recognizing simple arm movements basing upon a minimal set of parameters computed by a proper feature extraction procedure. This procedure uses just the measurements on the arm orientation, given by the Euler Angles, for reconstructing the underlying system dynamics generating the movement of the arm, in order to express it in terms of *Correlation Dimensions*, corresponding to the three Euler Angles of the arm. The latter information is then used by a classifier, i.e., a linear Support Vector Machine, in order to recognize the arm movement. To gather measurements of the Euler Angles of the arm, we exploited the exercises performed by jugglers whose wrist was equipped with a relatively tiny and cheap sensor. Juggling results particularly suitable since it has uses beyond hobby and entertainment. The use of juggling for scientific purposes is not new. Claude E. Shannon faced several studies on the subject and formulated some theorems [1]. More recently, a system, based on vision information, for classifying some movements, e.g., ball grasps [19], was presented. Juggling is a fruitful application domain to study and recognize human arm movement as will be discussed in the next sections.

To the best of our knowledge, there is only one work [18] where arm movement recognition is based on the orientation information of the arm. Nevertheless, the system uses a threshold-based IF-THEN classification algorithm, instead of the usual and most powerful machine learning techniques. Regarding to the specific case of juggling, there is only a previous work [19] that analyzes the dynamics of juggler's movement, using the only visual information. Therefore, this is the first work where the recognition of juggling movement is performed using the only information extracted from the reconstruction of the dynamics that has generated the movement. In the following, the paper is structured as follows: In Sect. 8.2, is introduced the definition of Correlation Dimension; In Sect. 8.3, the proposed approach is described, while in Sect. 8.4 the experiments performed are illustrated; Finally, conclusions are drawn in Sect. 8.5.

8.2   Correlation Dimension

Nonlinear Dynamics provides effective methods for the *model reconstruction* of the time series, namely for reconstructing the underlying dynamic system that yielded the time series.

Let $t(s)$, with $s = 1, \ldots, N$ be a time series, the *method of delays* [9] allows the model reconstruction of the time series in the following way. Starting from the time series, a dataset $S = \{T(s) : T(s) = [t(s), t(s-1), \ldots, t(s - M + 1)]\}$ in a M-dimensional space can be created. The *Takens-Mañé embedding theorem* [17, 21] states that if M is large, the manifold $\mathcal{T}$, so generated, is a faithful reconstruction of the attractor $\mathcal{U}$ of the dynamic system that created the time series. Hence the study of $\mathcal{T}$ allows retrieving all information on $\mathcal{U}$.

Fig. 8.1 The log-log plot on
α-Euler Angle Time series

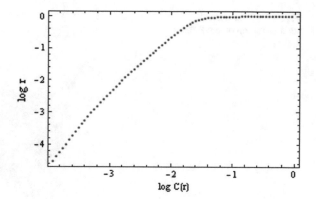

The dimension is the most important geometrical characteristic of an attractor. However, there is not an univocal definition of the dimension [5]. In the rest of the paper, the *Correlation Dimension* is used for defining the attractor dimension, since it can be easily computed.

The *Correlation Dimension* [12] of set S is defined in the following way. The *Correlation Integral* $C(\rho)$ of set S is given by

$$C(\rho) = \lim_{P \to \infty} \frac{2}{P(P-1)} \sum_{i=1}^{P} \sum_{j=i+1}^{P} \Theta(\|T_j - T_i\| \leq \rho) \qquad (8.1)$$

where $P = N - M$ and Θ is the Heaviside function that is 1 if the condition is fulfilled, 0 otherwise.

The Correlation Dimension D of S is the limit, if exists, of the Correlation Integral, namely:

$$D = \lim_{\rho \to 0} \frac{\ln(C(\rho))}{\ln(\rho)} \qquad (8.2)$$

The most popular method to estimate Correlation Dimension is the *Grassberger-Procaccia algorithm* [12]. As shown in Fig. 8.1, the algorithm resides in plotting $\ln(C(\rho))$ versus $\ln(\rho)$; the value of Correlation Dimension is given by the slope of the linear part of the curve. For more details on the algorithm and its software implementation, the reader can refer to [4].

8.3   Juggler's Arm Movement Recognition

The recognition of juggling movements is founded on the assumption that it can classify the movements extracting the single orientation information of a juggler arm. In physics, any orientation can be represented by composing the three elemental rotations, called *Euler Angles* [16], indicated by α, β, γ (see Fig. 8.2). They provide a

Fig. 8.2 Geometrical
representation of the Euler
Angles

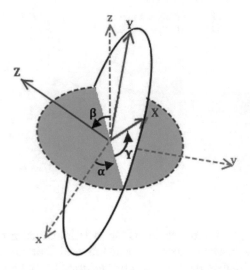

sequence of three elemental rotations to the coordinate system axes. However, Euler
angle orientation representation suffers by the so-called *gymbel lock*, i.e., in some
cases the third rotation cannot be performed after having carried out the first two.
For this reason, some recent trackers, e.g., Kinect [23], prefer to represent the rigid
body orientation in terms of *quaternions* [22].

For measuring the arm orientation we asked the juggler, in his performance, to put
on his wrist the *Colibri* inertial tracker by Trivisio Prototyping GmbH. The model
of the tracker, shown in Fig. 8.3, represents the body orientation in term of Euler
Angles, with a refresh time of 5 ms.

Hence, the juggler, during his exercise, generates three different time series, each
measuring an Euler Angle, sampled every 5 ms.

8.3.1   Feature Extraction and Classification Phases

The input of feature extraction procedure are the α-Euler, β-Euler, γ-Euler time series
generated by the tracker. The output are the Dimension correlations each associated
to a single Euler time series. Each juggling movement, considered in the experimen-
tation, has a duration of 30 s and therefore the tracker produces for each Euler Angle
6000 samples.

For each time series, the feature extraction procedure carries out the dynamics
reconstruction by means of the method of delays (see Sect. 8.2). Then, it is computed,
as output, the Correlation Dimension for each time series, namely $\mathcal{D}^{\alpha}, \mathcal{D}^{\beta}, \mathcal{D}^{\gamma}$.

As shown in Fig. 8.4, the feature representation in terms of Correlation Dimensions produces a linear separation between data. Hence, performing the classification
by the use of a linear classifier, e.g., a Linear Support Vector Machine (*Linear SVM*)
[8, 14] is quite appropriate.

Fig. 8.3 The Trivisio
Colibri

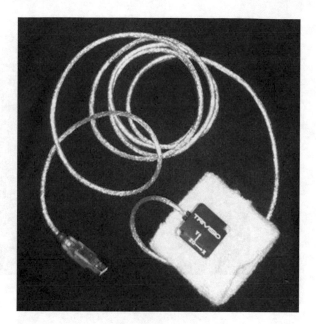

Fig. 8.4 Juggler
Movements represented by
α, β, γ Euler Angles
Correlation Dimensions. The
blue and the *purple circles*
indicate juggler movements
with three and four balls,
respectively. The Correlation
Dimensions, $\mathcal{D}^{\alpha}, \mathcal{D}^{\beta}, \mathcal{D}^{\gamma}$,
associated to Euler Angles
α, β, γ are denoted by x, y, z,
respectively

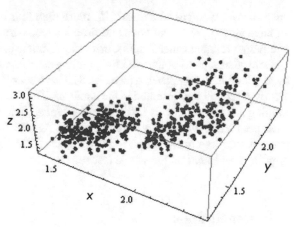

8.4 Experimental Results

For the validation of the proposed system, two diverse elementary juggling movements have been considered. To this purpose, it is necessary to remark that in juggling, elementary movements involving both hands, can be classified in two groups,
related to the number of balls involved in the exercise. If the movement involves
an odd number of balls, it belongs to the former category; otherwise to the latter.
Therefore, the proposed system has been validated considering juggling movements
with three and four balls. Exercises with two balls have been discarded involving

Fig. 8.5 Juggling movements with four balls

only an arm; whereas, the ones with more than four balls cannot be performed by an amateur juggler for an adequate time for reconstructing juggling dynamics and computing reliable Euler Angle Correlation Dimension estimates.

Being that said, a database of 500 juggling exercises of duration of 30 s, performed by jugglers[1] of different being (see Fig. 8.5) have been collected. Exercises with 3 and 4 balls are equally represented in the database. The database was randomly divided in training and test sets, with 150 and 350 exercises, respectively. Many Linear SVMs have been trained by using different values for SVM parameters and the best one was selected by cross-validation [13]. Linear SVM has misclassified just a single juggling movement on the whole test set.

8.5   Conclusions

The paper has described a system for recognizing the juggling movements by using the single arm orientation information expressed in terms of Euler Angles. The proposed approach implies that each juggler wears on his wrist a tracker that expresses the juggler arm orientation, during the exercise, in terms of three Euler Angle time series. The proposed system has two stages, i.e., the feature extractor and the classifier. The first module uses the method of delay for the dynamic system reconstruction that has produced the arm movements and then, computes three Euler Angle

---

[1]The juggler in the figure is Francesco Esposito, the second paper author, who gives his consent to show his image in the paper.

Correlation Dimensions. In the second stage, classification is performed by a Linear SVM. Experimental results show that the system can discriminate, making only one mistake, juggling exercises with three balls from the ones with four balls.

In the next future we plan to apply the system in the rehabilitation, for characterizing the movement dynamics, of people suffered by strokes.

Acknowledgements The research was developed when Francesco Esposito was at the Department of Science and Technology, University of Naples Parthenope, as B. Sc. student in Computer Science. Francesco Camastra and Antonino Staiano were funded by *Sostegno alla ricerca individuale per il triennio 2015–17* project of University of Naples Parthenope. This study was exempt from ethical approval procedures since it involved health subjects which volunteered their participation. Informed consents were signed by each participant after they were debriefed on the experimental protocol, the aims of the study and the procedures.

References

1. Beek, P.J., Lewbel, A.: The science of juggling. Sci. Am. **273**(5), 92–97 (1995)
2. Biswas, D., Cranny, A., Gupta, N., Maharatna, K., Achner, J., Klemke, J., Jöbges, M., Ortmann, S.: Recognizing upper limb movements with wrist worn inertial sensors using k-means clustering classification. Hum. Mov. Sci. **40**, 59–76 (2015)
3. Botia, J., Villa, A., Palma, J.: Ambient assisted living system for in-home monitoring of healthy independent elders. Expert Syst. Appl. **39**(9), 8136–8148 (2012)
4. Camastra, F., Esposito, F., Staiano, A.: Linear SVM-based recognition of elementary juggling movements using correlation dimension of Euler angles of a single arm. Neural Comput. Appl. 1–9 (to appear)
5. Camastra, F., Staiano, A.: Intrinsic dimension estimation: advances and open problems. Inf. Sci. **328**, 26–41 (2016)
6. Chernbumroong, S., Cang, S., Atkins, A., Yu, H.: Elderly activities recognition and classification for applications in assisted living. Expert Syst. Appl. **40**(5), 1662–1674 (2013)
7. Chernbumroong, S., Cang, S., Yu, H.: A practical multi-sensor activity recognition system for home-based care. Decis. Support Syst. **66**, 61–70 (2014)
8. Cortes, C., Vapnik, V.: Support vector networks. Mach. Learn. **20**, 273–297 (1995)
9. Eckmann, J.P., Ruelle, D.: Ergodic theory of chaos and strange attractors. Rev. Modern Phys. **57**(3), 617–656 (1985)
10. Field, M., Stirling, D., Pan, Z., Ros, M., Naghdy, F.: Recognizing human motions through mixture modeling of inertial data. Pattern Recogn. **48**(8), 2394–2406 (2015)
11. Fuentes, D., Gonzalez-Abril, L., Angulo, C., Ortega, J.: Online motion recognition using an accelerometer in a mobile device. Expert Syst. Appl. **39**(3), 2461–2465 (2012)
12. Grassberger, P., Procaccia, I.: Measuring the strangeness of strange attractors. Phys. D **9**, 189–208 (1983)
13. Hastie, T., Tibshirani, R., Friedman, R.: The Elements of Statistical Learning. Springer (2001)
14. Joachim, T.: Making large-scale SVM learning practical. In: Advances in Kernel Methods-Support Vector Learning, pp. 169–184. MIT Press (1999)
15. Kober, J., Graham, R., Mistry, M.: Playing catch and juggling with a humanoid robot. In: Proceedings IEEE-RAS International Conference on Humanoid Robots, pp. 875–881. IEEE (2012)
16. Landau, L., Lifshitz, E.M.: Course in Theoretical Physics, vol. 1. Butterworth-Heinemann, Mechanics (1996)
17. Mañé, R.: On the dimension of compact invariant sets of certain nonlinear maps. In: Dynamical Systems and Turbulence, Warwick 1980, pp. 230–242. Springer (1981)

18. Mazomenos, E.B., Biswas, D., Cranny, A., Rajan, A., Maharatna, K., Achner, J., Klemke, J., Jöbges, M., Ortmann, S., Langendörfer, P.: Detecting elementary arm movements by tracking upper limb joint angles with MARG sensors. IEEE J. Biomed. Health Inf. **20**(4), 1088–1099 (2016)
19. North, B., Blake, A., Isard, M., Rittscher, J.: Learning and classification of complex dynamics. IEEE Trans. Pattern Anal. Mach. Intell. **22**(9), 1016–1034 (2000)
20. Ordóõez, F., Iglesias, J., de Toledo, P., Ledezma, A., Sanchis, A.: Online activity recognition using evolving classifiers. Expert Syst. Appl. **40**(4), 1248–1255 (2013)
21. Takens, F.: Detecting strange attractors in turbulence. In: Dynamical Systems and Turbulence, Warwick 1980, pp. 366–381. Springer (1981)
22. Zhang, F.: Quaternions and matrices of quaternions. Linear Algebra Appl. **251**, 21–57 (1997)
23. Zhang, Z.: Microsoft kinect sensor and its effect. IEEE Trans. Multimedia **19**(2), 4–10 (2012)

Chapter 9
Cortical Phase Transitions as an Effect of Topology of Neural Network

Ilenia Apicella, Silvia Scarpetta and Antonio de Candia

Abstract Understanding the emerging of cortical dynamical state, its functional role, and its relationship with network topology, is one of the most interesting open questions in computational neuroscience. Spontaneous cortical dynamics often shows spontaneous fluctuations with UP/DOWN alternations and critical avalanches which resemble the critical fluctuations of a system posed near a non-equilibrium noise-induced phase transition. A model with structured connectivity and dynamical attractors has been shown to sustain two different dynamic states and a phase transition with critical behaviour is observed. We investigate here which are the features of the connectivity which permit the emergence of the phase transition and the large fluctuations near the critical line. We start from the original connectivity, that comes from the learning of the spatiotemporal patterns, and we shuffle the presynaptic units, leaving unchanged both the postsynaptic units and the value of the connections. The original structured network has a large clustering coefficient, since it has more directed connections which cooperate to activate a precise order of neurons, respect to randomized network. When we shuffle the connections we reduce the clustering coefficient and we destroy the spatiotemporal pattern attractors. We observe that the phase transition is gradually destroyed when we increase the ratio of shuffled connections, and already at a shuffling ratio of 70% both the phase transition and its critical features disappear.

I. Apicella · S. Scarpetta (✉)
Department of Physics "E.R. Caianiello", University of Salerno, Fisciano, SA, Italy
e-mail: sscarpetta@unisa.it

S. Scarpetta
INFN Salerno Group, Fisciano, SA, Italy

A. de Candia
Department of Physics, University of Naples "Federico II", Naples, Italy

A. de Candia
INFN Naples Branch, Complesso Universitario Monte S. Angelo, Naples, Italy

A. de Candia
CNR-SPIN, Naples Branch, Naples, Italy

© Springer International Publishing AG 2018
A. Esposito et al. (eds.), *Multidisciplinary Approaches to Neural Computing*,
Smart Innovation, Systems and Technologies 69,
DOI 10.1007/978-3-319-56904-8_9

9.1 Introduction

Thanks to recent experimental techniques, which allow to record the activity of many neurons simultaneously (both in-vivo and in-vitro), it is easier studying the complex collective dynamics emerging in highly connected networks of neurons, such as cortical networks. Spontaneous cortical activity can show critical collective features, such as the alternation between DOWN states of network quiescence and UP states of neural depolarization, observed in different system and conditions (both in-vitro [1–3] and in-vivo during slow-wave sleep, anesthesia and quiet walking [4, 5]).

Many recent works confirm the idea that brain operates close to a critical point (or close to a spinodal point), at which information progressing is optimized [6–8], so that it is interesting to investigate the role of criticality on cognitive activities or brain diseases and so on.

In this paper we focus on the presence of phase transition between UP and DOWN states; it should be important to emphasize that the power laws of avalanches size and duration distributions in our model has been shown to agree with experimental data of critical exponents of size and time avalanches distributions [9].

Many experiments both in-vitro [10, 11] and in-vivo [12–15] have demonstrated that cortical spontaneous activity occurs in precise spatio-temporal patterns. In this paper we study the spontaneous cortical dynamics of a neural network, in which a phase transition between replay and not-replay of stored spatiotemporal patterns emerges.

We call "UP state" the regime in which we have high firing rate with the replay of one of stored patterns, index of high correlated activity. Instead the "DOWN state" is the regime of quiescence without replay of pattern. Between this two states, a critical regime exists, with the alternation of UP and DOWN states. In this regime there is an intermitted replay of spatiotemporal pattern.

Noise level and strength of connections are the control parameters we change during the investigation, then we calculate firing rate and normalized variance, defined in next section.

As we shall see later an high firing rate doesn't necessarily imply an UP state, because if we change the topology of network we don't always observe retrieval pattern when firing rate is high and we can have uncorrelated Poissonian activity even at high rates. This change of topology consists of shuffling the connections thanks to a shuffling procedure discussing below. This procedure allows us to study the spontaneous dynamics of network with different position of connections, keeping unchanged their values, getting very different and interesting regimes of activity.

9.2 The Model

In order to simulate the spontaneous activity of a slice of brain cortex, we have a network of N spiking neurons, modeled as LIF (Leaky Integrate-and-Fire) units and represented by SRM (Spike Response Model) of Gerstner [16], in presence of a

Poissonian noise distribution. Neurons are connected by a sparse connectivity with the possibility to shuffle a fraction of the connections, in order to understand the role of topology in spontaneous cortical dynamics.

If we label with index i the postsynaptic neuron and with the index j the presynaptic one, when the neuron i does not fire, the postsynaptic membrane potential is:

$$u_i(t) = \sum_j \sum_{t_j < t_j < t} J_{ij}(e^{-(t-t_j)/\tau_m} - e^{-(t-t_j)/\tau_s}) + \sum_{\hat{t}_i < \hat{t}_i < t} J_i(e^{-(t-\hat{t}_i)/\tau_m} - e^{-(t-\hat{t}_i)/\tau_s}) \quad (9.1)$$

The Eq. (9.1) has two contributions: the first one is related to the connections strength, because J_{ij} is the connection strength between pre- and postsynaptic neurons; the second contribution is related to the noise of network, because $\hat{J}_i$ is extracted from a Gaussian distribution with mean 0 and standard deviation $\sigma = \sqrt{\alpha/\rho \sum_j J_{ij}^2}$, where α is the "noise level" of the network and $\rho = 1$ ms $^{-1}$ is the rate of Poissonian distribution $P(t) \propto e^{-\rho t}$. In the Eq. (9.1) τ_m is the characteristic time of membrane (in this paper $\tau_m = 10$ ms), τ_s is the characteristic time of synapse (in this paper $\tau_s = 5$ ms), t_j are the spike times of neuron j, $\hat{t}_i$ are the times of noise events releasing a random charge at some point of membrane of neuron i.

The (9.1) is the solution of a differential equation, describing a RC circuit, because in LIF model each unit has a membrane capacity C and a resistance R in parallel, so that we have for neuron i:

$$\frac{du_i(t)}{dt} = -\frac{u_i(t)}{\tau_m} + \frac{I_i(t)}{C} \quad (9.2)$$

with $\tau_m = RC$ and $I_i(t)$ is the input current, given by $I_i(t) = \sum_j \sum_{t_j < t_j < t} \frac{Q_{ij}}{\tau_s} e^{-(t-t_j)/\tau_s} + \sum_{\hat{t}_i < \hat{t}_i < t} \frac{\hat{Q}_i}{\tau_s} e^{-(t-\hat{t}_i)/\tau_s}$, where Q_{ij} is the total charge released at the synapse between neuron i and j and $\hat{Q}_i$ is a random charge released at some point of the membrane of neuron i. Q_{ij} and $\hat{Q}_i$ are related to J_{ij} and $\hat{J}_i$ respectively by the relations: $J_{ij} = \frac{Q_{ij}}{C(1-\frac{\tau_s}{\tau_m})}$ and $\hat{J}_i = \frac{\hat{Q}_i}{C(1-\frac{\tau_s}{\tau_m})}$. When the membrane potential $u_i(t)$ reaches the threshold θ, the unit emits a spike and then $u_i(t)$ is reset to zero, its resting value.

So far we have described the single unit. Since we want to investigate the effects of topology on network dynamics we build a structured connectivity that gives rise to a complex dynamics with a rich phases space, and then we shuffle the connections to check if crucial changes happen also in the dynamics. In the next section we will talk about the process of creation of the network connectivity, thanks to the "learning" and "pruning" procedures, and then we will describe the shuffling procedure.

9.2.1 Learning and Pruning Procedures

We set synapse strengths J_{ij} at the beginning with the "learning" procedure, inspired by STDP (Spike Timing Dependent Plasticity). Then, during the simulation we hold fixed J_{ij}, i.e. we don't use short term plasticity for sake of simplicity. Note that the sign of J_{ij} represent the type of synapse: if $J_{ij} < 0$ the synapse is inhibitory, while if $J_{ij} > 0$ it is excitatory. With this procedure we store $\mu = 1, 2, \ldots, P$ phase-coded patterns in the network connections, i.e. periodic ordered trains of spikes t_i^{μ} with period T^{μ} and with one spike per neuron and per cycle. Because of such periodic spikes train, the strength of connection J_{ij} changes:

$$\delta J_{ij} = H_i \sum_{n=-\infty}^{\infty} A(t_i^{\mu} - t_j^{\mu} + nT^{\mu}) \tag{9.3}$$

where $A(\tau)$ is a function of time, called "learning window", inspired to STDP, with $\tau = t_i^{\mu} - t_j^{\mu} + nT^{\mu}$. t_j^{μ} and t_i^{μ} are pre- and postsynaptic spikes time in pattern μ respectively, H_i is a constant that sets the strength of the connections, depending on the postsynaptic neuron. This learning procedure assures the balance between excitation and inhibition, i.e. $\sum_i J_{ij} = 0$.

To take into account the heterogeneity of neurons, we use two values of H_i: H_0 for "normal" neurons and $H_i = 3H_0$ for "leader" neurons, i.e. neurons that with higher incoming connection strengths amplify activity initiated by noise. In other words, leaders are neurons the ones which fire more than others, and they give rise to a cue able to initiate the short collective replay. They are chosen as a fraction of 3% of neurons with consecutive phases, for each pattern μ.

To improve the model's biologically plausibility, we delete some connections to make the connectivity sparse. With the "pruning" procedure we cut a fraction f_{prune}^+ (70%) of positive (excitatory) connections with the lowest value and a fraction $f_{prune}^{-,i}$ (depending on postsynaptic neuron) of negative (inhibitory) connections with the lowest absolute value. As it happens before pruning, also after pruning still there is a balance of positive and negative connections affering each postsynaptic unit, i.e. $\sum_i J_{ij} = 0$. In this way, only a part of connections survives: 27% of $N(N-1)$ connections are negative, 12% of $N(N-1)$ connections are positive, the other ones are equal to zero. In such a way we get a structured and sparse connectivity.

9.2.2 Shuffling Procedure

We investigate the effects of shuffling procedure on network dynamics. We start from the structured and sparse connectivity coming from learning and pruning procedures, and we apply a shuffling procedure. In such a way the network topology changes, but the strength of connections is preserved. Let's consider the connection J_{ij}, picked up randomly. Given the postsynaptic neuron i, we change the presynaptic neuron

j with another one k, chosen randomly among other neurons of the network. We use the strength of J_{ij} for the new connection J_{ik}, i.e. $J_{ik} = J_{ij}$ and then we put the old connection J_{ij} to zero. We repeat this procedure for a fraction of connections or all the connections of the network. Not only the strength of connections remains the same, but also the balance between inhibitory and excitatory connections entering each unit is preserved. In this way we have the possibility to investigate the spontaneous dynamics of the same model but with different network topology, from a structured connectivity to a random connectivity, with the same value of connections.

The key parameters are: the noise level α, the strength of connection H_0, expressed in units of the threshold θ of the neurons, and the fraction of connections we change cs, that is the ratio between the number of times we make the shuffling procedure and the number of connections, so we can have different situations from $cs = 0$ (structured connectivity) to $cs = 1$ (when all the connections are shuffled). For each cs we want to investigate, we change the value of α and H_0, and we calculate the spiking rate and the normalized variance. The normalized variance is defined as $\hat{\sigma} = \frac{\sigma}{<r>}$, where $r = \frac{n_{tot}}{N\Delta}$ is the rate, with n_{tot} total number of spikes in the time interval Δ, and $\sigma = N\Delta <r^2> - N\Delta <r>^2$ is the variance. Explaining r, the normalized variance can be written as $\hat{\sigma} = \frac{<n_{tot}^2> - <n_{tot}>^2}{<n_{tot}>}$. Note that if neurons are uncorrelated and Poissonian then $<n_{tot}^2> - <n_{tot}>^2 = <n_{tot}>$. As a consequence the normalized variance is equal to 1. Therefore if normalized variance is different from 1, this means that or (1) neurons are not uncorrelated or (2) each unit is not Poissonian. To understand the importance of topology in this dynamics we calculate the clustering coefficient of the network. The clustering coefficient (C) of a node of the network is a measure of the number of edges that exist between its nearest neighbors [17]. It is defined as

$$\bar{C} = \frac{\sum_{i=1}^{N} \left[\sum_{j,k \in \Delta(i)} \Gamma(j \to k) \right]}{\sum_{i=1}^{N} z_i(z_i - 1)} \tag{9.4}$$

where $\Delta(i)$ is the set of nodes j such that there is a connection from i to j, z_i is the number of nodes in $\Delta(i)$, and $\Gamma(j \to k)$ is one if there is a connection from j to k, zero otherwise.

Structured network has more clusters of directed connections which cooperate to activate a precise order of neurons, respect with randomized network. A dense local clustering coefficient we observe in the structured network (cs close or equal to 0) remembers the regular topology of a network, while the clustering coefficient is close to 0, from a particular value of cs close to 0.7 up to $cs = 1$ (random network). We calculate the normalized clustering coefficient, defined as $C = \frac{\bar{C} - C_1}{C_0 - C_1}$, where C_0 is clustering for zero shuffling, and C_1 is for randomized network and $\bar{C}$ is the clustering coefficient of other intermediate cases of networks.

So we have $C = 1$ for $cs = 0$, i.e. a network with structured connectivity and $C = 0$ for $cs = 1$, i.e. a random connectivity (Fig. 9.1a). In the next section we will show the main results we have got, focusing our attention in particular on three dif-

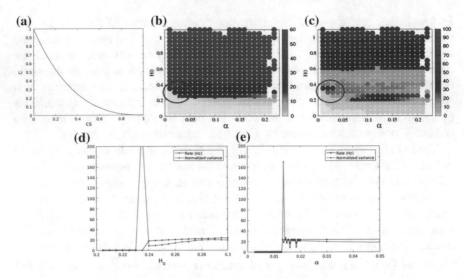

Fig. 9.1 **a** Changes of Topology. Normalized clustering coefficient $C = (\bar{C} - C_0)/(C_1 - C_0)$, where C_0 is clustering for zero shuffling, and C_1 is for randomized network, as a function of shuffled connections cs. From a value of cs close to 0.7 the clustering coefficient approaches to 0, as happens for a random network. A non-equilibrium phase transition occurs in the spontaneous dynamics of a network of $N = 3000$ neurons with structured connectivity ($cs = 0$). The rate (**b**) and the normalized variance (**c**) are shown as a function of noise intensity α and synaptic strength factor H_0. One can see a sharp transition from a region of Poissonian quiescence, with low rate and normalized variance close to 1 (*yellow points* in figure **b** and **c**) to a region of correlated high rate activity (*red points* in figure **b** and *magenta points* in figure **c**), with high values of both rate and normalized variance. Between them there is an intermediate region in which the rate gradually grows and the normalized variance has a peak, index of a transition (in *blue circle*). **d** Rate and normalized variance are shown as a function of H_0 for fixed $\alpha = 0.03$ ms^{-1}. Note that the transition between qiescence state and high correlated activity occurs when the rate grows and the normalized variance has a peak, for a particular value of connection strength, corresponding to the region in *blue circle* of figure **b** and **c**. **e** Rate and normalized variance are shown as function of α for fixed value of $H_0 = 0.3$. Note that also in this direction there is an abrupt growing of rate and a peak of normalized variance

ferent cases of topology: $cs = 0$ (structured connectivity coming from the learning), $cs = 1$ (random connectivity, when all the connections are shuffled) and $cs = 0.63$ (an intermediate case, when only 63% of connections are shuffled). We will observe three completely different behaviors.

9.3 Results

In order to study the spontaneous dynamics, we calculate the firing rate (number of spike per neuron and time interval) and the normalized variance (defined above) changing the value of noise level α and strength of connections H_0. For a network with structured connectivity ($cs = 0$) results are shown in Fig. 9.1. In this case we observe a transition from a regime of quiescence characterized by both low values

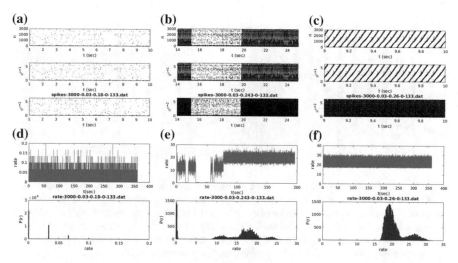

Fig. 9.2 Raster plot is the graphic in which we have the index neuron on the y-axis and the firing time on the x-axis. For each point (α, H_0) of the rate diagram we want to investigate, in figure **a**, **b** and **c** we show three raster plot, in the first one (*upper*) the neurons are ordered according to their own number, in the second one (in the *middle*) are ordered according to the pattern 1 and in the third one (*bottom*) according to the pattern 2. **a** The raster plot shows a quiescence regime in which few neurons fire and without a scheme, for $\alpha = 0.03$ ms^{-1} and $H_0 = 0.18$. **b** An intermitted reactivation of pattern 1, for $\alpha = 0.03$ ms^{-1} and $H_0 = 0.243$. This values of α and H_0 don't allow the permanence of pattern. **c** We have the permanence of retrieval pattern 1, for $\alpha = 0.03$ ms^{-1} and $H_0 = 0.26$. In this model the two patterns can't be retrieve at the same time, so that if we observe the perfect pattern 1, when neurons are ordered according to pattern 2 we observe a lot of neurons fire, because of high firing rate, but in disordered way. In figure **d**, **e** and **f** we show the rate to time (*upper*) and the distribution of rate (*bottom*) of the same point of figure **a**, **b** and **c** respectively. **d** The quiescence regime is shown, few neurons fire during the investigated time interval, indeed the distribution of rate has one peak at low value of rate, close to 0. **e** The bimodal activity and the alternation of up and down states are shown. We observe in *upper figure* the alternation of high and low values of firing rate in time. In the *bottom*, the distribution of rate has two peaks, one close to 0 and another one at higher value of rate. **f** An UP state. The rate always is high and the distribution of rate is similar to a Gaussian distribution with the mean value at high value of rate (about 20)

of rate and normalized variance to a regime of high correlated activity, characterized by high value of both rate and normalized variance. Between these two regime there is the transition region (in blue circle) when the transition occurs. Indeed in figure (d) and (e) we have respectively rate and normalized variance in function of H_0 for fixed value of $\alpha = 0.03$ ms^{-1} and in function of α for fixed value of $H_0 = 0.3$. The peak of normalized variance in correspondence of a growth of rate is the sign that a phase transition occurs. While in the region of high activity (red points in Fig. 9.1b) there is a replay of one of stored pattern (Fig. 9.2c, f), in the region with low rate (yellow points in Fig. 9.1b) there is uncorrelated Poissonian activity (see Fig. 9.2a, d). Between this two regions there is an interval of parameters where both the high-rate and low-rate states are metastable and the system switches between the two states (Fig. 9.2b, f). Even if this is not an equilibrium phase transition, but a dynamical one,

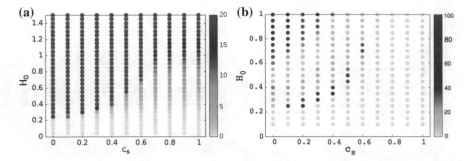

Fig. 9.3 Rate (**a**) and normalized variance (**b**) as a function of fraction of shuffled connections cs, at $\alpha = 0.05$ ms^{-1} and at different values of H_0. When cs increases the transition goes to upper values of H_0, until to a particular value of cs close to 0.7 where the transition ends. Indeed before this value of cs, for each cs we observe the gradually growth of rate and an abrupt increase of normalized variance following by low values again. At upper values of $cs = 0.7$ the normalized variance is always close to 1 (as we have said) even if the rate increases with H_0

this is similar to a first order transition since effects of hysteresis have been observed by preliminary investigations (not shown).

The raster plot (Fig. 9.2) confirms the idea that a phase transition between a high rate replay regime and a quiescence regime occurs. In raster plots we can see which neuron and when fires. We show three raster plot in order to point out the three different behaviors of the network dynamics: (A) quiescence state, that we call DOWN state, for $\alpha = 0.03$ ms^{-1} and $H_0 = 0.18$, in which few neurons fires, (B) a critical behaviors in which there is an intermitted reactivation of one of stored patterns, for $\alpha = 0.03$ ms^{-1} and $H_0 = 0.243$ (C) high correlated activity for $\alpha = 0.03$ ms^{-1} and $H_0 = 0.26$, in which neurons fire, retrieving perfectly one of stored patterns.

The dynamics of this neural network with structured connectivity is radically altered when its topology changes when we apply the shuffling procedure. It should be emphasized that the strength of connections are the same, even if we shuffle them.

In Fig. 9.3 the rate and normalized variance are shown as function of strength of connection H_0 and fraction of shuffled connections cs, for fixed value of noise level $\alpha = 0.05$ ms^{-1}. This two figures explain how the transition between the two different regimes (from quiescence-DOWN state to correlated activity-UP state) moves to higher values of H_0 as long as cs increases, until it disappears for a particular value of cs close to 0.7. Indeed for cs larger than this particular cs the normalized variance always is close to 1, index of dynamics dominated by Poissonian noise, so the transition is ended.

In particular we analyze two cases, different from the previous one of structured connectivity ($cs = 0$): the case near the end of phase transition, choosing $cs = 0.63$ (63% of connections are shuffled) and the case in which the network has a random connectivity ($cs = 1$, i.e. all the connections are shuffled).

In Fig. 9.4 we underline the end of phase transition when we shuffle all the connections. The sign of the end of this transition is the disappearing of peak in normalized variance and the gradual growth (very smooth) of firing rate when H_0 increases.

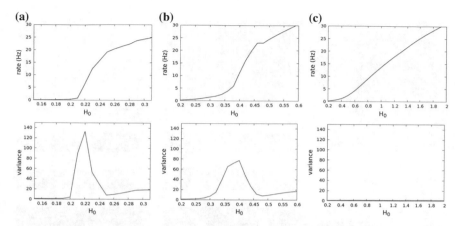

Fig. 9.4 Phase transition ends when we shuffle all the connections, keeping unchanged their values. Three cases are analyzed: **a** $cs = 0$, structured connectivity. The rate increases abruptly and the normalized variance has a peak in correspondence of this increasing of rate. **b** $cs = 0.63$, 63% of connections are shuffled. The increasing of rate is less abrupt than the figure **a** and the normalized variance has a smaller peak moved to higher values of H_0, like as the transitions moves to higher values of strength of connections but doesn't disappear. **c** $cs = 1$, all the connections are shuffled, but their values don't change. The firing rate increases very smoothly with the increasing of strength of connections, while the peak of normalized variance (seen in previous figures) disappears. The normalized always is close to 1, index of absence of the transition

The intermediate case, $cs = 0.63$ is very interesting. The phase transition occurs at high value of H_0 and at low noise, while at high noise there is a region of high rate but uncorrelated activity (normalized variance is close to 1). In particular at fixed value of $H_0 = 0.8$ (Fig. 9.5), for high value of noise ($\alpha = 0.2$ ms^{-1}) we have high value of rate but low value of normalized variance, close to 1, like in the case $cs = 1$ (see later). Indeed the raster plot shows a dynamics in which neurons fire a lot, but without a particular order (see Fig. 9.5b, d). For low value of noise ($\alpha = 0.02$ ms^{-1}), the situation is similar to the case of $cs = 0$, because in this case we observe a transition region with a peak of normalized variance. Indeed for this point the raster plot shows a perfect retrieval pattern (see Fig. 9.5a, c). The phase transition moves toward higher values of H_0 and lower values of noise level, when we increase the fraction cs of shuffled connections.

We repeat the same investigation for $cs = 1$. $cs = 1$ means that we shuffle all the connection, getting a completely random connectivity (while the set of connection's strengths are preserved). In this case the figures of firing rate and normalized variance are completely different from the case of $cs = 0$ (structured connectivity) and $cs = 0.63$ (intermediate case). While the rate shows a gradually growth (coming from high values of level noise and strength of connections), the normalized variance is always close to 1, even when the rate is high. It is the sign of a dynamics dominated by Poissonian noise. Indeed the raster plot doesn't show a particular scheme, even if the rate is always high (see Fig. 9.6c).

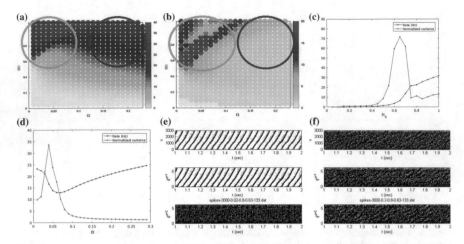

Fig. 9.5 The study of spontaneous dynamics for a network with $cs = 0.63$ (only a fraction of 63% of connections are shuffled). **a** The firing rate shows a gradually growth when we increase the strength of connections, but for high value of noise (in the *blue circle*) the normalized variance **b** is close to 1, similar to the case of $cs = 1$, while for low values of noise (in the *green circle*) the normalized variance shows a peak, similar to the case of $cs = 0$ where we observe a phase transition. **b** Normalized variance diagram, for the same values of α and H_0 of the firing rate diagram. **c** Rate and normalized variance in function of H_0 for fixed value of $\alpha = 0.02$ ms^{-1}. We observe a peak of normalized variance when the rate abruptly increases. **d** Rate and normalized variance in function of α for fixed value of $H_0 = 0.8$. We note a peak of normalized variance at value of noise close to 0.04 ms^{-1}, while for high value of α the normalized variance is always close to 1 even if the rate is high. **e** Raster plot of a network with $cs = 0.63$ and $\alpha = 0.02$ ms^{-1} and $H_0 = 0.8$ (*point in the blue circle* of panel **a** and **b**. In this case we observe the perfect retrieval stored pattern with high firing rate. **f** Raster plot of a network with $cs = 0.63$ and $\alpha = 0.2$ ms^{-1} and $H_0 = 0.8$ (*point in green circle* of panel **a** and **b**). Even if the firing rate is high, the neurons fire a lot, but without a precise scheme

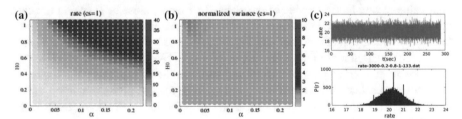

Fig. 9.6 The investigation of firing rate and normalized variance for $cs = 1$ (random connectivity) changing noise level α and strength of connections H_0 gradually. **a** Rate in function of α and H_0. We note a gradually growth of value of rate for high values of α and H_0, but this doesn't mean a transition occurs. Indeed in figure **b** we observe values of normalized variance always close to 1, index of dynamics dominated by Poissonian noise. **c** Rate to time (*upper*) and rate distribution (*bottom*) for a network with random connectivity ($cs = 1$), with strength of connection $H_0 = 0.8$ and noise level $\alpha = 0.2$ ms^{-1}, i.e. a *red point* in panel **a**. The rate is always high during the investigated time interval and the rate distribution is similar to a Gaussian distribution with a peak at rate close to 20. In the raster plot we don't observe a particular scheme according to which neurons fire

9.4   Conclusion

With this work we have seen how, in a structured network with a phase transition from quiescence state to a replay state, the spontaneous dynamics changes when we shuffle the connections. The balance between excitation and inhibition are preserved during shuffling. The clustering coefficient decreases when the fraction of shuffled connections increases. The clustering coefficient reaches the value that we observe for a completely randomized network when the shuffling fraction is about $cs = 0.7$. This is also the value of cs at which the phase transition seems to disappear. We can conclude that the transition and the rich dynamics in phase-space we observe when the network has a structured connectivity is crucially related to the network topology, induced by the learning procedure. The peak in the normalized variance that we observe near the transition is the signature of a system with high fluctuations as it is observed near a second order transition or near the spinodals of a first order transition. Further investigations are in progress to understand the order of the phase transition.

References

1. Plenz, D., Kitai, S.: Up and down states in striatal medium spiny neurons simultaneously recorded with spontaneous activity in fast-spiking interneurons studied in cortex-striatum-substantia nigra organotypic cultures. J Neurosci. **18**(1), 266–83 (1998)
2. Cossart, R., Aronov, D., Yuste, R.: Attractor dynamics of network UP states in the neocortex. Nature **423**, 283–288 (2003). doi:10.1038/nature01614
3. Shu, Y., Hasenstaub, A., McCormick, D.A.: Turning on and off recurrent balanced cortical activity. Nature **423**, 288–293 (2003). doi:10.1038/nature01616
4. Petersen, C., Hahn, T., Mehta, M., Grinvald, A., Sakmann, B.: Interaction of sensory responses with spontaneous depolarization in layer 2/3 barrel cortex. PNAS **100**, 13638–13643 (2003). doi:10.1073/pnas.2235811100
5. Luczak, A., Barth, P., Marguet, S.L., Buzski, G., Harris, K.D.: Sequential structure of neocortical spontaneous activity in vivo. PNAS **104**, 347–352 (2007). doi:10.1073/pnas.0605643104
6. Kinouchi, O., Copelli, M.: Optimal dynamical range of excitable networks at criticality. Nat. Phys. **2**, 348–352 (2006). doi:10.1038/nphys289
7. Deco, G., Jirsa, V.K., McIntosh, A.R.: Resting brains never rest: computational insights into potential cognitive architectures. Trends Neurosci. **36**, 268–274 (2013). doi:10.1016/j.tins.2013.03.001
8. Shew, W.L., Plenz, D.: The functional benefits of criticality in the cortex. Neuroscientist **19**, 88–100 (2013). doi:10.1177/1073858412445487
9. Scarpetta, S., de Candia, A.: Neural avalanches at the critical point between replay and non-replay of spatiotemporal patterns. PLoS One (2013). doi:10.1371/journal.pone.0064162
10. MacLean, J.N., Watson, B.O., Aaron, G.B., Yuste, R.: Internal dynamics determine the cortical response to thalamic stimulation. Neuron **48**, 811–823 (2005). doi:10.1016/j.neuron.2005.09.035
11. Lau P-M., Bi G-Q.: Synaptic mechanisms of persistent reverberatory activity in neuronal networks. Proc. Nat. Acad. Sci. USA **102**, 10333–10338 (2005). doi:10.1073/pnas.0500717102
12. Ji, D., Wilson, M.A.: Coordinated memory replay in the visual cortex and hippocampus during sleep. Nat. Neurosci. **10**, 100–107. doi:10.1038/nn1825 (2007)

13. Feng, H., Caporale, N., Yang, D.: Reverberation of recent visual experience in spontaneous cortical waves. Neuron **60**, 321–327 (2008). doi:10.1016/j.neuron.2008.08.026
14. Luczak, A., MacLean, J.: Default activity patterns at the neocortical microcircuit level. Front Integr. Neurosci. **6**, 30 (2012). doi:10.3389/fnint.2012.00030
15. Ribeiro, T.L., Ribeiro, S., Copelli, M.: Repertoires of spike avalanches are modulated by behavior and novelty. Front. Neural Circuits (2016). doi:10.3389/fncir.2016.00016
16. Gerstner, W., Kistler, W.: Spiking Neuron Models: Single Neurons, Populations, Plasticity. Cambridge University Press, Cambridge (2002)
17. Watts, S.: Collective dynamics of 'small-world' networks. Nature **393**, 440–442 (1998). doi:10.1038/30918

Chapter 10
Human Fall Detection by Using an Innovative Floor Acoustic Sensor

Diego Droghini, Emanuele Principi, Stefano Squartini, Paolo Olivetti
and Francesco Piazza

Abstract Supporting people in their homes is an important issue both for ethical and practical reasons. Indeed, in the recent years, the scientific community devoted particular attention to detecting human falls, since the first cause of death for elderly people is due to the consequences of a fall. In this paper, we propose a human fall classification system based on an innovative floor acoustic sensor able to capture the acoustic waves transmitted through the floor. The algorithm employed is able to discriminate human falls from non falls and it is based on Mel-Frequency Cepstral Coefficients and a two class Support Vector Machine. The dataset employed for performance evaluation is composed by falls of a human mimicking doll, everyday objects and everyday noises. The obtained results show that the proposed solution is suitable for human fall detection in realistic scenarios, allowing to guarantee a 0% miss probability at very low false positive rates.

Keywords Floor acoustic sensor · Human fall detection · Ambient assisted living · Support Vector Machine

D. Droghini (✉) · E. Principi · S. Squartini · P. Olivetti · F. Piazza
Department of Information Engineering, Università Politecnica delle Marche,
Via Brecce Bianche, 60131 Ancona, Italy
e-mail: d.droghini@pm.univpm.it
URL: http://www.univpm.it

E. Principi
e-mail: e.principi@univpm.it

S. Squartini
e-mail: s.squartini@univpm.it

P. Olivetti
e-mail: pa.olo.83@hotmail.it

F. Piazza
e-mail: f.piazza@univpm.it

© Springer International Publishing AG 2018   97
A. Esposito et al. (eds.), *Multidisciplinary Approaches to Neural Computing*,
Smart Innovation, Systems and Technologies 69,
DOI 10.1007/978-3-319-56904-8_10

10.1 Introduction

The decreasing birth rate [1] and the contemporary increase of the life expectancy at birth [4] in the majority of industrialized countries have been generating new challenges in the assistance of the elderly. The scientific community, companies and governments are trying to face them by investing in the development of efficient healthcare systems and solutions. The direction taken goes towards the development of smart home capable of taking care of the inhabitants by supporting and monitoring them in their daily actions [8, 14]. Since falls are one of the main cause of death for the elderly [11], several efforts have been devoted to the development of algorithms for automatically detecting these events.

10.1.1 Related Work

There are two broad categories into which the fall recognition systems are divided: the first comprises systems based on wearable sensors, the second the ones based on environmental sensors. The former are mainly represented by accelerometers and their main disadvantage is the obtrusiveness due to the necessity of continuously wearing a device. Examples of environmental sensors are microphones, radar Doppler, floor vibration sensors, pressure sensors, and cameras [17], and their main disadvantages are the limited covered area and high installation costs.

Recently, several works appeared in the literature that use audio signals. In these works, techniques originally developed in the speech processing community are often employed with promising results. In [19] the authors used a single far field microphone, then a Support Vector Machine trained with GMM supervectors is used to classify audio segments into falls and various types of noise. Others researchers have improved the results by employing multi-channel algorithms: in [16], the authors used a source separation technique to remove the possible interferences from background sound sources. A one class support vector machine (OCSVM) is then applied and the Mel-Frequency Cepstral Coefficients (MFCC) features from non-fall sounds are employed to construct the OCSVM data description model that distinguishes fall from non-fall sounds. In [9], the authors developed an acoustic fall detection system which consists of a circular microphone array that captures the sounds in a room.

Differently, fall detection systems based on wearable devices usually rely on accelerometers. Being portable devices, they are battery powered and they cannot be use while recharging. Additional disadvantages are that they can be forgotten by the user and that it may be deemed annoying. In related works, the authors exploit the information of the falling body acceleration identifying if there has been a fall if the acceleration values exceed a threshold. In particular, the authors in [3] employ tri-axial accelerometer sensors mounted on the trunk and on the thighs to determine the peak accelerations recorded during different types of falls. In [5], the system consists of a motion sensor network deployed on the ceiling to monitor motion and an electronic patch worn by the subjects to identify them and detect falls. More complex algorithms have been used in [12] where several machine learning techniques have been employed on data coming from tri-axial devices (accelerometer, gyroscope, and magnetometer/compass).

10.1.2 Contribution

The fall classification system presented in this paper uses an innovative sensor which belongs to the first category, i.e., it is placed on the environment. It is composed of a microphone located inside a resonant cavity that amplifies the floor vibrations captured with the aid of a membrane placed in contact with it. As shown in [13], this configuration makes the sensor particularly suitable to capture signals generated by fall events and at the same less sensitive to environmental noises. As a result, simplified algorithmic pipelines can be employed, and the computational complexity can be reduced, since it is possible to lower the sampling rates and a good performance can be achieved with a single sensor. This means that the cost of the system is lower compared to solutions that require multiple microphone and a more complicated acquisition and signal processing hardware [9].

The classification algorithm is based on a previous work by some of the authors [13]. It employs both low-level features, i.e., MFCCs [7], and high-level features, i.e., Gaussian Mean Supervectors which are then used by a Support Vector Machine to distinguish falls from no-falls. Despite MFCCs were originally developed for speech and speaker recognition tasks, they have been successfully applied also for acoustic event classification [18] and fall detection [20]. Differently from [13], where the algorithm discriminated falls of general objects, here the focus is specifically on the classification of human falls. The classifier, thus, has been designed to discriminate between two classes: human falls and generic sound events. In order to assess the performance of the approach, the dataset presented in [13] has been augmented with instances of everyday sounds (speech, footsteps, etc.), thus making the task more challenging. As a reference, results employing the original dataset [13] are also reported.

The outline of the paper is the following: Sect. 10.2 describes the proposed acoustic sensor, while Sect. 10.3 outlines the main characteristics of the algorithm. Section 10.4 presents the acoustic fall events dataset. The experiments conducted to assess the performance of the system are described in Sect. 10.5. Section 10.6 concludes the paper and proposes future developments.

10.2 The Floor Acoustic Sensor

As aforementioned, the Floor Acoustic Sensor (FAS, Fig. 10.1) is composed of a membrane that goes in direct contact with the floor. The waves transmitted through the ground are amplified by an inner container where the microphone is located.

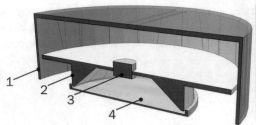 

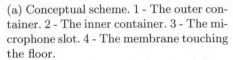

(a) Conceptual scheme. 1 - The outer container. 2 - The inner container. 3 - The microphone slot. 4 - The membrane touching the floor.

(b) Prototype used during the recordings.

Fig. 10.1 The Floor Acoustic Sensor

This microphone is characterized by an hypercardiod directivity pattern, thus it has been oriented so that the maximum gain is towards floor. Moreover, an outer enclosure covers all the sensor and between the two enclosures can be put a sound-absorbing material to minimize the possibility that environmental sounds reach the microphone.

10.3 The Fall Classification Algorithm

The classification algorithm is composed of two main parts. The first is the features extraction phase where we have extract the Mel Frequency Cepstral Coefficients from all audio files that comprise the dataset. For doing this, the feature extraction pipeline is the same used in [13]. In particular, the signal is segmented in frames 16 ms long overlapped by 8 ms, the parameter α of the pre-emphasis filter has been set to 0.97 and the number of filters which compose the filterbank has been set to 29. At the end, after the Discrete Cosine Transform, 13 statics coefficients are extracted that, together with their first and second derivatives, form the final feature vector of a signal.

The classification phase is similar to that one adopted in [13]: first it uses a mixture of Gaussians (GMM), trained on a large corpus of audio events with the Expectation Maximization algorithm to model the acoustic space (Universal Background Model, UBM). Then, for each audio segment, the Maximum a Posteriori (MAP) algorithm is used to calculate the Gaussian Mean Supervector (GMS) from the MFCCs. In contrast with the previous work [13], where we used a multi-class approach, here we employ a binary SVM to discriminate the class "fall" from "rest" which allows to distinguish human falls from the other types of sounds. In addition, the class decision is usually performed by evaluating the sign of the SVM discriminative function, which ultimately consists in deciding whether in example belongs to a class by setting a threshold equal to zero. However, in a human fall classification task, it is important to minimize the probability of missing a fall event, i.e., false negatives. In order to push the system towards this direction, we decided to consider the entire value of the SVM discriminative function, and then to set an appropriate threshold in order to minimize the occurrence of false negatives.

10.4   Dataset

The dataset has been created by the authors and is an enlarged version of the one used in [13]. It includes audio events corresponding to falls of several objects recorded in different conditions.[1] The fall events have been recorded in an isolated room particularly suitable for the propagation of acoustic waves through the floor. Each fall has been registered with the FAS placed on the ground and an aerial microphone (the same AKG 400 BL included in the floor sensor) located above the FAS, on a table 80 cm high. A Presonus AudioBox 44VSL sound card connected to a laptop has been used to record signals.

The sample rate has been set to 44100 kHz while the bit-depth at 32 bit. The dataset is composed of falls of different daily use objects, i.e., a ball, a metal basket, a book, a metal fork, a plastic chair, and a bag (Fig. 10.2a). Human falls have been simulated by employing the "Rescue Randy" doll[2] (Fig. 10.2b), a professional equipment employed in water rescues. The same type of doll has been used in many other work related to fall detection [2, 20] and is a well established practice in addressing this type of tasks. The doll has been dropped from upright position and from a chair, both forward and backward, for a total of 44 events, all included in the "fall" class. In order to further stress the system, a 20 min long recording session was carried out in which 2 persons have produced everyday noises such as talking, walking, dragging chairs, and playing with the ball. Then the track obtained in this session has been divided in 665 sub-tracks. The lengths of these sub-tracks have been randomly generated with Gaussian distribution. Mean value and standard deviation of this distribution have been calculated based on the lengths of the other files that form the dataset. In order to compare the algorithm with previous works [13], results obtained on the original dataset are also reported.

In addition to the clean dataset, a noisy version has been created to assess the performance in noisy condition. The noisy dataset consists in a musical background recorded with both sensors and digitally added to the clean events.

---

[1] The dataset is available at the following URL: http://www.a3lab.dii.univpm.it/research/fasdataset.
[2] http://www.simulaids.com/1475.htm.

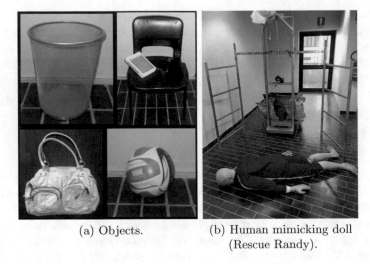

(a) Objects.   (b) Human mimicking doll
                                    (Rescue Randy).

Fig. 10.2 Equipment employed for creating the fall events dataset

10.5 Experiments

In this section, the experimental procedure is firstly discussed and then the algorithm performance is presented. In the experiments, the signals of the dataset described above have been downsampled to 8 kHz and the bit depth has been reduced to 16 bit. Both the choice of the sampling frequency and the choice of MFCCs are justified by the analysis performed in a previous work by the authors [13], where it was shown that the signals recorded with the FAS have the majority of the energy concentrated at low frequencies (below 1 kHz). Indeed, the mel scale used in the feature extraction pipeline has a higher resolution at low frequencies, that allows to better describe the portion of the spectrum where the majority of the energy resides. In addition, the algorithm has been evaluated using two features extraction pipelines:

- the first is the same described in the Sect. 10.3 and will be denoted as STD;
- the second pipeline does not include the pre-emphasis filter and will be denoted with NOPRE.

The experiments have been conducted with a 4-fold cross-validation strategy and a three-way data split in three different operating conditions:

- matched, where the training, validation and test sets share the same acoustic condition, i.e., clean or noisy;
- mismatched, where the training set is composed of clean signals while the validation and test sets are composed of noisy signals;
- multicondition, where the training, validation and test sets contain both clean and noisy data. In this case the sets have been divided so that they contain 1/3 of clean data and 2/3 of noisy data.

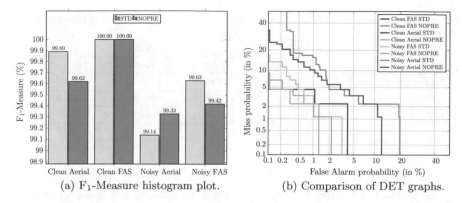

(a) F_1-Measure histogram plot.   (b) Comparison of DET graphs.

Fig. 10.3 Fall classification performance in matched condition with the dataset comprising every-day noises

The tests have been conducted also without using the signals of everyday noises, i.e., using the same dataset employed in [13]. The performance is evaluated in terms of F_1-Measure per class and the values have then been averaged to obtain a single performance metric. To better describe the algorithm behavior, we have used the Detection Error Trade-off (DET) curve, as defined in [10]. This graph allows evaluating the false alarm probability when miss probability is equal to 0 (henceforward named as FPM0), which is particularly relevant in a fall classification task.

10.5.1 *Matched*

Figure 10.3 shows the results obtained in the matched condition case study using the enlarged version of the dataset. In Fig. 10.3a, the FAS achieves higher F_1-Measures both in clean and noisy conditions. In the first case, the floor sensor exceeds the F_1-Measure of the aerial microphone obtained with the standard MFCC pipeline by 0.11%. In the noisy condition case study, the FAS with STD features achieves an F_1-Measure greater than 0.3% with respect to the aerial microphone with NOPRE features.

The DET curves are shown in Fig. 10.3b. Note that the lines relative to the FAS in clean condition are both absent. This is because independently from the threshold, either the miss probability or the false alarm probability is 0 and the DET curves assume values towards minus infinite (in logarithmic scale). This means that the FPM0 are 0 for both these configurations, while the lower FPM0 for the aerial microphone in clean condition is 1.3%. Regarding the tests with noisy signals, the performance difference between the two sensors increases, since the FPM0 is equal to 2.2% with the FAS (NOPRE) and is equal to 12.5% with the aerial sensor (STD).

Table 10.1 Fall classification performance in matched condition with the dataset excluding every-day noises

		STD		NOPRE	
		F_1-Measure	FPM0	F_1-Measure	FPM0
Clean	Aerial	96.29	9.85	93.11	4.95
	FAS	98.81	0.00	100.00	0.00
Noisy	Aerial	97.22	43.00	96.92	73.00
	FAS	99.63	6.05	99.62	9.85

Table 10.1 summarises the results obtained with the dataset version presented in [13], which does not comprise everyday noises. As it can be observed, with respect to the previous results, the performance decreases in all tests and for both the F_1-Measure and the FPM0, although the gap between the FAS and the aerial microphone increases.

10.5.2 Mismatched

The mismatched condition is the most difficult case for the classifier. As expected, the performance decrease for both sensors. Focusing on Fig. 10.4a, the best F_1-Measure is obtained with the NOPRE pipeline for both the FAS and aerial microphone, and is equal to 99.14% and 98.43% respectively. A consistent performance difference between the two sensors can be observed in Fig. 10.4b, where the smallest FPM0 for the FAS, obtained with NOPRE, is around 13% while for the aerial one is 95%, this time obtained with STD.

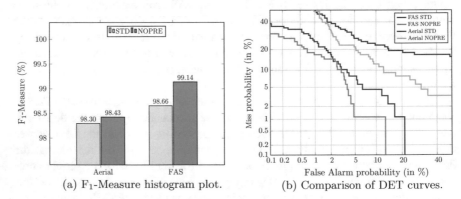

(a) F_1-Measure histogram plot.   (b) Comparison of DET curves.

Fig. 10.4 Fall classification performance in mismatched condition with the dataset comprising everyday noises

Table 10.2 Fall classification performance in mismatched condition (a) and multicondition (b) with the dataset excluding everyday noises. F_1 denotes the F_1-Measure

(a)					(b)				
%	STD		NOPRE		%	STD		NOPRE	
	F_1	FPM0	F_1	FPM0		F_1	FPM0	F_1	FPM0
Aerial	96.20	87.50	95.44	76.00	Aerial	96.80	16.50	96.78	31.4
FAS	97.80	34.50	98.11	12.50	FAS	99.62	0.00	99.43	0.00

The results of the mismatched experiments obtained with the dataset without everyday noises are reported in Table 10.2a. Regarding the F_1-Measure, a performance decrease can be observed compared to the results in Fig. 10.4a. Differently, the aerial microphone FPM0 improves, but it is again below the FPM0 achieved by the FAS with NOPRE MFCCs (12.5%).

10.5.3   Multicondition

Figure 10.5 shows the results in the multicondition case. The FAS superiority is confirmed, since it achieves an F_1-Measure equal to 99.78% regardless the feature extraction pipeline, while the aerial sensor obtains an F_1-Measure equal to 99.19% with the STD pipeline.

Regarding the DET plot (Fig. 10.5b), an FPM0 equal to 2.9% is achieved by the FAS, but differently from the previous cases, with the STD feature. The aerial microphone achieves an FPM0 equal to 9.8% with the NOPRE pipeline.

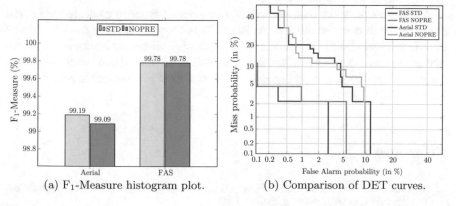

(a) F_1-Measure histogram plot.   (b) Comparison of DET curves.

Fig. 10.5 Fall classification performance in multicondition with the dataset comprising everyday noises

In the last table (Table 10.2b) are shown the result of the multicondition tests obtained by using the smaller dataset. Again there is an overall decrease for the F_1-Measure, greater for the aerial microphone, while the FAS exhibiting a FPM0 equal to 0% contrary to the increasing FPM0 for the aerial sensor.

10.6 Conclusion

In this work, a human fall classification system based on an innovative Floor Acoustic Sensor has been proposed. The sensor operates similarly to stethoscopes, with a microphone embedded in a resonant enclosure and a membrane in contact with the floor that captures the acoustic waves resulting from a fall. The classification algorithm extracts MFCC features from the signal acquired with the FAS, and then discriminates a fall from a generic event by using GMM supervectors and an SVM classifier. Differently from previous works [13], here we specifically addressed the human fall classification task by designing the classifier to discriminate human falls from other events.

The performance of the system has been evaluated on a corpus containing recordings of several events: falls of a human mimicking doll, falls of common objects and everyday noises (speech, footsteps, etc.). In order to assess the performance of the solution in adverse acoustic conditions, a noisy version of the dataset has been created. The experiments have been performed in three operating conditions: matched, mismatched and multicondition, and the performance has been evaluated in terms of average F_1-Measure and false alarm probability when the miss probability is equal to 0. The superiority of the FAS resulted evident in all the addressed conditions, in particular with an F_1-Measure equal to 100% and an FP0 equal to 0% in clean matched conditions, and an F_1-Measure equal to 99.14% and an FP0 equal to 13% in noisy mismatched conditions.

As future works, the datasets will be expanded in order to evaluate the system performance in case the falls occurs in a different room respect to the FAS one or in different scenario as falls occurs in presence of furniture. In addition, the detection of the time boundaries of the fall event will be addressed and techniques originally developed for enhancing speech will be evaluated to increase the robustness to acoustic distortions [6, 15].

References

1. Eurostat Statistic Explained: Fertility Statistics. http://ec.europa.eu/eurostat/statistics-explained/index.php/Fertility_statistics (2016)
2. Alwan, M., Rajendran, P.J., Kell, S., Mack, D., Dalal, S., Wolfe, M., Felder, R.: A smart and passive floor-vibration based fall detector for elderly. Proc. Inf. Commun. Technol. **1**, 1003–1007 (2006)

3. Bourke, A., O'Brien, J., Lyons, G.: Evaluation of a threshold-based tri-axial accelerometer fall detection algorithm. Gait Posture **26**(2), 194–199 (2007)
4. Carone, G., Costello, D.: Can europe afford to grow old? Finance Dev. **43**(3), 28–31 (2006)
5. Charlon, Y., Fourty, N., Bourennane, W., Campo, E.: Design and evaluation of a device worn for fall detection and localization: application for the continuous monitoring of risks incurred by dependents in an Alzheimer's care unit. Expert Syst. Appl. **40**(18), 7316–7330 (2013)
6. Cifani, S., Principi, E., Rocchi, C., Squartini, S., Piazza, F.: A multichannel noise reduction front-end based on psychoacoustics for robust speech recognition in highly noisy environments. In: Proceedings of Hands-Free Speech Communication and Microphone Arrays (HSCMA), pp. 172–175. Trento, Italy, 6–8 May 2008
7. Davis, S.B., Mermelstein, P.: Comparison of parametric representations for monosyllabic word recognition in continuously spoken sentences. IEEE Trans. Acoust. Speech Signal Process. **28**(4), 357–366 (1980)
8. Dawadi, P., Cook, D., Schmitter-Edgecombe, M.: Automated cognitive health assessment from smart home-based behavior data. IEEE J. Biomed. Health Inf. **20**(4), 1188–1194 (2016)
9. Li, Y., Ho, K., Popescu, M.: A microphone array system for automatic fall detection. IEEE Trans. Biomed. Eng. **59**(5), 1291–1301 (2012)
10. Martin, A., Doddington, G., Kamm, T., Ordowski, M., Przybocki, M.: The DET curve in assessment of detection task performance. In: Proceedings of the European Conference on Speech Communication and Technology, pp. 1895–1898 (1997)
11. Mubashir, M., Shao, L., Seed, L.: A survey on fall detection: principles and approaches. Neurocomputing **100**, 144–152 (2013)
12. Özdemir, A., Barshan, B.: Detecting falls with wearable sensors using machine learning techniques. Sensors **14**(6), 10691–10708 (2014)
13. Principi, E., Droghini, D., Squartini, S., Olivetti, P., Piazza, F.: Acoustic cues from the floor: a new approach for fall classification. Expert Syst. Appl. **60**, 51–61 (2016)
14. Principi, E., Squartini, S., Bonfigli, R., Ferroni, G., Piazza, F.: An integrated system for voice command recognition and emergency detection based on audio signals. Expert Syst. Appl. **42**(13), 5668–5683 (2015)
15. Rotili, R., Cifani, S., Principi, E., Squartini, S., Piazza, F.: A robust iterative inverse filtering approach for speech dereverberation in presence of disturbances. In: Proceedings of IEEE Asia Pacific Conference on Circuits and Systems (APCCAS), pp. 434–437. Macao, China, 30 Nov–3 Dec 2008
16. Salman Khan, M., Yu, M., Feng, P., Wang, L., Chambers, J.: An unsupervised acoustic fall detection system using source separation for sound interference suppression. Signal Process. **110**, 199–210 (2015)
17. Stone, E., Skubic, M.: Fall detection in homes of older adults using the Microsoft Kinect. IEEE J. Biomed. Health Inform. **19**(1), 290–301 (2015)
18. Temko, A., Nadeu, C.: Classification of acoustic events using SVM-based clustering schemes. Pattern Recogn. **39**(4), 682–694 (2006)
19. Zhuang, X., Huang, J., Potamianos, G., Hasegawa-Johnson, M.: Acoustic fall detection using Gaussian mixture models and GMM supervectors. In: Proceedings of ICASSP, pp. 69–72. Taipei, Taiwan, 19–24 Apr 2009
20. Zigel, Y., Litvak, D., Gannot, I.: A method for automatic fall detection of elderly people using floor vibrations and sound proof of concept on human mimicking doll falls. IEEE Trans. Biomed. Eng. **56**(12), 2858–2867 (2009)

Chapter 11
An Improved Hilbert-Huang Transform for Non-linear and Time-Variant Signals

Cesare Alippi, Wen Qi and Manuel Roveri

Abstract Learning in non-stationary/evolving environments requires methods able to process and deal with non-stationary streams. In this paper we propose a novel algorithm providing a time-frequency decomposition of time-variant signals. Outcoming signals can be used to identify anomalous events/patterns or extract features associated with the time-variance of the signal, precious information for any consequent learning action. The paper extends the Hilbert-Huang Transform notoriously used to deal with time-variant signals by introducing (i) a new Empirical Mode Decomposition that identifies the number of frequency modes of the signal and (ii) an extension of the Hilbert Transform that eliminates negative frequency-values in the time-frequency spectrum. The effectiveness of the proposed Transform has been tested on both synthetic and real time-variant signals acquired by a real-world intelligent system for landslide monitoring.

Keywords Hilbert-Huang Transform · Learning in non-stationary environment · Time-frequency analysis

C. Alippi (✉) · W. Qi · M. Roveri
Politecnico di Milano, Milan, Italy
e-mail: cesare.alippi@polimi.it

W. Qi
e-mail: wen.qi@polimi.it

M. Roveri
e-mail: manuel.roveri@polimi.it

C. Alippi
Università della Svizzera Italiana, Lugano, Switzerland

© Springer International Publishing AG 2018
A. Esposito et al. (eds.), *Multidisciplinary Approaches to Neural Computing*,
Smart Innovation, Systems and Technologies 69,
DOI 10.1007/978-3-319-56904-8_11

11.1 Introduction

In recent years learning in non-stationary/evolving environments is becoming a hot research topic [4]. Research aspires at addressing the problem of learning in those scenarios where the process generating the data changes/evolves over time either through the probability density function generating the i.i.d. inputs (no stationarity) or the equations ruling the dynamical system (time-variance). Addressing this problem is fundamental in a plethora of applications since such changes affect the effectiveness of the existing application relying on acquired data which, rapidly, might become obsolete. When time-variance occurs adaptive solutions, e.g., those based on active or passive approaches, must be considered to adapt the application and track the evolution over time [1]. Relevant application scenarios requesting learning in non-stationary environments are those for which a data-driven application is designed and sensors are present, as it happens in the monitoring and control of industry systems, critical infrastructures, smart grids, water distribution networks, and natural or physical environments.

Learning in non-stationary/evolving environments often requires the acquisition and processing of signals from the system under inspection, e.g., in order to extract features. However, most of the traditional signal processing techniques present in the literature assume signals to be time-invariant [3, 6, 12, 13]. In order to weaken this hypothesis, Wavelet analysis and Wigner-Ville distribution function [6] have been proposed: they do not require the time-invariance assumption but request signals to be linear. Recently, a new method called Hilbert-Huang Transform (HHT) has been introduced in the literature to process non-linear time-variant signals [8]. The idea of this method is to decompose the input signal into a time-frequency representation so as to identify the occurrence of events or, more in general, information of interest. This transform can be also considered part of a change detection mechanism designed to analyse changes in stationarity/time variance.

The standard HHT requires the execution of two different phases: the Empirical Mode Decomposition (EMD) and the Hilbert Spectral Analysis (HSA). In more detail, EMD aims at decomposing the input signal Y into various components representing a complete and orthogonal basis of the original signals (by means of repeated sifting mechanism and computing the average of upper and lower envelops). The outcome of the EMD is a set of functions $\{IMF_1; \ldots ; IMF_n; R\}$, where $IMF_1; \ldots ; IMF_n$ are the n Intrinsic Mode Functions (IMFs) representing the different oscillatory modes of the signal, while R accounts for the linear component representing the trend of Y. Subsequently, the HSA phase activates to provide the time-frequency spectrum of IMFs by means of the Hilbert Transform (HT) [16] and marginal spectrum processing [8]. The HT computes the change of phase angle θ of $IMFs$ to obtain n Instantaneous Frequency (IF) components, i.e., $\{IF_1, \ldots, IF_n\}$, namely the time-frequency spectrum.

Even if the HHT showed to be particularly effective in many application scenarios, e.g., fault detection [5, 11] and medical diagnosis [14], it suffers from two major problems as pointed out in several papers [9, 10]:

- the difficulty in identifying the proper number n of *IMF*s;
- the possible presence of negative frequency values in the time-frequency spectrum due to finite precisions in the computation of HT.

Both problems negatively affect the overall quality of signal decomposition, hence reducing the effectiveness of the envisaged applications relying on processed signals. In order to fix those problems, some works are recently proposed that only mitigate but not solve them. A critical review of such related literature is presented in Sect. 11.2.

The aim of this paper is to propose an extension of the HHT transform, called *HHT-C*, able to conclusively solve both aforementioned problems. The key elements of the proposed *HHT-C* pass through: an improved EMD algorithm (EMD-C) that automatically identifies the proper number n of IMFs and an extension of the HT algorithm (HT-NS) to eliminate negative frequency-values in the time-frequency spectrum. The effectiveness of the proposed *HHT-C* has been tested on both synthetic and real datasets.

The paper is organized as follows. Section 11.2 critically reviews the related literature. Section 11.3 details the proposed *HHT-C*, and Sect. 11.4 describes experimental results.

11.2 Analysis of the Literature

None of the solutions present in the related literature [7, 9, 15] jointly addresses the problems emerged in the previous section.

Among the solutions meant to address the first problem, namely identifying the proper number n of *IMF*s, a modified version of the HHT is proposed in [8]. There, the idea is to set a constraint on the standard deviation between two subsequent IMFs (i.e., two IMFs that have been generated in the sequel). Unfortunately, this often results in the generation of an unnecessary number of IMFs, as shown in [9]. A novel 'S-number' EMD method is proposed in [9]. There, the idea is to fix the number of sifting operations by means of the user-defined parameter S-number introduced to reduce the redundant IMFs. However, this method induces a fixed number n of IMFs once the length of the original signal Y is evaluated. The solution proposed in [15] relies instead on an ad-hoc figure of merit, which depends on the upper and lower envelope curves, to control the number of IMFs. Unfortunately, the residual R might still contain frequency components as also outlined in [15].

The second problem, namely removing the possible presence of negative frequency values in the time-frequency spectrum, is rarely investigated, though very important. Huang [7] proposes a generalized zero-crossing algorithm whose goal is to find all the zero-crossing and extreme points of *IMF*s and replacing them with average values. However, the computed IF curves might be characterized by unusual and not-smooth behaviours as shown in [7]. Rilling et al. [15] introduces a different method to compute IFs (HT-R) that adopts a normalizing frequency mechanism to generate each IF curve. Unfortunately, this solution does not guarantee IFs not to contain negative values [15].

11.3 The Improved Hilbert-Huang Transform

In this section we introduce HHT-C, the proposed extension for the Hilbert-Huang Transform designed to address the problems of the HHT highlighted in Sect. 11.1. As presented in Fig. 11.1, HHT-C is characterized by:

- a novel Empirical Mode Decomposition algorithm (EMD-C) able to identify the proper number n of IMFs to be considered;
- a Negative-Suppression Hilbert Transform (HT-NS) able to remove negative frequency values in the time-frequency spectrum.

11.3.1 The New Empirical Mode Decomposition Algorithm

The proposed EMD-C algorithm, which is detailed in Algorithm 1, represents an extension of the traditional EMD algorithm into two main directions: (i) a new IMF is created when the magnitude of the vector representing the mean value between the upper and lower envelope of the processed signal is below a user-defined

Fig. 11.1 The workflow of the proposed HHT-C. The input signal Y is sifted into the n IMFs and the residual R according to the novel Empirical Mode Decomposition algorithm (EMD-C). Subsequently, the computed IMFs are transformed into the time-frequency spectrum (IFs set) through the proposed Negative-Suppression Hilbert Transform (HT-NS)

threshold. As described in [8], this condition is requested when a new IMF is created; (ii) the iterative process of computing IMFs ends when the residual R does not contain anymore frequency components. In fact, as described in [8], the process of creating IMFs ends when all the frequency modes have been extended by the original signal.

In more detail, each IMF is created as follows (line 7–12). The upper U_e and lower L_e envelope curves of the processed signal as well as the mean signal M between U_e and L_e are computed (line 8 and 9, respectively). Afterwards, M is subtracted from the original signal to get residual R (line 11). According to the definition of IMFs [8], a new IMF can be created when the magnitude of M is close to zero, i.e., $|M| \cong 0$. This is verified by requesting the inner product $\sigma_1 = M \cdot M$ to be smaller than a user-defined parameter θ_1 (in our simulations $\theta_1 = 3E^{-5}$). When $\sigma_1 < \theta_1$ (line 7), residual R as computed in line 11 represents the new IMF to be included into the IMF library $IMFs$ (line 15) that is initially empty (line 5). The number n of stored IMFs increases accordingly (line 13). Otherwise, the processing at line 8–11 iterates.

Once a new IMF has been added to $IMFs$, the new residual R between the original signal Y and the sum of all the n previously stored $IMFs$ is computed (line 16). When R does not contain any more frequency modes, the algorithm terminates [8]. In order to test the ending condition a linear regression $\hat{R}$ on signal R (line 17) is fitted and the sum of squared error σ_2 between R and $\hat{R}$ (line 18) is computed. When σ_2 is less than an user-defined parameter θ_2, the process of generating $IMFs$ terminates and $IMFs$ contains the final IMFs (in our simulations $\theta_2 = 6E - 2$). Otherwise, R still contains frequency components and additional $IMFs$ must be extracted through the iteration of lines 6–18.

An additional condition (line 4) verifies that the number n of stored IMFs does not exceed the maximum number of IMFs MAX_{IMF} as defined in [17], i.e., $MAX_{IMF} = \lfloor log_2(N) \rfloor$. When this condition does not hold, the whole process of generating IMFs is repeated by using a larger value of θ_1 as per line 19 (where $\eta = 1E - 6$ in our simulation).

According to the orthogonal condition of IMFs, once IMFs have been computed in line 4–21, we have to verify that the inner product between any two IMFs i, j, i.e., $< IMF_i, IMF_j > = \sum \frac{|IMF_i \cdot IMF_j|}{(IMF_i)^2 \cdot (IMF_j)^2}, i < j, i, j = 1, 2, \ldots, n$, is close to zero. As proposed in [17], to simplify the computation of this orthogonal condition, we rely on the analysis of correlation coefficients (line 22), i.e., $\sigma_3^{i,j} = \frac{covariance(IMF_i, IMF_j)}{SD(IMF_i) \cdot SD(IMF_j)}, i < j, i, j = 1, 2, \ldots, n$. When $\sigma_3^{i,j}$ is larger than the given parameter θ_3, IMF_i and IMF_j are added together and the sum is stored in IMF_i of $IMFs$ (line 24), while the zero vector is assigned to IMF_j (line 25). In our simulations we set $\theta_3 = 0.3$.

ALGORITHM 1: Improved Empirical Mode Decomposition (EMD-C)

1 Input $Y = \{y(t), t = 1, \ldots, N\}$, $\theta_1, \theta_2, \theta_2, \eta$;
2 $\sigma_1 = 1, \sigma_2 = 0, R = Y, MAX_{IMF} = \lfloor log_2(N) \rfloor$;
3 $n = MAX_{IMF} + 1$;
4 **while** $(n > MAX_{IMF})$ **do**
5    $n = 0$, insert blank space $IMFs = \{\}$;
6   **while** $(\sigma_2 < \theta_2)$ **do**
7   **while** $(\sigma_1 > \theta_1)$ **do**
8   Compute upper U_e and lower L_e envelopes of R;
9   Compute mean signal $M = (U_e + L_e)/2$;
10       $\sigma_1 = M \cdot M$;
11       $R = R - M$;
12   **end**
13     $n = n + 1$;
14     $IMF_n = R$;
15     $IMFs = IMFs \bigcup IMF_n$
16     $R = Y - \sum_{i=1}^{n} IMF_i$;
17   Fit a linear regression model $\hat{R}$ on R;
18     $\sigma_2 = 1/N * (\hat{R} - R)^2$;
19   **end**
20   $\theta_1 = \theta_1 + \eta$;
21 **end**
22   Compute $\sigma_3^{i,j} = \frac{covariance(IMF_i, IMF_j)}{SD(IMF_i) \cdot SD(IMF_j)}, i < j, i, j = 1, 2, \ldots, n$;
23 **if** $(\sigma_3^{i,j} > \theta_3)$ **do**
24   $IMF_i = IMF_i + IMF_j$
25   $IMF_j = [\underline{0}]$;
26 **else**
27   Output $IMFs$
28 **end**

11.3.2 Negative-Suppression Hilbert Transform

The second step of the proposed HHT-C is the novel Negative-Suppression Hilbert Transform (HT-NS), an extension of the HT algorithm able to eliminate the presence of negative frequency values in the IF curves.

HT-NS relies on the basic concept of HT, which is meant to analyse the change of IMFs phase angle $\theta(t)$ to compute the IF as described in [16], i.e., $IF(t) = \frac{d(\theta(t))}{2\pi dt} = \frac{IMF(t+1) - IMF(t)}{2\pi dt}, t = 1, 2, \ldots, N$, where $IF(t)$ is the IF curve with time and IMF is an IMF generated by the EMD. In some time points, the difference between $IMF(t + 1)$ and $IMF(t)$ may be a negative value, hence generating negative frequency values in $IF(t)$.

To avoid negative values, we propose the following algorithm. IF curves $IF_1, \ldots,$ IF_n are computed by applying the standard HT to IMFs. Then, for each curve IF_i, $i = 1, \ldots, n$, the algorithm verifies whether negative values are present or not and, if needed, correct them as follows:

$$IF_i(t) = \begin{cases} 0 & \text{if } IF_i(t) < 0 \text{ and } t = 1 \\ IF_i(t-1) & \text{if } IF_i(t) < 0 \text{ and } t > 1 \\ IF_i(t) & \text{if } IF_i(t) \geq 0 \end{cases} \tag{11.1}$$

with $t = 1, \ldots, N$. This simple mechanism allows to guarantee the absence of negative frequency values in the IF curves.

11.4   Experimental Results

The aim of this section is to assess the performances of the proposed HHT-C by it comparing with the standard HHT and the HHT-R algorithm (being composed by EMD-R and HT). As a relevant tested we considered both simulated signals and a time-variant signal acquired by an intelligent monitoring system for landslide forecasting deployed on the Italian Alps [2].

The figures of merit we considered to compare HHT-C, HHT and HHT-R are the following: (1) the number n of IMFs; (2) the largest magnitudes MAX_{IP} of the generated IMFs; (3) the sum of squared difference σ_2 between the residual R and its linear regression $\hat{R}$, name $\sigma_2 = 1/N(\hat{R} - R)^2$; and (4) the largest amount of negative values C_{IF} present in the computed IFs.

The synthetic experiment refers to the following non-linear and time-variant problem

$$Y_s(t) = \sin(6\pi t) + \cos(33\pi t) + \sin(77\pi t) + t,$$

$N = 500$ samples have been extracted with a sampling frequency equal $1000 \, Hz$.

A comparison among HHT-C, HHT and HHT-R is shown in Table 11.1. As expected, the proposed HHT-C outperforms the other methods in terms of MAX_{IP} and C_{IF} meaning that (1) the generated IMFs guarantee the condition on the mean signal M described in Sect. 11.3.1; (2) no frequency values are present in generated IFs. HHT-H provides the lowest σ_2 value at the expenses of a larger number of generated IMFs and at larger MAX_{IP} and C_{IF}. In addition, Table 11.1 details the average values of the IFs, i.e., AVG_{IF1}, AVG_{IF2}, etc., showing the high ability of HHT-C in correctly modelling the 3 oscillatory modes present in the original signal.

The real dataset refers to data acquired by a clinometer sensor, as shown in Fig. 11.2. Here the sampling period is 1 h, while the data length is 512 samples. Experimental results are shown in Table 11.1. HHT-C, HHT and HHT-R decomposes 5, 8 and 7 IMFs, respectively. More specifically, MAX_{IP} and σ_2 reveal the

Table 11.1 Synthetic and real experiments: the results

Experiement		HHT-C	HHT-H	HHT-R
Synthetic	AVG_{IF1}	38.5029	38.4995	38.5124
	AVG_{IF2}	16.4934	16.4921	16.5245
	AVG_{IF3}	2.9884	2.9819	3.6652
	AVG_{IF4}	–	1.6303	0.1645
	AVG_{IF5}	–	1.0243	–
	AVG_{IF6}	–	0.2501	–
	n	3	6	4
	MAX_{IP}	0.0026	0.1411	0.0169
	σ_2	0.0637	1.8971×10^{-27}	2.3116×10^3
	C_{IF}	0	626	453
Real-world	n	5	8	7
	MAX_{IP}	0.1002	0.1195	0.1049
	σ_2	7.2992×10^{-8}	1.8483×10^{-5}	7.2403×10^{-5}
	C_{IF}	0	85	79

Fig. 11.2 The considered real-world dataset: clinometer measurements acquired by an intelligent monitoring system for landslide forecasting deployed on the Italian Alps

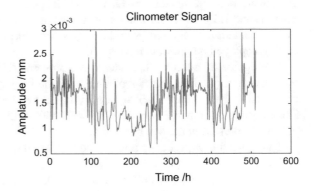

advantage of HHT-C in orthogonality and sifting residual. In addition, the values of C_{IF} show the ability of the proposed HHT-C to remove all the negative frequency values from the IFs, while HHT and HHT-R do not.

11.5 Conclusions

The aim of the paper was to introduce an extension of HHT, called HHT-C, able to address the two problems of the traditional HHT, i.e., finding out the proper number n of *IMF*s and removing negative frequency values in the time-frequency spectrum. The proposed HHT-C consists of two steps: a novel EMD algorithm to identify the correct number of IMFs and Negative-Suppression Hilbert Transform (HT-NS) able to guarantee the absence of negative values from IFs spectrum. The effectiveness of the proposed HHT-C has been tested on both synthetic and real-data coming from an intelligent monitoring system. Moreover, experimental results show that the proposed HHT-C outperforms the solutions present in the literature.

Hence, the proposed HHT-C could be considered as a precious tool for the analysis of time-variant signals in solutions and mechanisms meant to operate in the field of learning in non-stationary/evolving environments.

References

1. Alippi, C.: Intelligence for Embedded Systems. Springer (2014)
2. Alippi, C., Camplani, R., Galperti, C., Marullo, A., Roveri, M.: A high-frequency sampling monitoring system for environmental and structural applications. ACM Trans. Sensor Netw. (TOSN) **9**(4), 41 (2013)
3. Diks, C.: Nonlinear Time Series Analysis.: Methods and Applications, vol. 4. World Scientific (1999)
4. Ditzler, G., Roveri, M., Alippi, C., Polikar, R.: Learning in nonstationary environments: a survey. IEEE Comput. Intell. Mag. **10**(4), 12–25 (2015)
5. Esfahani, E.T., Wang, S., Sundararajan, V.: Multisensor wireless system for eccentricity and bearing fault detection in induction motors. IEEE/ASME Trans. Mechatron. **19**(3), 818–826 (2014)
6. Gröchenig, K.: Foundations of Time-frequency Analysis. Springer Science & Business Media (2013)
7. Huang, N.E.: Computing frequency by using generalized zero-crossing applied to intrinsic mode functions (2006)
8. Huang, N.E., Shen, Z., Long, S.R., Wu, M.C., Shih, H.H., Zheng, Q., Yen, N.C., Tung, C.C., Liu, H.H.: The empirical mode decomposition and the Hilbert spectrum for nonlinear and non-stationary time series analysis. In: Proceedings of the Royal Society of London A: Mathematical, Physical and Engineering Sciences, vol. 454, pp. 903–995. The Royal Society (1998)
9. Huang, N.E., Wu, M.L.C., Long, S.R., Shen, S.S., Qu, W., Gloersen, P., Fan, K.L.: A confidence limit for the empirical mode decomposition and Hilbert spectral analysis. In: Proceedings of the Royal Society of London A: Mathematical, Physical and Engineering Sciences. vol. 459, pp. 2317–2345. The Royal Society (2003)
10. Huang, N.E.: Hilbert-Huang Transform and Its Applications, vol. 16. World Scientific (2014)
11. Huang, S.D., Cao, G.Z., He, Z.Y., Pan, J., Duan, J.A., Qian, Q.Q.: Nonlinear modeling of the inverse force function for the planar switched reluctance motor using sparse least squares support vector machines. IEEE Trans. Ind. Inform. **11**(3), 591–600 (2015)
12. Kantz, H., Schreiber, T.: Nonlinear Time Series Analysis, vol. 7, Cambridge University Press (2004)
13. Kuo, S.M., Lee, B.H., Tian, W.: Real-Time Digital Signal Processing: Fundamentals, Implementations and Applications. Wiley (2013)
14. Park, J.W., Chu, M.K., Kim, J.M., Park, S.G., Cho, S.J.: Analysis of trigger factors in episodic migraineurs using a smartphone headache diary applications. PLoS One **11**(2), e0149577 (2016)
15. Rilling, G., Flandrin, P., Gonçalves, P.: On empirical mode decomposition and its algorithms. In: Proceedings of IEEE-EURASIP Workshop on Nonlinear Signal and Image Processing NSIP-03 (2003)
16. Schreier, P.J., Scharf, L.L.: Statistical Signal Processing of Complex-valued Data. The Theory of Improper and Noncircular Signals. Cambridge University Press (2010)
17. Wu, Z., Huang, N.E.: Ensemble empirical mode decomposition: a noise-assisted data analysis method. Adv. Adapt. Data Anal. **1**(01), 1–41 (2009)

Chapter 12
Privacy-Preserving Data Mining for Distributed Medical Scenarios

Simone Scardapane, Rosa Altilio, Valentina Ciccarelli, Aurelio Uncini and Massimo Panella

Abstract In this paper, we consider the application of data mining methods in medical contexts, wherein the data to be analysed (e.g. records from different patients) is distributed among multiple clinical parties. Although inference procedures could provide meaningful medical information (such as optimal clustering of the subjects), each party is forbidden to disclose its local dataset to a centralized location, due to privacy concerns over sensible portions of the dataset. To this end, we propose a general framework enabling the parties involved to perform (in a decentralized fashion) any data mining procedure relying solely on the Euclidean distance among patterns, including kernel methods, spectral clustering, and so on. Specifically, the problem is recast as a decentralized matrix completion problem, whose proposed solution does not require the presence of a centralized coordinator, and full privacy of the original data can be ensured by the use of different strategies, including random multiplicative updates for secure computation of distances. Experimental results support our proposal as an efficient tool for performing clustering and classification in distributed medical contexts. As an example, on the known Pima Indians Diabetes dataset, we obtain a Rand-Index for clustering of 0.52 against 0.54 of the (unfeasible) centralized solution, while on the Parkinson speech database we increase from 0.45 to 0.50.

Keywords Distributed learning · Biomedicine · Kernel methods · Spectral clustering · Privacy

S. Scardapane · R. Altilio · V. Ciccarelli · A. Uncini · M. Panella (✉)
Department of Information Engineering, Electronics and Telecommunications (DIET),
University of Rome "La Sapienza", Via Eudossiana 18, 00184 Rome, Italy
e-mail: massimo.panella@uniroma1.it

S. Scardapane
e-mail: simone.scardapane@uniroma1.it

R. Altilio
e-mail: rosa.altilio@uniroma1.it

A. Uncini
e-mail: aurelio.uncini@uniroma1.it

© Springer International Publishing AG 2018
A. Esposito et al. (eds.), *Multidisciplinary Approaches to Neural Computing*,
Smart Innovation, Systems and Technologies 69,
DOI 10.1007/978-3-319-56904-8_12

12.1 Introduction

Health care and biomedicine are two of the most prolific areas for the application of data mining methods [15], with successful implementations ranging from clustering of patients to rule extraction for expert systems, automatic diagnosis, and many others. In this paper, we are concerned with one particular aspect of medical scenarios, which hinders the use of standard machine learning techniques in practice. Specifically, many medical databases are *distributed* in nature [13], i.e. different parties may possess separate records on the process to be analysed. As an example, consider the problem of training a classifier to perform automatic diagnosis of a specific disorder (e.g. a cancer), starting from a set of standardized medical measurements. In this case, different hospitals have access to historical training data relative to disjoint patients, and it would be highly beneficial to collect these separate sources in order to train an effective classifier. At the same time, however, releasing medical data to a centralized location (to perform training) generally goes against a number of privacy concerns on sensible information, being subject to privacy attacks even if identifiers are removed before releasing it [2]. So the question becomes, is it possible to perform inference in a decentralized fashion (i.e., without the need for a central coordinator), and without requiring the exchange of training data?

In the literature, this is known as the problem of 'distributed machine learning', and many algorithms have been proposed to train specific classes of neural networks models, subject to the constraints detailed above. These include algorithms for distributed training of support vector machines (SVMs) [4, 9], random-weights networks [10, 11], kernel ridge regression [7], and many others, also considering computing energy constraints [1]. Our aim in this paper is instead more general, and starts from the known fact that a large number of learning techniques depend on the input data only through the computation of pairwise Euclidean distances among points. Examples of methods belonging to this category include kernel algorithms (e.g. SVMs), spectral clustering, k-means, and many others. Thus, instead of solving the original distributed learning problem, we can focus on the equivalent problem of completing in a distributed fashion the full matrix of Euclidean distances (EDM). Recasting the problem in this way allows us to leverage over a large number of works on matrix completion and EDM completion [6], especially in the distributed setting [3].

Particularly, we consider a distributed gradient-descent algorithm to this end, originally proposed for SVM inference over networks [3]. The proposed algorithm consists of two iterative steps, which are performed locally by every party (*agent*) in the network. First, each agent performs a single gradient descent step with respect to a locally defined cost function. Then, the new estimate of the EDM is averaged with respect to the estimates of other agents connected to it, and the process is repeated until convergence. Due to the way in which information is propagated, this kind of iterative techniques go under the general name of 'diffusion' strategies [8]. Additionally, we reduce the computational complexity by the exploitation of the specific structure of the EDM, by operating on a suitable factorization of the original matrix.

Once all the agents have access to the global estimate of the EDM, many data mining techniques can be applied directly (e.g. spectral clustering [14]), or by simple in-network operations (e.g. SVMs), as we discuss subsequently. Additionally, if there is the need of applying more than one technique, the same estimate can be reused for all of them, making the framework particularly useful whenever data must be used in an 'exploratory' fashion, without a particular predefined objective in mind. In order to show the applicability of the framework, we present experimental results for clustering of three well-known medical databases, showing that the solutions obtained are comparable to that of a fully centralized implementation.

The rest of the paper is organized as follows. In Sect. 12.2 we provide our algorithm for EDM completion. Since this requires the exchange of a small portion of the dataset, we present in the subsequent section two efficient methods to ensure privacy preservation. Then, a set of experimental evaluations are provided in Sect. 12.4, followed by some concluding remarks in Sect. 12.5.

12.2   Proposed Framework

Consider the application of a data mining procedure on a dataset of N examples $S = \left(\mathbf{x}_i\right)_{i=1}^{N} \in \mathbb{R}^d$, e.g. vectors to be suitably clustered. In a supervised setting, they can also be supplemented by additional labels. We assume that the dataset S is not available on a centralized location. Instead, it is partitioned over L agents (e.g. hospitals), such that the kth agent has access to a dataset S_k and $\bigcup_{k=1}^{L} S_k = S$. For generality, we can fully describe the connectivity between the agents in the form of an $L \times L$ connectivity matrix $\mathbf{C}$, where $C_{ij} \neq 0$ if and only if agents i and j are connected. In this paper, we assume that the network is connected (i.e., every agent can be reached from any other agent with a finite number of steps), and undirected (i.e., $\mathbf{C}$ is symmetric). Based on what we stated previously, we also assume that no coordinating entity is available, and communication is possible only if two agents are directly connected.

Suppose that the data mining procedure depends on the inputs $\mathbf{x}_i$ only through the computation of Euclidean distances among them (such as in the case of kernel methods). In this case, the overall distributed data mining procedure can be recast as the distributed computation of the Euclidean distance matrix (EDM) $\mathbf{E}$, where $E_{ij} = \left\|\mathbf{x}_i - \mathbf{x}_j\right\|_2$. For unsupervised problems, knowledge of this matrix is generally enough to solve the overall problem. In the supervised case, we would instead be left with a distributed optimization problem where only labels are distributed, which can be solved efficiently (see [3] for a fuller treatment on this aspect). To formalize this equivalent problem, we note that with a proper rearrangement of patterns, the global EDM $\mathbf{E}$ can always be expressed as:

$$
\mathbf{E} = \begin{bmatrix} \mathbf{E}_1 & ? & ? \\ ? & \ddots & ? \\ ? & ? & \mathbf{E}_L \end{bmatrix}, \tag{12.1}
$$

where $\mathbf{E}_k$ denotes the EDM computed only from the patterns in S_k. This structure implies that the sampling set is not random, and makes non-trivial the problem of completing $\mathbf{E}$ solely from the knowledge of the local matrices. At the opposite, the idea of exchanging the entire local datasets between nodes is unfeasible because of the amount of data which would need to be shared. Starting from these considerations, based on [3] we propose the following distributed procedure:

1. **Patterns exchange**: every agent exchanges a fraction p of the available S_k with its neighbours. This is necessary so that the agents can increase the number of known entries in their local matrices. How to ensure privacy in this step is described in the following section.
2. **Local EDM computation**: each agent computes, using its original dataset and the data received from its neighbours, an incomplete approximation $\hat{\mathbf{E}}_k \in \mathbb{R}^{N \times N}$ of the real EDM matrix $\mathbf{E}$.
3. **Entries exchange**: the agents exchange a sample of their local EDMs $\hat{\mathbf{E}}_k$ with their neighbours (similarly to step 1).
4. **Distributed EDM completion**: the agents complete the estimate $\tilde{\mathbf{E}}$ of the global EDM using the strategy detailed next.

To formalize this last step, define a local matrix $\boldsymbol{\Omega}_k$ as:

$$
\boldsymbol{\Omega}_k = \begin{cases} 1 & \text{if } \hat{E}_{ij} \neq 0 \\ 0 & \text{otherwise} \end{cases}. \tag{12.2}
$$

We aim at finding a matrix $\tilde{\mathbf{E}}$ such that the following (joint) cost function is minimized:

$$
\min_{\tilde{\mathbf{E}} \in \text{EDM}(N)} \sum_{k=1}^{L} J_k(\tilde{\mathbf{E}}) = \sum_{k=1}^{L} \left\| \boldsymbol{\Omega}_k \circ \left(\hat{\mathbf{E}}_k - \tilde{\mathbf{E}} \right) \right\|_F^2, \tag{12.3}
$$

where $\circ$ denotes the Hadamard product, and $\text{EDM}(N)$ is the set of EDMs of size $N \times N$. To solve problem (12.3) in a fully decentralized fashion, we use the algorithm introduced in [3], which in turn derives from the framework of diffusion adaptation (DA) for optimization [8] and on previous works on EDM completion [6]. In particular, we approximate the objective function in Eq. (12.3) by:

$$
J_k(\mathbf{V}) = \left\| \boldsymbol{\Omega}_k \circ \left[\hat{\mathbf{E}}_k - \kappa \left(\mathbf{V}\mathbf{V}^{\mathrm{T}} \right) \right] \right\|_F^2 \quad k = 1, \ldots, L, \tag{12.4}
$$

where $\kappa(\cdot)$ is the Schoenberg mapping, which maps every positive semidefinite (PSD) matrix to an EDM, given by:

$$\kappa(\mathbf{E}) = \mathrm{diag}(\mathbf{E})\mathbf{1}^{\mathrm{T}} + \mathbf{1}\mathrm{diag}(\mathbf{E})^{\mathrm{T}} - 2\mathbf{E}, \tag{12.5}$$

such that $\mathrm{diag}(\mathbf{E})$ extracts the main diagonal of $\mathbf{E}$ as a column vector, and we also exploits the known fact that any PSD matrix $\mathbf{D}$ with rank r admits a factorization $\{\mathbf{D} = \mathbf{V}\mathbf{V}^{\mathrm{T}}\}$, where $\mathbf{V} \in \mathbb{R}_*^{N \times r} = \{\mathbf{V} \in \mathbb{R}^{N \times r} : \det(\mathbf{V}^{\mathrm{T}}\mathbf{V}) \neq 0\}$. This allows to strongly reduce the computational cost of our algorithm, as the objective function is now formulated only in terms of the low-rank factor $\mathbf{V}$. The diffusion gradient descent for the distributed completion of the EDM is then defined by an alternation of updating and diffusion equations in the form of [3]:

1. **Initialization**: All the agents initialize the local matrices $\mathbf{V}_k$ as random $N \times r$ matrices.
2. **Update of V**: At time n, the kth agent updates the local matrix $\mathbf{V}_k$ using a gradient descent step with respect to its local cost function:

$$\tilde{\mathbf{V}}_k[n+1] = \mathbf{V}_k[n] - \eta_k[n]\nabla_{\mathbf{V}_k}J_k(\mathbf{V}). \tag{12.6}$$

where $\eta_k[n]$ is a positive step-size. It is straightforward to show that the gradient of the cost function is given by:

$$\nabla_{\mathbf{V}_k}J_k(\mathbf{V}) = \kappa^*\left\{\boldsymbol{\Omega}_k \circ \right.$$
$$\left. \circ\left(\kappa\left(\mathbf{V}_k[n]\,\mathbf{V}_k^{\mathrm{T}}[n]\right) - \hat{\mathbf{E}}_k\right)\right\}\mathbf{V}_k[n], \tag{12.7}$$

where $\kappa^*(\mathbf{A}) = 2\left[\mathrm{diag}(\mathbf{A}\mathbf{1}] - \mathbf{A}\right)$ is the adjoint operator of $\kappa(\cdot)$.
3. **Diffusion**: In order to propagate information over the network, the updated matrices are combined according to the mixing weights $\mathbf{C} \in \mathbb{R}^{L \times L}$:

$$\mathbf{V}_k[n+1] = \sum_{i=1}^{L} C_{ki}\tilde{\mathbf{V}}_i[n+1]. \tag{12.8}$$

where $C_{ki} > 0$ if and only if agents k and i are connected, in order to send information only through neighbours.

The above process is repeated for a maximum of T iterations to ensure convergence (see [8] for a general introduction on DA algorithms).

12.3 Techniques for Privacy Preservation

The algorithm in the previous section is extremely general, but its efficient implementation requires the distributed computation of a small subset of distances (step 1 in the algorithm). In this section, we show two techniques which are able to preserve privacy (i.e., avoid the exchange of the original data), during this phase.

The first is the random projection-based technique developed in [5]. Suppose that both agents agree on a projection matrix $\mathbf{R} \in \mathbb{R}^{m \times d}$, with $m < d$, such that each entry R_{ij} is independent and chosen from a normal distribution with mean zero and variance σ^2. We have the following lemma:

Lemma 1 *Given two input patterns* $\mathbf{x}_i, \mathbf{x}_j$, *and the respective projections:*

$$\mathbf{u}_i = \frac{1}{\sqrt{m\sigma}} \mathbf{R} \mathbf{x}_i, \text{ and } \mathbf{u}_j = \frac{1}{\sqrt{m\sigma}} \mathbf{R} \mathbf{x}_j, \tag{12.9}$$

we have that:

$$\mathbb{E}\left\{\mathbf{u}_i^T \mathbf{u}_j\right\} = \mathbf{x}_i^T \mathbf{x}_j. \tag{12.10}$$

Proof See [5, Lemma 5.2].

In light of Lemma 1, exchanging the projected patterns instead of the original ones allows to preserve, on average, their product. A thorough investigation on the privacy-preservation guarantees of this protocol can be found in [5]. Additionally, we can observe that this protocol provides a reduction on the communication requirements of the application, since it effectively reduces the dimensionality of the patterns to be exchanged by a factor m/d.

The second technique is the k-anonymity presented in [12]. In this case, we assume that the pattern $\mathbf{x}_i$ is composed by both *quasi-identifier* fields (e.g., age) and *sensible* fields (e.g., diagnosis). We say that a dataset is k-anonymous if, for any pattern, there exist at least $k-1$ other patterns with the same quasi-identifiers. It is possible to preserve k-anonymity by performing what is called "generalization" on the dataset [12], wherein the quasi-identifiers are binned in a set of Q predefined bins, and only the information on the corresponding bins is included in the dataset. Different values for Q correspond to different privacy values for k, with an inverse relation [12]. In this paper, we only wish to analyse the influence of this operation on our framework. For this reason, we choose to perform generalization artificially on the full dataset, while a decentralized implementation would require a sophisticated procedure going outside the scope of the paper.

12.4 Experimental Results

12.4.1 Experimental Setup

In this section, we evaluate the performance of the proposed algorithm for decentralized spectral clustering [14] with the privacy-preserving protocols described in Sect. 12.3. Note that spectral clustering can be achieved directly with the use of the

Table 12.1 Detailed description of each dataset

Dataset	Features	Instances	Classes
Pima Indians Diabetes	8	769	2
Breast Cancer Wisconsin	32	569	2
Parkinson Speech	26	1040	6

EDM, so no additional distributed step is necessary after completing the matrix. We consider three different (medical) public datasets available on the UCI repository,[1] a schematic description of which is given in Table 12.1. The number of attributes is always greater than three and depends on the specific features of the dataset. In all cases, for clustering the optimal solution is known beforehand for testing purpose. Below we add some additional information on each dataset.

- *Pima Indians Diabetes Dataset*: It is a classification dataset composed by 768 instances. The task is to identify whenever the tests are positive for diabetes or negative. Eight attributes are used for this purpose.
- *Breast Cancer Wisconsin Dataset*: It is a binary classification dataset of 569 instances composed by 32 attributes. The features describe the characteristics of the cell nuclei present in the image. The task is to identify the correct diagnosis (M = malignant, B = benign).
- *Parkinson Speech Dataset*: The dataset contains data of 20 Parkinson's Disease patients (PD) and 20 healthy subjects for which multiple types of sound recording are taken. Globally 1040 instances composed by 26 attributes are used to identify the correct type of sound recording (6 in total).

Five different runs of simulation are performed for each dataset which is preventively normalized between -1 and 1 before the experiments and randomly partitioned among the agents. A network of 7 agents is considered, where every pair of nodes is connected with a fixed probability $p = 0.5$ according to the so-called "Erdos-Rènyi model". The only requirement is that the graph is connected. We compare the following strategies:

- *Centralized:* this simulates the case where a dataset is collected beforehand on a centralized location (for comparison).
- *No-privacy:* the dataset is used with no privacy protocol applied to the data;
- *Randomization protocol:* the privacy of the data in step 1 is preserved by computing the distance on the projected patterns according to (12.9); parameter d is chosen in $k = [2, \ldots, 8]$ to minimize RMSE;
- *K-anonymity:* the privacy of the data is preserved by generalization on the quasi-identifiers of the dataset. We use 4 bins for each quasi-identifier.

---

[1]https://archive.ics.uci.edu/ml/datasets.html.

All experiments are carried out using MATLAB R2013b on a machine with Intel Core i5 processor with a CPU @ 3.00 GHz with 16 GB of RAM. All parameters of the algorithms are set according to [3].

12.4.2 Results and Discussion

We begin by evaluating the results of the framework with the randomization procedure. Three quality indexes are computed for both the privacy-preserving protocols and the privacy-free algorithm, namely the Rand Index, the Falks-Mallows index (F-M Index) and the F-measure. All of the indexes range in [0, 1], with 1 indicating a perfect correlation between the true label of the cluster and the output of the clustering algorithm, and 0 the perfect negative correlation. In Table 12.2 we report the mean and the standard deviation of each quality index averaged over 10 k-means evaluations and over the different agents in the distributed case. The best result for each index is highlighted in bold. The results of the three approaches are reasonably aligned; for all of the datasets they are very similar and in some cases the algorithm with privacy-preservation outperforms the traditional one. For evaluating the k-anonymity, we use the Pima Indians Diabetes Dataset described in Sect. 12.4, where the first and the eighth feature, that are respectively the number of pregnancies and the age of the subject, are used as quasi-identifiers. In Table 12.3 we computed the three quality indexes for the k-anonymity protocol, the randomization and the no-privacy transformation strategy. As shown in Table 12.3, we can obtain a comparable performance with respect to the privacy-free algorithm, additionally in the k-anonymity protocol the results are even better.

Table 12.2 Experimental results for the randomization. We show the average and the standard deviation of the indexes. Best results for each algorithm are highlighted in bold

Dataset	Algorithm	F-Measure	Rand-Index	F-M Index
Pima Indians Diabetes	Centralized	0.511 ± 0.000	$\mathbf{0.542 \pm 0.005}$	0.721 ± 0.014
	No-privacy	$\mathbf{0.682 \pm 0.106}$	0.505 ± 0.000	0.711 ± 0.000
	Randomization	0.679 ± 0.050	0.523 ± 0.004	$\mathbf{0.723 \pm 0.003}$
Breast Cancer Wisconsin	Centralized	$\mathbf{0.785 \pm 0.000}$	0.543 ± 0.004	0.728 ± 0.004
	No-privacy	0.772 ± 0.330	0.624 ± 0.169	0.779 ± 0.086
	Randomization	0.609 ± 0.389	$\mathbf{0.682 \pm 0.106}$	$\mathbf{0.815 \pm 0.041}$
Parkinson Speech	Centralized	0.665 ± 0.001	0.450 ± 0.000	0.705 ± 0.000
	No-privacy	$\mathbf{0.674 \pm 0.0.047}$	$\mathbf{0.504 \pm 0.000}$	$\mathbf{0.710 \pm 0.000}$
	Randomization	0.672 ± 0.027	0.501 ± 0.003	0.708 ± 0.002

Table 12.3 Experimental results for the k-anonymity. We show the average and the standard deviation of the F-Index, Rand Index, F-M Index for the Randomization, k-anonimity and privacy-free protocols. Best results for each algorithm are highlighted in bold

Dataset	Algorithm	F-Measure	Rand-Index	F-M Index
Pima Indians Diabetes	No-privacy	0.682 ± 0.106	0.505 ± 0.000	0.711 ± 0.000
	Randomization	0.679 ± 0.050	0.523 ± 0.004	0.723 ± 0.003
	K-anonymity	**0.779 ± 0.167**	**0.561 ± 0.031**	**0.749 ± 0.021**

12.5   Conclusion

In this paper, we presented a general framework for performing distributed data mining procedures on medical scenarios. The algorithms rely on the distributed computation of a matrix of distances, which is obtained via an innovative gradient descent procedure. Preliminary results on a clustering application show the feasibility of the approach, which is able to reach almost-optimal performance with respect to a fully centralized implementation. Additionally, we have investigated two different techniques allowing to preserve privacy even during the exchange of patterns among agents. Future research direction will involve designing more efficient procedures for the distributed computation of the EDM, together with an analysis of the different customizations of the framework for multiple algorithms.

References

1. Baccarelli, E., Cordeschi, N., Mei, A., Panella, M., Shojafar, M., Stefa, J.: Energy-efficient dynamic traffic offloading and reconfiguration of networked data centers for big data stream mobile computing: review, challenges, and a case study. IEEE Netw. **30**(2), 54–61 (2016)
2. Clifton, C., Kantarcioglu, M., Vaidya, J., Lin, X., Zhu, M.Y.: Tools for privacy preserving distributed data mining. ACM SiGKDD Explor. Newsl. **4**(2), 28–34 (2002)
3. Fierimonte, R., Scardapane, S., Uncini, A., Panella, M.: Fully decentralized semi-supervised learning via privacy-preserving matrix completion. IEEE Trans. Neural Netw. Learn. Syst. (2016) in press. doi:10.1109/TNNLS.2016.2597444
4. Forero, P.A., Cano, A., Giannakis, G.B.: Consensus-based distributed support vector machines. JMLR **11**, 1663–1707 (2010)
5. Liu, K., Kargupta, H., Ryan, J.: Random projection-based multiplicative data perturbation for privacy preserving distributed data mining. IEEE Trans. Knowl. Data Eng. **18**(1), 92–106 (2006)
6. Mishra, B., Meyer, G., Sepulchre, R.: Low-rank optimization for distance matrix completion. In: 2011 50th IEEE Conference on Decision and Control and European Control Conference (CDC-ECC'11), pp. 4455–4460. IEEE (2011)
7. Predd, J.B., Kulkarni, S.R., Poor, H.V.: Distributed learning in wireless sensor networks. IEEE Signal Process. Mag. **23**(4), 56–69 (2006)
8. Sayed, A.H.: Adaptive networks. Proc. IEEE **102**(4), 460–497 (2014)
9. Scardapane, S., Fierimonte, R., Di Lorenzo, P., Panella, M., Uncini, A.: Distributed semi-supervised support vector machines. Neural Netw. **80**, 43–52 (2016)

10. Scardapane, S., Wang, D., Panella, M.: A decentralized training algorithm for echo state networks in distributed big data applications. Neural Netw. **78**, 65–74 (2016)
11. Scardapane, S., Wang, D., Panella, M., Uncini, A.: Distributed learning for random vector functional-link networks. Inf. Sci. **301**, 271–284 (2015)
12. Sweeney, L.: k-anonymity: A model for protecting privacy. Int. J. of Uncertainty, Fuzziness and Knowledge-Based Systems **10**(05), 557–570 (2002)
13. Vieira-Marques, P.M., Robles, S., Cucurull, J., Navarro, G., et al.: Secure integration of distributed medical data using mobile agents. IEEE Intelligent Systems **21**(6), 47–54 (2006)
14. Von Luxburg, U.: A tutorial on spectral clustering. Statistics and computing **17**(4), 395–416 (2007)
15. Yoo, I., Alafaireet, P., Marinov, M., Pena-Hernandez, K., Gopidi, R., Chang, J.F., Hua, L.: Data mining in healthcare and biomedicine: a survey of the literature. J. Med. Syst. **36**(4), 2431–2448 (2012)

Chapter 13
Rule Base Reduction Using Conflicting and Reinforcement Measures

Luca Anzilli and Silvio Giove

Abstract In this paper we present an innovative procedure to reduce the number of rules in a Mamdani rule-based fuzzy systems. First of all, we extend the similarity measure or degree between antecedent and consequent of two rules. Subsequently, we use the similarity degree to compute two new measures of conflicting and reinforcement between fuzzy rules. We apply these conflicting and reinforcement measures to suitably reduce the number of rules. Namely, we merge two rules together if they are redundant, i.e. if both antecedent and consequence are similar together, repeating this operation until no similar rules exist, obtaining a reduced set of rules. Again, we remove from the reduced set the rule with conflict with other, i.e. if antecedent are similar and consequence not; among the two, we remove the one characterized by higher average conflict with all the rules in the reduced set.

Keywords Fuzzy systems · Rule base reduction · Rule base simplification · Conflicting and reinforcement measures

13.1 Introduction

The number of rules in a fuzzy system (FIS, Fuzzy Inference System) exponentially increases with the number of the input variables and the number of the linguistic values that these inputs can take (antecedent fuzzy terms) [1, 2]. Several approaches for reducing fuzzy rule base have been proposed using different techniques such as interpolation methods, orthogonal transformation methods, clustering techniques [3–8]. A typical tool to perform model simplification is merging similar fuzzy sets and rules using similarity measures [9–14].

L. Anzilli (✉)
Department of Management, Economics, Mathematics and Statistics,
University of Salento, Lecce, Italy
e-mail: luca.anzilli@unisalento.it

S. Giove
Department of Economics, University Ca' Foscari of Venice, Venice, Italy
e-mail: sgiove@unive.it

© Springer International Publishing AG 2018   129
A. Esposito et al. (eds.), *Multidisciplinary Approaches to Neural Computing*,
Smart Innovation, Systems and Technologies 69,
DOI 10.1007/978-3-319-56904-8_13

In this paper we propose a new procedure for simplifying rule-based fuzzy systems. Starting from similarity measures we introduce two new measures of conflicting and reinforcement between fuzzy sets. Then we develop a simplification methodology using the introduced conflicting and reinforcement measures to merge similar rules and to remove redundant rules from the rule set.

The paper is organized as follows. In Sect. 13.2 we briefly review the basic notions of fuzzy systems. In Sect. 13.3 we define conflicting and reinforcement measures. In Sect. 13.4 we present the merging methodology and, finally, in Sect. 13.5 we illustrate our rule-base reduction method.

13.2 Fuzzy Systems

The *knowledge* of a FIS can be obtained from available data using some optimization tool as a neural approach, or by direct elicitation from one or a group of Experts. In the latter case, the Experts represent their knowledge by defining a set of inferential rules. The input variables are processed by these rules to generate an appropriate output.

In the case of a FIS with n input variables, $x_1, \ldots, x_n$ and a single output y (miso fuzzy system, [2]) every rule has the form

$$R_i : \quad \text{IF } x_1 \text{ is } A_{i,1} \text{ and }, \ldots, \text{ and } x_n \text{ is } A_{i,n} \text{ THEN } y \text{ is } B_i \qquad i = 1, \ldots, N$$

where $A_{i,j}$ is a fuzzy sets of universe space X_j and B_i is a fuzzy set of universe space Y, and N is the number of rules. The fuzzy set $A_{i,j}$ is the *linguistic label* associated with j-th antecedent in the i-th rule and B_i is the linguistic label associated with the consequent in the i-th rule. We recall that a linguistic label can be easily represented by a fuzzy set [15]. The rule i, R_i, can be represented by the ordered couple $R_i = \left(\bigcap_{j=1}^{n} A_{i,j}(x_j), B_i \right)$, being $A_{i,j}(x_j)$ the j-th component of the antecedent and B_i the consequent, $i = 1, \ldots, N$, and $\bigcap$ is the conjunction operator. Every Rule R_i in the data base is characterized by a confidence degree e_i (or rule weight), with $e_i > 0$ (see [8, 16])). Rule weights can be applied to complete rules or only to the consequent part of the rules [16]. In the first case, the weight is used to modulate the activation degree of a rule, and in second to modulate the rule conclusion.

13.3 Conflicting and Supporting Rules

13.3.1 Similarity Measures Between Fuzzy Sets

We denote by $Sim(A, B)$ the similarity between fuzzy sets A and B with respect to a similarity measure Sim. Different similarity measures for fuzzy sets have been proposed in literature [17–23]. They can be classified into two main groups:

geometric and *set-theoretic* similarity measures. Axiomatic definitions of similarity can be found in [18, 23].

An example of a geometric similarity measure based on distance between two fuzzy sets is

$$Sim_1(A, B) = \frac{1}{1 + D(A, B)}$$

being $D(A, B)$ a suitable distance among the two fuzzy sets A and B.

An example of a similarity measure between two fuzzy sets, based on the set-theoretic operations of intersection and union, is (see [15])

$$Sim_2(A, B) = \frac{M(A \cap B)}{M(A \cup B)} \tag{13.1}$$

where

$$M(A) = \int_{-\infty}^{+\infty} A(x)\, dx. \tag{13.2}$$

is the size of fuzzy set A. Details for computing (13.1) are given in [4, 14].

13.3.2 Similarity Measures Between Rules

Let us consider a fuzzy system with n input variables ($=$ number of antecedents of each rule) and N rules

$$R_i = \left(\bigcap_{j=1}^{n} A_{i,j}(x_j), B_i \right), \qquad i = 1, \dots, N$$

being $A_{i,j}(x_j)$ the j-th component of the antecedent and B_i the consequent. Each Rule R_i in the data base will be characterized also by a confidence degree e_i.

Definition 13.1 Following measures will be considered (see [11, 14]):

(1) Similarity between the antecedent of two rules, R_k, R_ℓ $(k, \ell = 1, 2 \dots, N)$

$$\mu_{k,l} = Sim(Ant_k, Ant_\ell) = T_{j=1}^{n}\, Sim(A_{k,j}, A_{\ell,j})$$

where T is a t-norm (in [11, 14] $T = \min$);

(2) Similarity between the consequent of two rules, R_k, R_ℓ

$$v_{k,l} = Sim(Cons_k, Cons_\ell) = Sim(B_k, B_\ell).$$

13.3.3 Conflicting and Reinforcement Degrees

Definition 13.2 Based on the two above measures, $\mu_{k,\ell}$ and $\nu_{k,\ell}$, we can propose the following conflicting and reinforcing degrees:

(i) *Conflicting degree*, a measure of the conflict among a couple of rules:

$$c(k, \ell) = \mu_{k,\ell} \, (1 - \nu_{k,\ell}) f(e_k, e_\ell) ; \qquad\qquad (13.3)$$

(ii) *Reinforcement degree*, a measure of the agreement among a couple of rules:

$$r(k, \ell) = \mu_{k,\ell} \, \nu_{k,\ell} f(e_k, e_\ell) \qquad\qquad (13.4)$$

being $f(e_k, e_\ell)$ a suitable aggregation function, symmetric and idempotent, not decreasing in both its two arguments.

Proposition 13.1 *Conflicting and reinforcement degrees satisfy the following properties: for any $k, \ell = 1, 2 \ldots, N$*

(i) $0 \le c(k, \ell) \le 1, \, 0 \le r(k, \ell) \le 1$
(ii) $c(k, \ell) = c(\ell, k), \, r(k, \ell) = r(\ell, k)$
(iii) $c(k, k) = 0, \, r(k, k) = f(e_k, e_k) = e_k$
(iv) $c(k, \ell) = 1 \implies r(k, \ell) = 0$
(v) $r(k, \ell) = 1 \implies c(k, \ell) = 0$
(vi) $0 \le c(k, \ell) + r(k, \ell) \le \min\{\mu_{k,\ell}, f(e_k, e_\ell)\} \le 1$
(vii) *if the aggregation function f is such that $f(e_k, e_\ell) = 0$ only if $c_k = 0$ or $c_\ell = 0$, that is the only annihilator element (see [24]) of f is 0, then:*

$$\mu_{k,\ell} > 0 \implies c(k, \ell) + r(k, \ell) > 0 .$$

Proof Property (ii) holds since similarity measure is symmetric. Property (iii) follows taking into account that $\mu_{k,k} = \nu_{k,k} = 1$ since $Sim(A, A) = 1$ (assuming that Sim is a normal similarity measure). Properties (vi) and (vii) follow from the relation

$$c(k, \ell) + r(k, \ell) = \mu_{k,\ell} \cdot f(e_k, e_\ell)$$

and observing that, since $e_\ell > 0$ for any ℓ, we have $f(e_k, e_\ell) > 0$. $\square$

We observe that both $c(k, \ell)$ and $r(k, \ell)$ can be equal to zero, but if one is close to one the other is close to zero.

13.4   Merging Methodology

13.4.1   Merging Fuzzy Sets

Different shape of membership functions exist in the specialized literature. Among them we recall trapezoidal, triangular, bell-shaped fuzzy number. A trapezoidal fuzzy number is defined[1] by the 4-ple $A = (a_1, a_2, a_3, a_4)$, with $a_1 < a_2 \leq a_3 < a_4$, and has membership function

$$A(x) = \begin{cases} 0 & x \leq a_1 \text{ or } x \geq a_4 \\ \dfrac{x - a_1}{a_2 - a_1} & a_1 \leq x \leq a_2 \\ 1 & a_2 \leq x \leq a_3 \\ \dfrac{a_4 - x}{a_4 - a_3} & a_3 \leq x \leq a_4 \end{cases}$$

More in general, if the (continuous) fuzzy number A is characterized by the membership $A(x)$, its α-cuts are by the intervals $A(\alpha) = \{x | A(x) \geq \alpha\} = [a_1(\alpha), a_2(\alpha)]$, with $\alpha \in [0, 1]$. The size $M(A)$ of fuzzy set A, as defined in (13.2), can be computed using α-cuts by

$$M(A) = \int_0^1 M(A(\alpha)) \, d\alpha$$

where $M(A(\alpha))$ is the size of α-cut $A(\alpha)$. We extend this concept by defining

$$M_l(A) = \int_0^1 M(A(\alpha)) \, l(\alpha) \, d\alpha$$

being $l(\alpha)$ a suitable *weighting* function, $l(\alpha) : [0, 1] \rightarrow [0, 1]$ (see [25, 26]). Then we introduce the following extended version of the similarity (13.1)

$$Sim_2^l(A, B) = \frac{M_l(A \cap B)}{M_l(A \cup B)} \, .$$

In order to compute the similarity, we observe that a discrete representation of A can be done through a finite subset of its α-cuts, see [15]. Particularly useful is an equally spaced grid for α, as $\alpha_i = \frac{i}{T}$, $i = 0, 1, .., T$, with step size $\frac{1}{T}$. For a discretized fuzzy number A we have

$$M_l(A) = \sum_{i=1}^{T} \int_{\alpha_{i-1}}^{\alpha_i} M(A(\alpha)) \, l(\alpha) \, d\alpha \approx \frac{1}{T} \sum_{i=1}^{T} M(A(\alpha_i)) \, l(\alpha_i) \, .$$

---

[1]A triangular fuzzy number is a sub-case of a trapezoidal one, with $a_2 = a_3$, while a *bell-shape* recalls a gaussian distribution.

Then, taking into account that $(A \cap B)(\alpha) = A(\alpha) \cap B(\alpha)$ and $(A \cup B)(\alpha) = A(\alpha) \cup B(\alpha)$, we get the following formulas

$$M_l(A \cap B) = \frac{1}{T} \sum_{i=1}^{T} M((A(\alpha_i) \cap B(\alpha_i)) \, l(\alpha_i)$$

$$M_l(A \cup B) = \frac{1}{T} \sum_{i=1}^{T} M((A(\alpha_i) \cup B(\alpha_i)) \, l(\alpha_i)$$

where $M((A(\alpha_i) \cap B(\alpha_i)) = \max\{min(a_2(\alpha_i), b_2(\alpha_i)) - max(a_1(\alpha_i), b_1(\alpha_i)), 0\}$ and $M((A(\alpha_i) \cup B(\alpha_i)) = M(A(\alpha_i)) + M(B(\alpha_i)) - M(A(\alpha_i) \cap B(\alpha_i))$. As a consequence, the similarity degree among two discretized fuzzy numbers A and B is given by

$$Sim_2^l(A, B) = \frac{\sum_{i=1}^{T} M((A(\alpha_i) \cap B(\alpha_i)) \, l(\alpha_i)}{\sum_{i=1}^{T} M((A(\alpha_i) \cup B(\alpha_i)) \, l(\alpha_i)} \, .$$

Two fuzzy sets A and B can be merged into a fuzzy set $C = \lambda A + (1 - \lambda) B$, where $0 < \lambda < 1$ is a suitably selected parameter. If $A = (a_1, a_2, a_3, a_4)$ and $B = (b_1, b_2, b_3, b_4)$ are trapezoidal fuzzy numbers, the merged (trapezoidal) fuzzy number $C = (c_1, c_2, c_3, c_4)$ is given by[2]

$$c_1 = \lambda a_1 + (1 - \lambda) b_1$$
$$c_2 = \lambda a_2 + (1 - \lambda) b_2$$
$$c_3 = \lambda a_3 + (1 - \lambda) b_3$$
$$c_4 = \lambda a_4 + (1 - \lambda) b_4 \, .$$

13.4.2 Merging Rules

Let us fix a pre-specified antecedent-similarity threshold $\bar{\mu} > 0$. If $\mu_{k,\ell} > \bar{\mu}$ then we can merge Rules R_k, R_ℓ into a single rule $R_{(k,\ell)}$ with confidence degree $e_{k,l}$ given by

$$e_{k,l} = h(c(k, \ell), r(k, \ell)) \cdot f(e_k, e_\ell) \, . \tag{13.5}$$

We require that function $h : [0, 1]^2 \to [0, 1]$ satisfy the following properties:

(i) $h(c, r)$ is decreasing with respect to c
(ii) $h(c, r)$ is increasing with respect to r

---

[2]Alternatively, in [9] the following merging procedure is proposed: if $A = (a_1, a_2, a_3, a_4)$ and $B = (b_1, b_2, b_3, b_4)$ are trapezoidal fuzzy numbers, the merged (trapezoidal) fuzzy number $C = (c_1, c_2, c_3, c_4)$ is obtained by $c_1 = \min(a_1, b_1), c_2 = \lambda_2 a_2 + (1 - \lambda_2) b_2, c_3 = \lambda_3 a_3 + (1 - \lambda_3) b_3, c_4 = \max(a_4, b_4)$, where $\lambda_2, \lambda_3 \in [0, 1]$.

(iii) $h(1, r) = 0$
(iv) $h(0, r) = r$.

Examples of functions $h(c, r)$ are:

(a) $h_1(c, r) = r(1 - c)$
(b) $h_2(c, r) = T(n(c), r)$, where T is a t-norm and n is a fuzzy complement (that is a decreasing function $n : [0, 1] \to [0, 1]$ such that $n(0) = 1$ and $n(1) = 0$).
    We observe that h_1 is a special case of h_2 with the product t-norm $T = T_P$ and the standard fuzzy complement $n(c) = 1 - c$.
(c) $h_3(c, r) = \frac{r}{c+r}$ (we may set $h_2(c, r) = 1$ if $c + r = 0$; note that for $\mu_{k,\ell} > \bar{\mu} > 0$ we have $c(k, \ell) + r(k, \ell) > 0$). We observe that $h_3(c, r) = v_{k,\ell}$.

13.5   Rule Base Reduction Method

We now present a reduction algorithm to perform a rule base simplification of a Mamdani fuzzy system [15]. The main idea is the following: if two rules have similar antecedents and similar consequents we merge them into a single rule; if two rules have similar antecedents but dissimilar consequents we remove the rule which have the greater conflict. Our method consists two steps.

First step. We merge two Rules characterized by a value of similarity $\mu_{k,\ell}$ greater than a pre-specified threshold $\bar{\mu}$ and a value $v_{k,\ell}$ greater than a pre-specified threshold $\bar{v}$ in a single rule. The antecedent (consequent) of the merged Rule will be obtained merging together the antecedents (consequents) of the two Rules using a suitable averaging operator. The confidence degree of the merged Rule will be increased in the case of reinforcement, but decreased in the case of conflicting. The merging algorithm thus will proceed considering every couple of Rules, selecting the most antecedent-similar couple and merge the two Rules into a single one, consequently modifying the aggregated confidence. The merging procedure will continue until two Rules with antecedent-similarity and consequent-similarity greater that a specified threshold exist in the data base.

Second step. We consider the reduced rule set $\mathscr{R}$. If two Rules R_k, R_ℓ have a value of similarity $\mu_{k,\ell}$ greater than a pre-specified threshold $\bar{\mu}$ then we compute the total conflicting degrees

$$c_k = \sum_{R_m \in \mathscr{R}} c(k, m), \qquad c_\ell = \sum_{R_m \in \mathscr{R}} c(k, m)$$

and remove the Rule having the greater conflict degree. The removing procedure will continue until two Rules with antecedent-similarity greater than $\bar{\mu}$ and different total conflicting degree exist in the data base.

The previous methodology can be formalized in the following algorithm:

1. calculate $\mu_{k,\ell}$ and $v_{k,\ell}$ for $k, \ell = 1, \dots, N$;
2. calculate the values of $c(k, \ell)$ and $r(k, \ell)$ for $k, \ell = 1, \dots, N$;
3. for each $k, \ell = 1, \dots, N, k \neq \ell$: if $\mu_{k,\ell} > \bar{\mu}$ and $v_{k,\ell} > \bar{v}$ then we merge R_k and R_ℓ and assign to merged Rule $R_{(k,\ell)}$ a confidence degree $e_{(k,\ell)}$ computed according to (13.5);
4. let $\mathscr{R} = \{R_1, \dots, R_p\}, p \leq N$, be the new (reduced) rule set;
5. for each $k, \ell = 1, \dots, p, k \neq \ell$: if $\mu_{k,\ell} > \bar{\mu}$ (and thus $v_{k,\ell} \leq \bar{v}$) then

   - if $c_k > c_\ell$ we remove R_k,
   - if $c_k < c_\ell$ we remove R_ℓ.

13.6 Conclusion

In this paper we proposed a novel methodology for Rule base reduction of Mamdani fuzzy systems based on conflicting and reinforcement measures. The reduction is achieved by merging antecedents and consequents of two Rules and assigning to the merged Rule an increased (decreased) confidence degree in the case of reinforcement (conflicting).

As a future development, we intend to investigate the properties of conflicting and reinforcement measures and, moreover, to apply the proposed simplification procedure to Takagi-Sugeno fuzzy systems.

References

1. Simon, D.: Design and rule base reduction of a fuzzy filter for the estimation of motor currents. Int. J. Approximate Reasoning **25**(2), 145–167 (2000)
2. Lazzerini, B., Marcelloni, F.: Reducing computation overhead in miso fuzzy systems. Fuzzy Sets Syst. **113**(3), 485–496 (2000)
3. Bellaaj, H., Ketata, R., Chtourou, M.: A new method for fuzzy rule base reduction. J. Intell. Fuzzy Syst. Appl. Eng. Technol. **25**(3), 605–613 (2013)
4. Tsekouras, G.E.: Fuzzy rule base simplification using multidimensional scaling and constrained optimization. Fuzzy Sets Syst. (2016) (to appear)
5. Baranyi, P., Kóczy, L.T., Gedeon, T.T.D.: A generalized concept for fuzzy rule interpolation. Fuzzy Syst. IEEE Trans. **12**(6), 820–837 (2004)
6. Wang, H., Kwong, S., Jin, Y., Wei, W., Man, K.-F.: Multi-objective hierarchical genetic algorithm for interpretable fuzzy rule-based knowledge extraction. Fuzzy Sets Syst. **149**(1), 149–186 (2005)
7. Nefti, S., Oussalah, M., Kaymak, U.: A new fuzzy set merging technique using inclusion-based fuzzy clustering. Fuzzy Syst. IEEE Trans. **16**(1), 145–161 (2008)
8. Riid, A., Rüstern, E.: Adaptability, interpretability and rule weights in fuzzy rule-based systems. Inf. Sci. **257**, 301–312 (2014)
9. Setnes, M., Babuška, R., Kaymak, U., van Nauta Lemke, H.R.: Similarity measures in fuzzy rule base simplification. Syst. Man Cybern. Part B Cybern. IEEE Trans. **28**(3), 376–386 (1998)

10. Chen, M.-Y., Linkens, D.A.: Rule-base self-generation and simplification for data-driven fuzzy models. Fuzzy Sets Syst. **142**, 243–265 (2004)
11. Jin, Y.: Fuzzy modeling of high-dimensional systems: complexity reduction and interpretability improvement. Fuzzy Syst. IEEE Trans. **8**(2), 212–221 (2000)
12. Chao, C.-T., Chen, Y.-J., Teng, C.-C.: Simplification of fuzzy-neural systems using similarity analysis. Syst. Man Cybern. Part B Cybern. IEEE Trans. **26**(2), 344–354 (1996)
13. Babuška, R., Setnes, M., Kaymak, U., van Nauta Lemke, H.R.: Rule base simplification with similarity measures. In: Proceedings of the Fifth IEEE International Conference on Fuzzy Systems, 1996, vol. 3, pp. 1642–1647. IEEE (1996)
14. Jin, Y., Von Seelen, W., Sendhoff, B.: On generating FC3 fuzzy rule systems from data using evolution strategies. Syst. Man Cybern. Part B Cybern. IEEE Trans. **29**(6), 829–845 (1999)
15. Dubois, D., Prade, H.: Fuzzy sets and systems: theory and applications, vol. 144. Academic Press (1980)
16. Nauck, D., Kruse, R.: How the learning of rule weights affects the interpretability of fuzzy systems. In: Fuzzy Systems Proceedings, 1998. IEEE World Congress on Computational Intelligence., The 1998 IEEE International Conference on, vol. 2, pp. 1235–1240. IEEE (1998)
17. Zadeh, L.A.: Similarity relations and fuzzy orderings. Inf. Sci. **3**(2), 177–200 (1971)
18. Wang, W.-J.: New similarity measures on fuzzy sets and on elements. Fuzzy Sets Syst. **85**(3), 305–309 (1997)
19. Johanyák, Z.C., Kovács, S.: Distance based similarity measures of fuzzy sets. In: Proceedings of SAMI, vol. 2005 (2005)
20. Beg, I., Ashraf, S.: Similarity measures for fuzzy sets. Appl. Comput. Math **8**(2), 192–202 (2009)
21. Deng, G., Jiang, Y., Fu, J.: Monotonic similarity measures between fuzzy sets and their relationship with entropy and inclusion measure. Fuzzy Sets and Syst. (2015)
22. Zwick, R., Carlstein, E., Budescu, D.V.: Measures of similarity among fuzzy concepts: a comparative analysis. Int. J. Approximate Reasoning **1**(2), 221–242 (1987)
23. Couso, I., Garrido, L., Sánchez, L.: Similarity and dissimilarity measures between fuzzy sets: A formal relational study. Inf. Sci. **229**, 122–141 (2013)
24. Grabisch, M., Marichal, J.-L., Mesiar, R., Pap, E.: Aggregation functions: means. Inf. Sci. **181**(1), 1–22 (2011)
25. Delgado, M., Vila, M.A., Voxman, W.: On a canonical representation of fuzzy numbers. Fuzzy Sets Syst. **93**(1), 125–135 (1998)
26. Facchinetti, G.: Ranking functions induced by weighted average of fuzzy numbers. Fuzzy Optim. Decis. Making **1**(3), 313–327 (2002)

Chapter 14
An Application of Internet Traffic Prediction with Deep Neural Network

Sanam Narejo and Eros Pasero

Abstract The advance knowledge of future traffic load is helpful for network service providers to optimize the network resource and to recover the demand criteria. This paper presents the task of internet traffic prediction with three different architectures of Deep Belief Network (DBN). The artificial neural network is created with the depth of 4 hidden layers in each model to learn the nonlinear hierarchal essence present in the time series of internet traffic data. The deep learning in the network is executed with unsupervised pretraining of the layers. The emphasis is given to the topology of DBN that achieves excellent prediction accuracy. The adopted approach provides accurate traffic predictions while simulating the traffic data patterns and stochastic elements, achieving 0.028 Root Mean Square Error (RMSE) value on the test data set. To validate our choice for hidden layer size selection, further more experiments were done for chaotic time series prediction.

Keywords Deep learning · Deep belief networks · Internet traffic prediction · Restricted boltzmann machine

14.1 Introduction

In this contemporary era, the world of internet is facing an increasing challenge to design more reliable and robust strategies for network communication. The rapid changes in the network topologies bring abundant changes in data traffic. Internet Service Provider (ISP) is an organization that provides services for accessing the internet. Figure 14.1 shows the infrastructure of an ISP. Quality of Service (QoS) is

S. Narejo (✉) · E. Pasero (✉)
Department of Electronics and Telecommunications, Politecnico Di Torino,
Torino, Italy
e-mail: sanam.narejo@polito.it

E. Pasero
e-mail: eros.pasero@polito.it

© Springer International Publishing AG 2018
A. Esposito et al. (eds.), *Multidisciplinary Approaches to Neural Computing*,
Smart Innovation, Systems and Technologies 69,
DOI 10.1007/978-3-319-56904-8_14

Fig. 14.1 Infrastructure of an
ISP

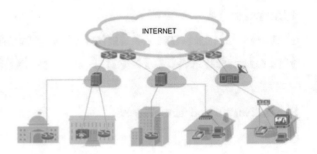

the concept in which error rates, transmission rates and other physical characteristics of the network can be measured, improved, and, to some extent, guaranteed in advance. Analysis of the network traffic load plays a crucial role in optimization of resources, design, management and control of network services [1]. The ATM and high speed networks provide integrated services of voice, video and data. These types of services need a good congestion control mechanism on the network while maintaining the maximum network utilization, optimizing the resource allocation such that the service provision imitates with quality of service constraints. Therefore, internet traffic prediction is very important for the progress of more proficient traffic engineering and anomaly detection tools. On the other hand, the models developed for future traffic forecasting do support the detection of anomalies residing in the data networks. Moreover, irregular amount of spam elements and other security attacks can be identified by comparing the actual traffic with the predictions of forecasting models.

Traffic modeling is fundamental to the network performance evaluation. Traffic measurement studies have demonstrated the nature of network load as nonlinear and self-similar [2, 3]. The transportation data is demonstrated as time series with nonlinear and chaotic characteristics [4, 5]. The traffic data in time series is found to be correlated over both short and long time scales. Hence, most of important correlations and bursting properties are often ignored when traffic is characterized using simple models. Consequently, our model follows the approach of deep Artificial Neural Network (ANN).

The purpose of our work is to address an approach for time series prediction by using the essential attributes of deep learning and its significant strategies for architectural topologies. The proposed models in our approach investigate the structure of internet traffic time series to grasp the accuracy in predictions. The models are based on stacking of Restricted Boltzmann Machines (RBMs) to create a deep architecture of neural network. The layers of an RBM are first trained in an unsupervised fashion. Each layer extracts the nonlinear representation of data. The abstractions of the top hidden layer form highly nonlinear hierarchal features as input set for predicting the network traffic load. In this manner, the estimated future traces of traffic load are predicted at the output layer which is trained in a supervised manner at the fine tuning step of the whole model.

14.2   Theoretical Background

The use of ANN is encapsulated in a wide area of applications on the grounds of its generalization capability to unforeseen situations. ANN is trained on a set of inputs to reach a specific target output using an appropriate learning algorithm until the outcomes of the network tend to match the given targets. This way the parameters of the network are selected, which are able to deduce the nonlinear relation between given inputs and outputs. To this extent, the trained model estimates the predictions on unseen input samples. In order to increase the efficiency of ANN, its computational capacity requires more neurons per layer or increasing the number of hidden layers. This results in the deep architecture of neural computational model, in which the intermediate layers or hidden layers perform as multiple layers of abstraction. Massive research is taking place as far as deep architectures of ANN are concerned.

Deep ANNs are achieving encouraging improvements over previous shallow ANN architectures [6]. Deep ANNs contain numerous levels of non-linearities depending on the depth of hidden layers. The deep hierarchical architecture allows them to efficiently represent highly nonlinear patterns and highly-varying functional abstractions. Although, it was not clear how to train such deep networks, as the random initialization of network parameters appears to often get stuck in poor solutions [7]. In [8] a learning algorithm was introduced that greedily trains, one a layer of network at a time. Deep architectures can be formed either by stacking autoencoders or Deep Belief Networks (DBNs) of RBMs [8].

DBN is the most common and effective approach among all deep learning models. DBN is formed by stacking of RBMs. RBM is a bipartite graphical, energy based, probabilistic model and it is interpreted as stochastic neural network. RBM relies on two layer structure comprising on visible and hidden nodes as shown in Fig. 14.2.

These nodes are binary units which means h and v $\in \{0, 1\}$. These nodes are conditionally independent given one another. Therefore, the probabilistic representation of these hidden and visible nodes with sigmoid activation function i.e. $\sigma(x) = 1/(1 + e^{-x})$ is given as in Eqs. (1) and (2). W_{ij} is a weight associated with the edge between units V_j and h_i. Whereas b_j and c_i are real valued bias terms associated with the jth visible and the ith hidden variable, respectively. Sampling in an RBM is obtained by running a Markov chain to convergence, using Gibbs

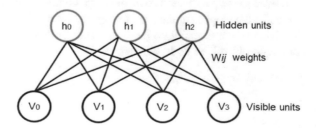

Fig. 14.2 Graphical structure of RBM with 3 hidden units and 4 visible units

sampling. The weight parameter is updated with the rate of change as shown in Eq. (3) [8].

$$p(h_{i=1}|v) = \sigma\left(\sum_{j=1}^{m} w_{ij}v_j + c_i\right) \qquad (1)$$

$$p(V_{j=1}|h) = \sigma\left(\sum_{i=1}^{n} w_{ij}h_i + b_j\right) \qquad (2)$$

$$\Delta wij = \in (\langle vjhi\rangle_{data} - \langle vjhi\rangle_{model}) \qquad (3)$$

14.3 Related Work

The literature review reveals the presence of various models available for time series analysis, prediction and forecasting. Specifically focusing on the internet traffic predictions, internet traffic has been empirically studied by conducting statistical analysis [9]. The traditional statistical approaches include Box-Jenkins, Autoregressive Moving Average (ARMA) and its variational models. In another research work, the traffic load is also modeled using filtering of non-stationary components and applying the nonlinear threshold autoregressive model [10]. These models are based on discrete linear stochastic processes. Therefore, these models only capture the linear correlation structure present in the time series irrespective of identifying any nonlinear patterns.

Most statistical methods are parametric models that need a great deal of statistical information, whereas ANNs are non-parametric, data-driven and self-adaptive models. The experts in [11] applied the nonlinear time series prediction models for internet traffic prediction and evaluated their effectiveness. These models included Radial Basis Function (RBF) and Support Vector Machine (SVM). A comparative study of Time Series Forecasting methods was conducted on the basis of TCP/IP traffic [12]. Their proposed neural assembled technique as an online forecasting system produced the lowest error as compared to other models.

Recent studies demonstrate the implementation of deep ANN models for time series prediction and forecasting [13–15]. The deep learning model is implemented for internet traffic prediction problem [16]. The stacked autoencoder approach is followed as deep architecture in above mentioned research problems. In [17] a DBN model is developed for the road transportation problem. However, to the best of our knowledge, the research on deep belief network and stacked RBMs for time series prediction is still in its initial stages.

14.4 Formulation of Proposed Models

As it is outlined in the introduction, deep learning is followed as the basic structure for developing a new neural predictive model for our work. In this work, we present three different architectures of the deep learning model. The proposed deep neural models perform the time series prediction of internet traffic data. The general construction of the input/output variables and structure for hidden layers of deep ANN for the prediction task is shown in Fig. 14.3.

14.4.1 Dataset Description

The internet traffic dataset is taken from the Time Series Data Library. The time series dataset consists of a total of 1657 samples recorded hourly from an ISP. The data show aggregated traffic in the United Kingdom academic network backbone from 2004 to 2005. The dataset contains two attributes, one is traffic load values recorded in bits and the other one is a particular time interval of each record. In order to increase the accuracy of the predictions the input set is further segmented as an hour, month, day, internet traffic at time t, t−1 and t−2. The number of lag terms included as input parameters are calculated by the autocorrelation function. Each input variable is normalized in the range of (0, 1). Among the normalized data, the first 1200 samples are chosen to train the model and the rest of the data is used as a test set.

14.4.2 Models Description

As mentioned in earlier sections, our task is to implement a deep ANN model which forecasts the traffic load for the next hour. The deep hierarchy of the model converts

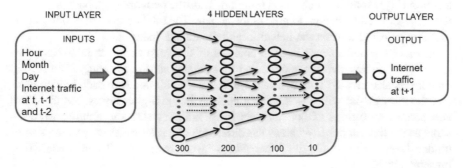

Fig. 14.3 The architecture of DBN based Model I

Table 14.1 Parameter
settings of DBN

Parameters	Value
Max iterations	100
Initial momentum	0.5
Final momentum	0.9
Learning rate	0.1
Batch size	5
Dropout rate	0.0
Transfer function	Sigmoid

the input feature set into nonlinear higher level abstractions to extract more crucial patterns and trends available in the data set. As can be seen in Fig. 14.3, the model is implemented with DBN of 4 hidden layers of size 300–200–100–10, an input layer of 6 neurons for input variables and an output layer of 1 neuron to predict future traffic load. One cannot choose the topology of the network, i.e. the depth of the network and the hidden layer size on an arbitrary basis. The researchers in [19] have given emphasis to three different topologies for model development while considering the width selection of hidden layers. This suggests that the preferable choice of model topology is keeping the width of the hidden layers either constant throughout the model or in increasing/decreasing order of size. Therefore, the architecture of our model is based on decreasing the hidden layer size from bottom to top. The number of neurons in the first hidden layer is calculated through Monte Carlo simulation from the range (100–600).

Each layer of the DBN is independently trained as an RBM with the sigmoid activation function. The preferred settings for the DBN are shown in Table 14.1. Initially the first hidden layer next to the input layer is trained in an unsupervised manner, framing an initial set of parameter values for the first layer of the neural network. Afterwards the outputs obtained from this layer are used as a new feature set which is given as input to the next level layer. The layers are added constructively and are trained independently one by one. This procedure is known as greedy layerwise pretraining. The procedure is repeated an arbitrary number of times in order to obtain more nonlinear representations. This provides the parameter initialization of the neural network. Once the layers of RBM in DBN have been trained, an output layer is added with a cost function. Thus the network is globally fine tuned with a supervised algorithm to predict the targets. The total number of 500 iterations was implemented to train the network with the added linear output layer.

In order to conduct comparative analysis of the deep architectural ANNs, two more DBN models were developed and investigated during the study. The second model was also created with constant width size of 300 nodes in each hidden layer and the third model with fewer hidden neurons in the higher layers, but the depth and parameter settings of both models for pretraining and fine tunning were the same as the first model I. Whereas the difference of model relies only in the size of hidden layers. The topology of the hidden layers for model II and model III is presented in Fig. 14.4.

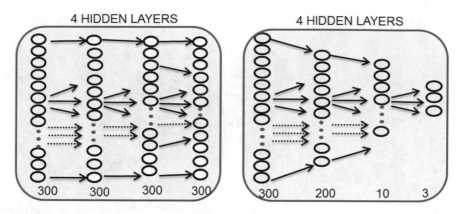

Fig. 14.4 Topological view of hidden layers for Models II and III

14.5 Results and Discussion

To identify the best topological structure of deep ANN for internet traffic prediction, the three different models of DBN were developed as discussed earlier. The performance measure of each deep ANN model is given in Table 14.2 in terms of Mean Square Error (MSE) and Root Mean Square Error (RMSE). All of the models achieve appreciable performance, but comparatively the first model produces more accurate predictions. This indicates that even for deeper architectures of ANN the proper selection of the number of hidden neurons on each layer is important, otherwise searching for the proper parameter space to initialize the network model may become difficult.

As it is seen in the Table 14.2, the values of performance measures specify the worst results obtained from model III. The test error is larger than the training error. This is because of the small size of the top hidden layers. There should be enough neurons in the top hidden layers to bring the training error closer to 0 as implemented in the model I. Consequently, due to unsupervised pretraining the test errors are lower than the training errors in the model I and model II, this observation indicates consistency with previous research [18, 19]. It is clearly visible from the measures of Table 14.2 values that, the estimated error values of model I and II are quite close to each other. This justifies the fact that with the appropriate hidden neurons in deeper networks, unsupervised pretraining works as a data-dependent regularizer.

Table 14.2 Performance measures of models I, II and III

Deep ANN (DBN)	RMSE		MSE		R
	Training	Test	Training	Test	Test
Model I (300–200–100–10)	0.0300	0.0286	8.9e-04	8.1e-04	0.987
Model II (300–300–300–300)	0.0319	0.0310	0.0010	9.5e-04	0.984
Model III (300–200–10–3)	0.0331	0.0427	0.0011	0.0018	0.976

Fig. 14.5 Internet traffic data

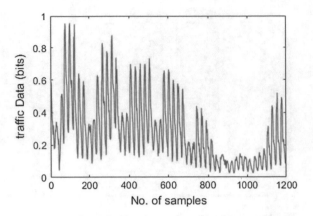

Fig. 14.6 Unsupervised
pretraining of layers

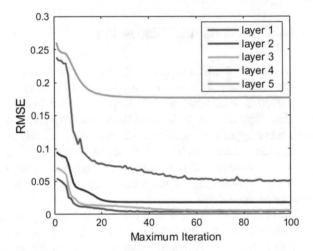

It is important to reiterate that the topology of the DBN implemented in the model I was found to perform better than the other two. Henceforth, the outcomes from this model are focused in this paper. Figure 14.5 presents the structure and nonlinear trend present in the internet traffic data series. Figure 14.6 illustrates the RMSE performance during the unsupervised pretraining of each layer, starting with the input layer.

The feature learning at multiple levels of abstractions allows a neural network to absorb complex function mappings of input distributions, with the exception of any hand-engineered features. Figure 14.7 presents the learnt features on each hidden layer of the model. It is evident that these learnt features are distributed representations of raw input.

Figure 14.8 presents the distribution of errors estimated by model I on the training data set in form of histogram. It can be seen that the histogram is symmetric, and the errors are mostly centered near 0. The predictions of traffic load

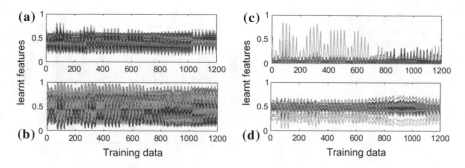

Fig. 14.7 Nonlinear representations as learnt features of hidden layers of Model I **a** h.1 **b** h.2 **c** h.3 **d** h.4

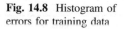

Fig. 14.8 Histogram of errors for training data

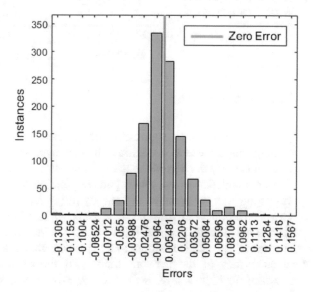

estimated by model I, along with the original traffic values are drawn in Fig. 14.9. The model resulted as RMSE of 0.0286 on test data for 457 samples. Thus the equivalent MAPE efficiency is 7.4% on test set. The regression value of R obtained on test data by model I is 0.98, as given in Table 14.2.

To validate our strategy for hidden layer width selection of deeper neural networks, we conducted two more experiments on Lorenz and Mackey-Glass chaotic time series. Two different models, one for Lorenz and other for Mackey-Glass were developed and trained on both series respectively. The hidden layer width of both the models is similar 600–200–100–10, 4 lag terms as input variables to input layer and 1 output. The results are shown in Table 14.3. These outcomes estimated from both models demonstrate that, our followed strategy for the topology of deep networks specifically DBNs is capable to produce outstanding predictions.

Fig. 14.9 Sixty samples of
predictions estimated by
model I (test data)

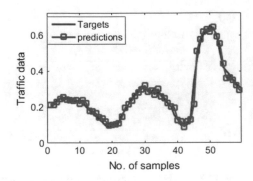

Table 14.3 Performance
measures for Lorenz and
Mackey-Glass

Deep ANN (DBN)	RMSE	
	Training	Test
Model LO	0.0016	0.0015
Model MC	0.0028	0.0023

14.6   Conclusion

The paper presents an application of internet traffic prediction with DBN, also
investigating the selection criteria of hidden layer width in deep neural networks. In
order to identify the suitable topological structure, three different prediction models
of deep architecture were developed. Depending on their comparative analysis, it
was found that the number of neurons in the hidden layers is crucial for deeper
networks. The model with the decreasing width of hidden layers, albeit with sufficient hidden units in higher layers were found to be a proficient predictive model.
The results indicate that Model I and II are capable to accurately capture the salient
characteristics of the network traffic. To further justify the feasibility of our
approach for hidden layer size strategy, experiments were performed for chaotic
time series prediction which also produced promising results. Further extending the
adopted model to produce long term forecasts is intended as a future work.

References

1. Feng, H., Shu, Y.: Study on network traffic prediction techniques. In: International
   Conference on Wireless Communications, Networking and Mobile Computing, vol. 2,
   pp. 1041–1044. IEEE (2005)
2. Park, K., Willinger, W. (eds.): Self-similar Network Traffic and Performance Evaluation,
   pp. 94–95. Wiley, New York (2000)
3. Floyd, S., Paxson, V.: Difficulties in simulating the internet. IEEE/ACM Trans. Netw.
   (TON) **9**(4), 392–403 (2001)

4. Shang, P., Li, X., Kamae, S.: Nonlinear analysis of traffic time series at different temporal scales. Phys. Lett. A **357**(4), 314–318 (2006)
5. Shang, P., Li, X., Kamae, S.: Chaotic analysis of traffic time series. Chaos, Solitons Fractals **25**(1), 121–128 (2005)
6. Romeu, P., Zamora-Martínez, F., Botella-Rocamora, P., Pardo, J.: Time-series forecasting of indoor temperature using pre-trained deep neural networks. In: International Conference on Artificial Neural Networks, pp. 451–458. Springer, Heidelberg (2013)
7. Bengio, Y., Lamblin, P., Popovici, D., Larochelle, H.: Greedy layer-wise training of deep networks. Adv. Neural. Inf. Process. Syst. **19**, 153 (2007)
8. Hinton, G.E., Osindero, S., Teh, Y.W.: A fast learning algorithm for deep belief nets. Neural Comput. **18**(7), 1527–1554 (2006)
9. Zhang, Z.L., Ribeiro, V.J., Moon, S., Diot, C.: Small-time scaling behaviors of Internet backbone traffic: an empirical study. In: Twenty-Second Annual Joint Conference of the IEEE Computer and Communications, vol. 3, pp. 1826–1836. IEEE (2003)
10. You, C., Chandra, K.: Time series models for internet data traffic. In: 1999 Conference on Local Computer Networks, LCN'99, pp. 164–171. IEEE (1999)
11. Hasegawa, M., Wu, G., Mizuni, M.: Applications of nonlinear prediction methods to the internet traffic. In: Circuits and Systems, 2001. ISCAS 2001. The 2001 IEEE International Symposium on Vol. 3, pp. 169–172. IEEE (2001)
12. Cortez, P., Rio, M., Rocha, M., Sousa, P.: Internet traffic forecasting using neural networks. In: International Joint Conference on Neural Network Proceedings, pp. 2635–2642. IEEE (2006)
13. Huang, W., Song, G., Hong, H., Xie, K.: Deep architecture for traffic flow prediction: deep belief networks with multitask learning. IEEE Trans. Intell. Transp. Syst. **15**(5), 2191–2201 (2014)
14. Liu, J.N., Hu, Y., You, J.J., Chan, P.W.: Deep neural network based feature representation for weather forecasting. In: International Conference on Artificial Intelligence (ICAI), p. 1. WorldComp (2014)
15. Dalto, M.: Deep neural networks for time series prediction with applications in ultra-short-term wind forecasting. In: IEEE ICIT (2015)
16. Oliveira, T.P., Barbar, J.S., Soares, A.S.: Multilayer Perceptron and Stacked Autoencoder for Internet Traffic Prediction. In: IFIP International Conference on Network and Parallel Computing, pp. 61–71. Springer, Heidelberg. (2014)
17. Huang, W., Hong, H., Li, M., Hu, W., Song, G., Xie, K.: Deep architecture for traffic flow prediction. In: International Conference on Advanced Data Mining and Applications pp. 165–176. Springer, Heidelberg (2013)
18. Larochelle, H., Bengio, Y., Louradour, J., Lamblin, P.: Exploring strategies for training deep neural networks. J. Mach. Learn. Res.pp. 1–40. (2009)
19. Bengio, Y.: Learning deep architectures for AI. Foundations and trends®. Mach. Learn. **2**(1), 1–127 (2009)

Chapter 15
Growing Curvilinear Component Analysis (GCCA) for Dimensionality Reduction of Nonstationary Data

Giansalvo Cirrincione, Vincenzo Randazzo and Eros Pasero

Abstract Dealing with time-varying high dimensional data is a big problem for real time pattern recognition. Only linear projections, like principal component analysis, are used in real time while nonlinear techniques need the whole database (offline). Their incremental variants do no work properly. The onCCA neural network addresses this problem; it is incremental and performs simultaneously the data quantization and projection by using the Curvilinear Component Analysis (CCA), a distance-preserving reduction technique. However, onCCA requires an initial architecture, provided by a small offline CCA. This paper presents a variant of onCCA, called growing CCA (GCCA), which has a self-organized incremental architecture adapting to the nonstationary data distribution. This is achieved by introducing the ideas of "seeds", pairs of neurons which colonize the input domain, and "bridge", a different kind of edge in the manifold graph, which signal the data nonstationarity. Some examples from artificial problems and a real application are given.

Keywords Dimensionality reduction · Curvilinear component analysis · Online algorithm · Neural network · Vector quantization · Projection · Seed · Bridge

G. Cirrincione
University of Picardie Jules Verne, Laboratory Testing Inc. (LTI), Amiens, France
e-mail: exin@u-picardie.fr

V. Randazzo (✉) · E. Pasero
Politecnico di Torino, DET, Turin, Italy
e-mail: vincenzo.randazzo@polito.it

E. Pasero
e-mail: eros.pasero@polito.it

© Springer International Publishing AG 2018   151
A. Esposito et al. (eds.), *Multidisciplinary Approaches to Neural Computing*,
Smart Innovation, Systems and Technologies 69,
DOI 10.1007/978-3-319-56904-8_15

15.1   Introduction

Data mining is ever increasingly facing the extraction of meaningful information
from big data (e.g. from internet), which are often very high dimensional. For both
visualization and automatic purposes, their dimensionality has to be reduced. This
is also important in order to learn the data manifold, which, in general, is lower
dimensional than the original data. Dimensionality reduction (DR) also mitigates
the curse of dimensionality: e.g., it helps classification, analysis and compression of
high-dimensional data. Most DR techniques work offline, i.e. they require a static
database (batch) of data, whose dimensionality is reduced. They can be divided into
linear and nonlinear techniques, the latter being in general slower, but more
accurate in real world scenarios. See for an overview [1]. However, the possibility
of using a DR technique working in real time is very important, because it allows
not only having a projection after only the presentation of few data, but also
tracking non-stationary data distributions (e.g. time-varying data manifolds). This
can be used, for example, for all applications of real time pattern recognition, where
the data reduction step plays a very important role: fault diagnosis, novelty
detection, intrusion detection for alarm systems, computer vision and scene analysis
and so on. Working in real time requires a data stream, a continuous input for the
DR algorithms, which are defined as on-line or, sometimes, incremental (synonym
for non-batch). They require, in general, data drawn from a stationary distribution.
The fastest algorithms are linear and use the Principal Component Analysis
(PCA) by means of linear neural networks, like the Generalized Hebbian Algorithm
(GHA [2] and the incremental PCA (candid covariance-free CCIPCA [3]). Nonlinear DR techniques are not suitable for online applications. Many efforts have
been tried in order to speed up these algorithms: updating the structure information
(graph), new data prediction, embedding updating. However, these incremental
versions (e.g. iterative LLE [4]) require too a cumbersome computational burden
and are useless in real time applications. Neural networks can also be used for data
projection. In general they are trained offline and used in real time (recall phase). In
this case, they work only for stationary data and can be better considered as implicit
models of the embedding. Radial basis functions and multilayer perceptrons work
well for this purpose (out-of-sample techniques). However, their adaptivity can be
exploited by either creating ad hoc architectures and error functions or using
self-organizing maps (SOM) and variants. The former comprises multilayer perceptrons trained on a precomputed Sammon's mapping or with a backpropagation
rule based on the Sammon's technique and an unsupervised architecture
(SAMANN [5]). These techniques require the stationarity of their training set. The
latter family of neural networks comprises the self-organizing feature maps
(SOM) and its incremental variants. SOM is inherently a feature mapper with fixed
topology (which is also its limit). Its variants have no topology (neural gas, NG [6])
or a variable topology and pave the way to pure incremental networks like growing
neural gas (GNG [7]). These networks, in conjunction with the Competitive
Hebbian Learning (CHL [8]), create a graph representing the manifold, which is the

first step for most DR techniques. NG plus CHL is called Topology representing network (TRN [9]). The approach is called TRNMap [10] if the DR technique is a multidimensional scaling (MDS); here the projection follows the graph estimation, which results in the impossibility to track changes in real time. If the graph is computed by GNG, then the DR can be computed by OVI-NG [11], if Euclidean distances are used, and GNLG-NG [12] if geodesic distances replace Euclidean distances. However, from the point of view of real time applications, only the former is interesting, because it estimates, in the same time, the graph updating and its projection. For data drawn from a nonstationary distribution, as it is the case for fault and prefault diagnosis and system modeling, the above cited techniques basically fail. For instance, the methods based on geodesic distances always require a connected graph. If the distribution changes abruptly (jump), they cannot track anymore. Recently, an ad hoc architecture has been proposed (onCCA [13]), which tracks nonstationarity by using an incremental quantization synchronously with a fast projection based on the Curvilinear Component Analysis (CCA [14]). It requires an initial architecture provided by a fast offline CCA.

The purpose of this paper is the presentation of an improved version of onCCA, here called growing CCA (GCCA), which, by using the new idea of seed, does not need an initial CCA architecture. It also uses the principle of bridges in order to detect changes in the data stream. After the presentation of the traditional (offline) CCA in Sects. 15.2 and 15.3 introduces the new algorithm and discusses both its basic ideas and the influence of its user-dependent parameters. Sect. 15.4 shows the results of a few simulations on artificial and real problems. Sect. 15.5 presents the conclusions.

15.2 The Curvilinear Component Analysis

One of the most important nonlinear techniques for dimensionality reduction is the Curvilinear Component Analysis (CCA [14]), which is a non-convex technique based on weighted distances. It derives from the Sammon mapping [1], but improves it because of its properties of unfolding data and extrapolation. CCA is a self-organizing neural network. It performs the quantization of a data training set (input space, say X) for estimating the corresponding non-linear projection into a lower dimensional space (latent space, say Y). Two weights are attached to each neuron. The first one has the dimensionality of the X space and is here called X-weight: it quantizes the input data. The second one, here called Y-weight, is placed in the latent space and represents the projection of the X-weight. In a sense, each neuron can be considered as a correspondence between a vector and its projection. The input vector quantization can be performed in several ways, by using, for instance, classical neural unsupervised techniques. The CCA projection, which is the core of the algorithm, works as follows. For each pair of different weight vectors in the X space (data space), a between-point distance D_{ij}, calculated as

$D_{ij} = \|x_i - x_j\|$. The objective is to constraint the distance L_{ij} of the associated Y-weights in the latent space, computed as $L_{ij} = \|y_i - y_j\|$, to be equal to D_{ij}. Obviously, this is possible only if all input data lay on a linear manifold. In order to face this problem, CCA defines a distance function, which, in its simplest form, is the following:

$$F_\lambda(L_{ij}) = \begin{cases} 0 & \text{if } \lambda < L_{ij} \\ 1 & \text{if } \lambda \geq L_{ij} \end{cases} \tag{15.1}$$

That is a step function for constraining only the under threshold between-point distances L_{ij}. In this way, the CCA favors short distances, which implies local distance preservation. For each pair i, j of N neurons, the CCA error function is given by:

$$E_{CCA} = \frac{1}{2} \sum_{i=1}^{N} E_{CCA}^i = \frac{1}{2} \sum_{i=1}^{N} \sum_{j=1}^{N} (D_{ij} - L_{ij})^2 F_\lambda(L_{ij}) \tag{15.2}$$

Defining as $y(j)$ the weight of the j-th projecting neuron, the stochastic gradient algorithm for minimizing (15.2) follows:

$$\mathbf{y}(j) \leftarrow \mathbf{y}(j) + \alpha(D_{ij} - L_{ij}) F_\lambda(L_{ij}) \frac{\mathbf{y}(j) - \mathbf{y}(i)}{L_{ij}} \tag{15.3}$$

where α is the learning rate.

15.3 The Growing CCA (GCCA)

The growing CCA is a neural network whose number of neurons is determined by the quantization of the input space. Each neuron has two weights: the first in the data space (X-weight) is used for representing the input distribution, the second in the latent space (Y-weight) yields the corresponding projection. The neurons are connect-ed by links which define the manifold topology. The original concepts are the idea of seed and bridge. The seed is a pair of neurons, which (except in the network initializa-tion) colonize the nonstationary input distribution, in the sense that they are the first neurons representing the change in data. Seeds are created by the neuron doubling explained in Fig. 15.1. The bridge is a qualitatively different link, which indicates a non-stationarity of the input. Hence, there are two types of links: edges, created by CHL, and bridges. Each neuron is equipped with a threshold which represents its receptive field in data space. It is estimated as the distance in X-space between the neuron and its farthest neighbor (neighbors are defined by the graph) and is used for determining the novelty of input data. GCCA is incremental both in the sense that it can increase or decrease (pruning by age) the number of

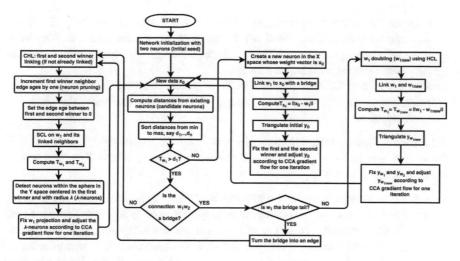

Fig. 15.1 The GCCA flowchart

neurons and the quantization and the projection work simultaneously. The learning rule is the soft competitive learning (SCL [13]) except in neuron doubling, which requires the hard competitive learning (HCL [13]). The projection is based on (15.3), which, as a consequence of the choice of (15.1), implies the idea of λ-neurons.

15.3.1 The Algorithm

The initial structure of GCCA is a seed, i.e. a pair of neurons. The X-weights are random. However, a good choice is the use of two randomly drawn inputs. The associated Y-weights can be chosen randomly, but it is better that one projection is the zero vector, for normalization purposes.

The basic iteration, represented in the flowchart of Fig. 15.1, starts at the presentation of a new data, say $x_0 \in X$. All neurons are sorted according to the Euclidean distances between x_0 and their X-weights. The neuron with the shortest distance (d_1) is the winner. If its distance is higher than the scalar threshold of the neuron (novelty test), a new neuron is created. Otherwise, there is a weight adaptation and a linking phase.

Neuron creation The X-weight vector is given by x_0. The winner and the new neuron are linked by a bridge (this link does not respect CHL). The new neuron threshold is d_1. The associated projection (Y-weight) in latent space requires two steps:

1. Determination of the initial values of the projection (y_0): a *triangulation* inspired by [15] is used, in which the winner and second winner projections are the centers of two circles (in the first two dimensions of the latent space), whose

radii are the distances in data space from the input data, respectively. There are two intersections and the initial two components are chosen as the farthest from the third winner projection. If the latent space is more than two-dimensional, the other components are chosen randomly.

2. One or several CCA iterations (15.3) in which the first and second winner projections are considered as fixed, in order to estimate the new y_0 (*extrapolation*).

Adaptation, linking and doubling If a new neuron is not created, it is checked if the winner, whose X-weight is x_{-1}, and the second winner, whose X-weight is x_{-2}, are connected by a bridge.

1. If there is no bridge, these two neurons are linked by an edge (whose age is set to zero) and the same age procedure as in [13] is used as follows. The age of all other links emanating from the *winner* is incremented by one; if a link age is greater than the *agemax* scalar parameter, it is eliminated. If a neuron remains without links, it is removed (*pruning*). X-weights are adapted by using SCL [13]: x_{-1} and its direct topological neighbors are moved towards x_0 by fractions α_1 and α_n, respectively, of the total distance

$$\Delta x_{-i} = \alpha_1 (x_0 - x_{-i}) \quad i = 1 \tag{15.4a}$$

$$\Delta x_{-i} = \alpha_1 (x_0 - x_{-i}) \quad \text{otherwise} \tag{15.4b}$$

and the thresholds of the winner and second winner are recomputed. Then the neurons whose Y-weights are within the sphere of radius λ centered in the *first winner* are determined, say λ-*neurons* (*topological constraint*). One or several CCA iterations (15.3), in which the first winner projection is fixed, are done for estimating the new projections of the λ-neurons (*interpolation*).

2. If it is a bridge, it is checked if the winner is the bridge tail; in this case step 1 is done and the bridge becomes an edge. Otherwise, a seed is created by means of the neuron doubling:

   (a) A virtual adaptation of the X-weight of the winner is estimated by HCL (only (15.4a) is used) and considered as the X-weight of a new neuron (doubling).
   (b) The winner and the new neuron are linked (age set to zero) and their thresholds are computed (it corresponds to their Euclidean distance).
   (c) The initial projection of the new neuron (Y-weight) is estimated by the same triangulation as before.
   (d) One or several CCA iterations (15.3) in which the projections of the two neurons of the bridge are considered as fixed, in order to estimate the final projection of the new neuron (*extrapolation*).

15.3.2 Considerations

The algorithm requires very few user-dependent parameters. They are needed for the CCA projection, the competitive learning and the pruning. The CCA projection requires the learning rate α and the λ parameter, which determines the choice of the neurons for the projection step. The selection of this parameter is very important, because a too small value could imply a collection of local projections without any coordination. Indeed, the accurate setting of λ is the way GCCA creates its global projection. Instead, the network is not very sensitive to the choice of the number of iterations for each projection. The neuron pruning requires setting the value of *edgemax*, i.e. the maximum value of the age before pruning: a too low value implies a smaller number of neurons. The constant learning parameters α_1 for the *first winner* (for CHL and HCL, see (15.4a)) and α_n: constant learning rate for the *first winner neighbors* (for CHL, see (15.4b)) are needed for the X-weight adaptation.

Bridges are fundamental in tracking nonstationary data. They are links between a neuron and a new data (new neuron). As a consequence, they point to the change in data. They have two basic characteristics: the length and the density. A long bridge, whose new neuron has doubled, represents an effective change in the input distribution; instead, if the new neuron has no edges, it represents an outlier. The density yields further insight in the time-varying distribution. In case of abrupt change in the input distribution (jump), there are a few long bridges. In case of smooth displacement of data, the density of bridges is proportional to the displacement speed of the distribution. In case of very slow displacement, only the border (frontier of the distribution domain) neurons win the competition and move in average in the direction of the displacement. The other neurons are static. Very slow displacement implies no bridges. Bridges appear only if the learning rate of SCL is not constant.

15.4 Examples

Two examples, showing a two-dimensional projection (for visualization) follow: the first one deals with a static unidimensional manifold embedded in the three-dimensional space, the second one, instead, with nonstationary data in a fault diagnosis.

All the simulations have been implemented on MATLAB®. The first deals with data drawn uniformly from a spiral distribution of 30,000 noiseless points (see Fig. 15.2 left). The parameters of GCCA are the following: $\rho = 0.07$, $\alpha_1 = 0.4$, $\alpha_n = 0.1$, agemax $= 2$, $\lambda_i = 20$, $\lambda_f = 0.6$, epochs$_i = 5$, epochs$_f = 1$, $\alpha_{cca} = 0.001$. The results are shown in Fig. 15.2 right after 30,000 instants. It can be deduced that the quantization spans the input domain uniformly and the projection unfolds data correctly.

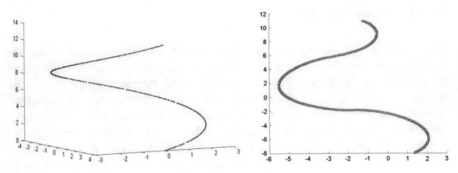

Fig. 15.2 3D-Spiral (*left*); 2D projection (*right*)

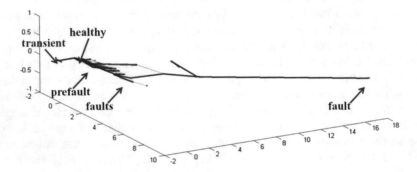

Fig. 15.3 GCCA edges and bridges for the bearing diagnostic experiment

The second example deals with a more challenging problem: data drawn from a dataset coming from the bearing failure diagnostic and prognostic platform [16], which provides access to accelerated bearing degradation tests. Here, the dataset contains 2155 5-dimensional vectors whose components correspond to statistical features extracted by measurements drawn from four vibration transducers installed in a kinematic chain of an electrical motor. In particular, this test deals with a nonstationary framework which evolves from the healthy state to a double fault occurring in the inner-race of a bearing and in the ball of another bearing. The parameters of GCCA are the following: $\rho = 0.01$, $\alpha_1 = 0.05$, $\alpha_n = 0.005$, age-max $= 2$, $\lambda_i = 20$, $\lambda_f = 0.6$, epochs$_i = 1$, epochs$_f = 1$, $\alpha_{cca} = 0.01$. The GCCA learns the chain behavior and tracks it, by adapting in real time the data projection. Figure 15.3 shows the motor life-cycle, from the initial transient phase, through the healthy state, towards, first, a prefault (characterized by an increasing bridges density), and, finally, the two faults which are clearly identified in the figure by the longer bridges.

15.5 Conclusion

The GCCA neural network is the only method able to track a nonstationary input distribution and to project it in a lower dimensional space. In a sense, GCCA learns a time-varying manifold. The algorithm is based on three key ideas: the first is the seed, a pair of neurons which colonizes (start of the new vectorization) a change in the input distribution domain; the second is the bridge, which not only allows the visualization of data changes, but also discriminates the outliers and yields the possibility (by its geometry and density) to infer more information about the nonstationarity; the third is the locality of the projection, given by the selection of the λ-*neurons* for the CCA iterations. The global coherence of the projection is obtained by modulating λ.

Future work will deal with the implementation in this network of other projection techniques, a deeper analysis of bridges and a minor change in the computation of the short distances for approximating the geodesic distances.

Acknowledgements This work has been partly supported by OPLON Italian MIUR project.

References

1. Van der Maaten, L., Postma, E., Van der Herik, H.: Dimensionality reduction: a comparative review. TiCC TR 2009-005, Delft University of Technology (2009)
2. Sanger, T.D.: Optimal Unsupervised Learning in a Single-Layer Neural Network. Neural Netw. **2**, 459–473 (1989)
3. Weng, J., Zhang, Y., Hwang, W.S.: Candid covariance-free incremental principal components analysis. IEEE Trans. Pattern Anal. Mach. Intell. **25**(8), 1034–1040 (2003)
4. Kouropteva, O., Okun, O., Pietikainen, M.: Incremental locally linear embedding. Pattern Recogn. **38**, 1764–1767 (2005)
5. De Ridder, D., Duin, R.: Sammon's mapping using neural networks: a comparison. Pattern Recogn. Lett. **18**, 1307–1316 (1997)
6. Martinetz, T., Schulten, K.: A "neural gas" network learns topologies. In: Artificial Neural Networks, pp. 397–402. Elsevier (1991)
7. Fritzke, B.: A growing neural gas network learns topologies. In: Advances in Neural Information Processing System, vol. 7, pp. 625–632. MIT Press (1995)
8. White, R.: Competitive hebbian learning: algorithm and demonstations. Neural Netw. **5**(2), 261–275 (1992)
9. Martinetz, T., Schulten, K.: Topology representing networks. Neural Netw. **7**(3), 507–522 (1994)
10. Vathy-Fogarassy, A., Kiss, A., Abonyi, J.: *Topology Representing Network Map—A New Tool for Visualization of High-Dimensional Data*, in Transactions on Computational Science I, Vol. 4750 of the series Lecture Notes in Computer Science pp. 61–84, 2008
11. Estevez, P., Figueroa, C.: Online data visualization using the neural gas network. Neural Netw. **19**, 923–934 (2006)
12. Estevez, P., Chong, A., Held, C., Perez, C.: Nonlinear projection using geodesic distances and the neural gas network. Lect. Notes Comput. Sci. **4131**, 464–473 (2006)

13. Cirrincione, G., Hérault, J., Randazzo, V.: The on-line curvilinear component analysis (onCCA) for real-time data reduction. In: International Joint Conference on Neural Networks (IJCNN), pp. 157–165 (2015)
14. Demartines, P., Hérault, J.: Curvilinear component analysis: a self-organizing neural network for nonlinear mapping of data sets. IEEE Trans. Neural Netw. **8**(1), 148–154 (1997)
15. Karbauskaitė, R., Dzemyda, G.: Multidimensional data projection algorithms saving calculations of distances. Inf. Technol. Control **35**(1), 57–61 (2006)
16. Nasa prognostic data repository. http://ti.arc.nasa.gov/tech/dash/pcoe/prognostic-data-repository

Chapter 16
Convolutional Neural Networks with 3-D Kernels for Voice Activity Detection in a Multiroom Environment

Paolo Vecchiotti, Fabio Vesperini, Emanuele Principi,
Stefano Squartini and Francesco Piazza

Abstract This paper focuses on employing Convolutional Neural Networks (CNN) with 3-D kernels for Voice Activity Detectors in multi-room domestic scenarios (mVAD). This technology is compared with the Multi Layer Perceptron (MLP) and interesting advancements are observed with respect to previous works of the authors. In order to approximate real-life scenarios, the DIRHA dataset is exploited. It has been recorded in a home environment by means of several microphones arranged in various rooms. Our study is composed by a multi-stage analysis focusing on the selection of the network size and the input microphones in relation with their number and position. Results are evaluated in terms of Speech Activity Detection error rate (SAD). The CNN-mVAD outperforms the other method with a significant solidity in terms of performance statistics, achieving in the best overall case a SAD equal to 7.0%.

16.1 Introduction

In the recent years, the research on automatic-assisted home environments has been an active area for study, with particular attention to the processing of audio signals [4, 15, 20]. Typically, to increase the quality of the audio signal and improve

P. Vecchiotti (✉) · F. Vesperini · E. Principi · S. Squartini · F. Piazza
Department of Information Engineering, Università Politecnica delle Marche,
Via Brecce Bianche, 60131 Ancona, Italy
e-mail: p.vecchiotti@pm.univpm.it

F. Vesperini
e-mail: f.vesperini@pm.univpm.it

E. Principi
e-mail: e.principi@univpm.it

S. Squartini
e-mail: s.squartini@univpm.it

F. Piazza
e-mail: f.piazza@univpm.it

© Springer International Publishing AG 2018   161
A. Esposito et al. (eds.), *Multidisciplinary Approaches to Neural Computing*,
Smart Innovation, Systems and Technologies 69,
DOI 10.1007/978-3-319-56904-8_16

the performance of the successive audio analysis stages in complex systems, pre-processing algorithms are employed [6, 9, 16]. Hence, the Voice Activity Detection (VAD) element is considered fundamental in these systems, since the speech signal exhaustively characterizes the human activity. In a multi-room domestic environment, Automatic Speech Recognition (ASR) engines can use the information of both the speech segments time boundaries and the room in which the speaker is located in order to improve the word recognition performance. In this context, completely data-driven approaches have been investigated in our precedent works, due to recent success of deep learning approaches, especially regarding the case of multiple audio signals [3, 21]. In this paper we focus on the use of three-dimensional kernels for Convolutional Neural Networks (CNN), taking advantage of an arrangement of the input data to the network rarely used in the audio field. Thus, due to speech signal degradation caused by background noise and reverberation, a multiple sensor (i.e., microphone arrays) deployment is necessary, leading to a rapid increase of data to process.

The state-of-the-art VADs require many processing-stages to obtain the final decision, including the computation of typical characteristics of the acoustic wave or signal statistical descriptors [5]. In recent times, promising VAD approaches take advantage of deep neural networks. A speech/non-speech model based on a Multi-Layer Perceptron (MLP) neural network is proposed in [1], while in [13] multiple features are feed to a Deep Belief Neural Network (DBN) to segment the signal in multichannel utterances. CNNs have been recently employed in VAD tasks [11, 18] with encouraging results. In [14], the authors use a CNN to relabel training examples for a feedforward neural network, obtaining relative reductions in equal error rate of up to 11.5%.

A multi-room domestic scenario requires the room localization and the time detection of speech events. For this purpose we propose the investigation of a 3-D Convolutional neural network (CNN-mVAD) for multichannel audio processing. A similar architecture employed in image classification was presented in [7] with remarkable performance. Our interest goes to the exploitation of the peculiarities of this technology compared to a typical neural network architecture, the Multi Layer Perceptron (MLP-mVAD). This paper contribution is on the choice of a CNN with 3-D kernels. They lead to the possibility of jointly processing simultaneous information from different audio channels, similarly to what occurs in image processing with RGB channels. In addition, CNNs are able to exploit the temporal evolution of the audio signal, and this is an useful feature for the VAD purpose [22].

The outline of the paper is as follows. A general description of the mVAD algorithm and the key features of the two different neural network classifiers are given in Sect. 16.2. Dataset information are presented in Sect. 16.3. In Sect. 16.4 experimental setup is described, then we report and discuss the achieved results for the different number of employed audio channels. Finally, Sect. 16.5 concludes this contribution.

16.2 Neural Network Multi-room VAD

The algorithm proposed in this work is suitable for speech detection in a n room context. In Fig. 16.1 the block diagram of the NN-mVAD is depicted. Initially, features are extracted from the input audio signals. Successively, the Neural Network adopts a multi-class strategy in order to perform the classification task. In particular, the NN has an output layer of $K = 2^n$ units, where e.g., $n = 2$ due to the chosen rooms in our case of study. This leads to 4 output classes, one for each condition of speech/non-speech in the 2 considered rooms. Due to softmax behaviour, the 4 classes mean the joint probability of the 4 different events. Marginalization is then applied, obtaining separated probabilities for each room. Finally, the outputs are processed by a threshold block and a hangover scheme, with the focus on handling isolated speech detections.

16.2.1 Feature Extraction

The feature extraction stage operates on signals sampled at 16 kHz. The used frame size is equal to 25 ms and the frame step is equal to 10 ms. For our purpose, we exploit *LogMel* as feature set, following results obtained for acoustic modelling and music structure analysis in [12, 19]. LogMel features are extracted from the spectrogram of an excerpt of the signal, where 40 mel-band filters are applied, taking the logarithm of the power spectrum for each band. The range of feature values is then standardized to have zero mean and unitary standard deviation. LogMel coefficients are correlated, being the result of a filtering process of the spectrum. On the contrary, features such as MFCCs, which are the result of a Discrete Cousine Transform of the LogMel features, are not correlated, due to spatial compression performed by DCT itself. The choice of LogMel as unique features set was motivated these aspects, differently from what we did in previous works. In particular, in the convolution process performed by kernels, we focus on features with intrinsic correlation and their ability to highlight repetitive patterns. Due to multi-channel approach, features are structured in a specific order. For the MLP case, features of the

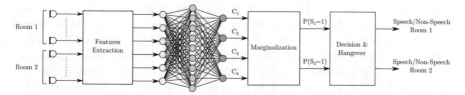

Fig. 16.1 Block diagram of the proposed Neural Network Multi-room VAD in a 2 rooms application. Several microphones can be exploited for each room, consisting in the multi-channel approach. Analysed classifiers are MLP and CNN

different microphones and rooms are concatenated for each frame, resulting in a vector. For the CNN case a temporal context is exploited [18]. In particular, considered the current feature vector $\mathbf{x}[t]$ at the frame index t and a context size equal to C, the feature vector $\mathbf{x}[t]$ is concatenated with the previous and successive feature vectors $\{\mathbf{x}[t-c], \ldots, \mathbf{x}[t-1], \mathbf{x}[t+1], \ldots, \mathbf{x}[t+c]\}$, with $c = 1, \ldots, \frac{(C-1)}{2}$. This procedure leads to a 2-D feature matrix for each microphone. Finally, the 2-D matrices are stacked together, resulting in a 3-D matrix, where the dimensions correspond to the length of the feature set, the context and the number of selected microphones.

16.2.2 Neural Network Classifier

The Convolutional Neural Network is compared to a Multi Layer Perceptron. In this section a brief description of the two networks is reported.

Multi-layer Perceptron. The MLP artificial neural network was introduced in 1986 [17]. The main element is the artificial neuron, consisting in an activation function applied to the sum of the weighted inputs. Neurons are then arranged in layers, with feed forward connections from one layer to the next. The supervised learning of the network makes use of the stochastic gradient descent with error back-propagation algorithm.

Convolutional Neural Network. CNN is a feed-forward artificial neural network inspired by the animal visual cortex, whose neurons elaborate overlapping regions of the visual field [8]. CNN consists in the arrangement of three different layers: convolutional layers, pooling layers and layers of neurons. The convolutional layer performs the mathematical operation of convolution between a multi-dimensional input and a fixed-size kernel. Successively, a non-linearity is applied element-wise. The pooling layers reduces the dimensions of the input matrix, in our case max-pooling is applied. This process deals with translations of the input patterns. Finally, one or more layer of fully connected neurons acts as a classifier, exactly as a MLP.

A particular attention goes to the 3-D convolutional kernel. It processes a 3-D input matrix, as depicted in Fig. 16.2. Convolution is performed along x and y axis. In z axis, the input matrix and the kernel have both the same number of layers. As result, for each z layer, a 2-D convolution is evaluated, leading to a number of *feature maps* equal to the number of layers. Finally, a 2-D output matrix is obtained by summing the feature maps in the z axis. As conclusion, for our case, the 3-D convolution process is suitable only for the first convolution layer of the CNN, successively, a 2-D convolution is performed in the following layers. Since we focus on audio application, it is interesting to analyse a signal excerpt which is extended in time. Thus, we make use of *strides* combined with frame context. In particular, this operation does not merge adjacent frames in order to obtain the input matrix, but it selects frames with a jump equal to the stride value.

Fig. 16.2 Convolution
process for a 3-D kernel
(*red*) over a 3-D input matrix
(*light blue*). The result is a
2-D matrix (*blue*)

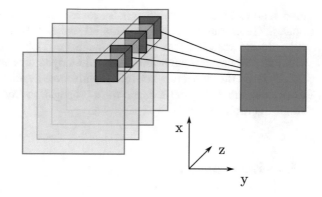

16.2.3   Marginalization

The joint probabilities of the two rooms are marginalized by summing the conditional
probabilities related to a specific room:

$$P(S_1 = 1) = P(S_1 = 1, S_2 = 0) + P(S_1 = 1, S_2 = 1), \qquad (16.1)$$
$$P(S_2 = 1) = P(S_1 = 0, S_2 = 1) + P(S_1 = 1, S_2 = 1). \qquad (16.2)$$

denoting with $S_i = 1$ the presence of speech in the room i and with $S_i = 0$ its
absence. A threshold is then applied to the marginalization probabilities, leading to
a binary signal. A smoothing step handles errors produced by the classifiers.

16.2.4   Decision and Hangover

We exploit a *hangover* technique, which relies on a counter. In particular, for two
consecutive speech frames the counter is set to a predefined value. On the contrary,
for each non-speech frame, the counter decreases by 1. If the counter is negative, the
actual frame is classified as non-speech. The value of the counter is set to $\eta = 8$.

16.3   DIRHA Dataset

The dataset we used for our experiments was provided by the DIRHA project [2], it
contains signals recorded in an apartment equipped with 40 microphones installed
on the walls and the ceiling of each room.[1] The whole dataset is composed of two
subsets called *Real* and *Simulated*, but we used only the latter since it contains more

---

[1] http://dirha.fbk.eu/simcorpora.

data, it is characterised by higher noise source rate and a wide variety of background
noises. The Simulated dataset counts 80 scenes 60 s long consisting in localized
acoustic and speech events in Italian language (23.6% of the total time), on which
different real background noise with random dynamics are superimposed. It is artificially built: the signals are convolved with some available measured room impulse
responses, simulating the acoustic wave propagation from the sound source to each
single microphone.

16.4   Experiments

The analysis of proposed mVADs relies on a two-stage strategy: a network size selection and a microphone combination selection. The experiments are conducted by
means of the k-fold cross-validation technique to reduce the performance variance.
In this case we choose $k = 10$, a validation set is also employed during the training,
thus, 64-8-8 scenes respectively compose the training, validation and test sets. The
performance has been evaluated using the false alarm rate (FA), the deletion rate
(Del) and the overall speech activity detection (SAD) defined as follows:

$$\text{Del} = \frac{N_{del}}{N_{sp}}, \quad \text{FA} = \frac{N_{fa}}{N_{nsp}}, \quad \text{SAD} = \frac{N_{fa} + \beta N_{del}}{N_{nsp} + \beta N_{sp}}, \tag{16.3}$$

where N_{del}, N_{fa}, N_{sp} and N_{nsp} are the total number of deletions, false alarms, speech
and non-speech frames, respectively. The term $\beta = N_{nsp}/N_{sp}$ acts as regulator term
for the class unbalancing. Two different GPU-based toolkits have been employed for
the experiments: a custom version of GPUMLib [10] for MLP-mVAD and *Keras*
(Theano-based)[2] for CNN-mVAD. The MLP networks were trained with a fixed
momentum of 0.9, learning rate equal to 0.01 and a Gaussian distribution with zero
mean and standard deviation of 0.1 for weight initialization. For the CNN networks
we used a fixed learning rate of 2.5×10^{-3} and a random weight initialization.

16.4.1   Results

In this section, the obtained results in terms of SAD are discussed and compared for
the two different architectures of neural network. The analysis of proposed mVADs
relies on a multi-stage strategy, where the best network size and microphone channel
combination are searched. The steps are the following:

1. one network per room, one microphone per room;
2. one network per two room: one, two and three microphone per room.

[2]http://keras.io/.

Regarding network size selection, MLP-mVAD network topologies are explored by means of 1 or 2 hidden layers with respectively 4, 8, 10, 15, 20, 25, 40 units per layer and all their combinations. For CNN-mVAD, due their greater number of hyperparameters and increased training time, a comprehensive grid search was not reasonable, thus we adopted a progressive strategy, based on intermediate results.

Concerning audio channels selection, we initially selected a subset of 9 microphones: 4 in the kitchen (i.e., K2L, K1R, K3C, KA5) and 5 in the living room (i.e., L1C, L2R, L3L, L4R, LA4). In the experiments with one microphone per room, we evaluated the performance for all of them, successively, in the following stages we analyse only combinations obtained with the best performing ones.

One network per room, one microphone per room (1R 1MxR). In this step we evaluated the performance considering two different VADs, one for the kitchen and one for the living room. In the network size selection, the best MLP-VAD resulted to have one layer with 10 units and 8 units respectively for the kitchen and the living room. In the second stage, the best performing microphone for the kitchen was the KA5, while for the living room the LA4: both of them are placed at the center of the room ceiling and the averaged SAD was equal to 12.5%. The two networks exploited for the CNN-VAD are reported in Table 16.1. As for MLP-VAD, best microphones are KA5 and LA4, with an average 9.9% SAD.

One network per two rooms, one microphone per room (2R 1MxR). From this step we started to evaluate the performance of properly mVAD, using both in training and in test audio channels coming from the two rooms. First of all we used only one channel per room: the best MLP-mVAD has one layer with 15 units and the audio captured by the pair KA5, LA4 (confirming the result of the previous step), leading to a SAD equal to 11.7%. For the CNN-mVAD, SAD equal to 9.3% is again obtained with the pair of microphones KA5 and LA4. CNN topology is reported in Table 16.1.

One network per two rooms, two microphones per room (2R 2MxR). We progressively introduced one more audio channel per room, primarily by repeating the network topology selection. Compared to the previous step, the best configuration for MLP-mVAD has only one hidden layer with 8 neurons. In the microphone selection, on the basis of the above analysis we evaluate the 12 combinations of double pairs of channels, achieving with the couple KA5, K1R (from the kitchen) and LA4, L2R (from the living room) an absolute improvement of −2.9% of SAD in respect to the case with one microphone per room. Settings of the CNN-mVAD are shown in Table 16.1. Again, best microphones are the same of the MLP-VAD: KA5, K1R, LA4, L2R. The resulting SAD is 8.1% (Fig. 16.3).

One network per two rooms, three microphones per room (2R 3MxR). In the last step experiments we evaluate the performance of mVADs that process three audio channels per room. For the MLP-mVAD the network topology remains the same as in the case with two microphones per room and the best result (SAD = 7.4%) is obtained with the combination K1R, K2L, KA5, L1C, L2R, LA4. The CNN-mVAD achieves 7.0% SAD with topology shown in Table 16.1, selected microphones are: K1R, K3C, KA5, L2R, L4R, LA4.

Table 16.1 Network topology parameter for CNN- and MLP-mVAD

CNN

		1R 1MxR		2R 1MxR	2R 2MxR	2R 3MxR
		Kitchen	Living room			
Input Params	Strides	8	10	8	8	8
	Context	17	23	25	23	23
First convolutional layer	N Kern	16	16	32	128	256
	Size	6 × 6	6 × 6	4 × 4	4 × 4	4 × 4
	Pooling	2 × 2	2 × 2	2 × 2	–	–
Second convolutional layer	N Kern	24	16	64	64	32
	Size	4 × 4	4 × 4	3 × 3	3 × 3	3 × 3
	Pooling	–	–	–	–	–
Third convolutional layer	N Kern	24	16	128	32	32
	Size	3 × 3	3 × 3	3 × 3	3 × 3	3 × 3
	Pooling	–	–	–	–	–
Fully connected layers	Num. of units	100	100	500	250	500
		20	20	100	100	100
SAD min (%)		9.0	10.7	9.3	8.1	7.0

MLP

Fully connected layers	Num. of units	10	15	10	8	8
		–	–	–	–	–
SAD min (%)		11.8	13.3	11.7	8.8	7.4

Fig. 16.3 Box-plot of the resulting SADs for the microphone selection experiments in the different steps. Evident is the improvement given by increasing the microphone number, and, for the CNN-mVAD, the related statistical robustness

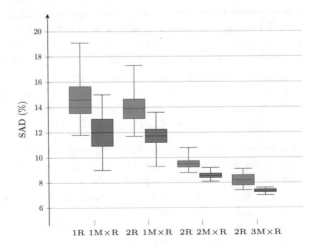

16.5 Conclusion

A neural network approach for voice activity detection in a domestic environment is presented in this paper, paying specific attention to a smart use of the input features. In particular, a multi-channel strategy is implemented, consisting in the usage of multiple microphones as input of the algorithm. Two networks are investigated, which are MLP and CNN. The latter was recently exploited for audio task, with remarkable results. Moreover, due to the suitable CNN structure for multi-channel investigation, we make use of 3-D convolutional kernel, whose dimensions correspond to frequency, time and used microphones. In detail, LogMel features are chosen to represent the frequency domain, in order to convolve the CNN kernel with correlated inputs. Time is explored by means of a temporal context plus strides, allowing the CNN to process an excerpt of the signal with duration about 2 s. Multi-channel features are stacked together, leading to a 3-D input matrix.

The optimization strategy consists in two steps, a network size selection and a microphone selection. Four different studies are conducted, which are a two network approach for single room VAD with only one microphone, and a unique network for the two rooms VAD, featuring one, two and three microphones. The latter achieves the best performance in terms of SAD, leading to 7.4% for MLP and 7.0% for CNN. A remarkable aspect of the CNN mVAD is the robustness to the microphone choice, with lower mean and standard deviation. The independence from the audio source positioning is an interesting applicative result. On the contrary, due to the dimension of the CNN, simulation time is considerably longer compared to MLP.

Future works will be oriented to the employment of raw audio data as input for the CNN, in order to exploit the network feature extraction capability.

References

1. Abad, A., Matos, M., Meinedo, H., Astudillo, R.F., Trancoso, I.: The L2F system for the EVALITA-2014 speech activity detection challenge in domestic environments. In: Proceedings of EVALITA, pp. 147–152 (2014)
2. Cristoforetti, L., Ravanelli, M., Omologo, M., Sosi, A., Abad, A., Hagmüller, M., Maragos, P.: The DIRHA simulated corpus. In: Proceedings of LREC, vol. 5. Reykjavik, Iceland (2014)
3. Ferroni, G., Bonfigli, R., Principi, E., Squartini, S., Piazza, F.: A deep neural network approach for voice activity detection in multi-room domestic scenarios. In: Proceedings of IJCNN, pp. 1–8. Killarney, Ireland (2015)
4. Gemmeke, J.F., Ons, B., Tessema, N., Van Hamme, H., van de Loo, J., De Pauw, G., Daelemans, W., Huyghe, J., Derboven, J., Vuegen, L., Van Den Broeck, B., Karsmakers, P., Vanrumste, B.: Self-taught assistive vocal interfaces: an overview of the ALADIN project. In: Proceedings of Interspeech, pp. 2039–2043. Lyon, France (2013)
5. Giannoulis, P., Tsiami, A., Rodomagoulakis, I., Katsamanis, A., Potamianos, G., Maragos, P.: The Athena-RC system for speech activity detection and speaker localization in the DIRHA smart home. In: Proceedings of HSCMA, 2014, pp. 167–171. Florence, Italy (2014)
6. Hussain, A., Chetouani, M., Squartini, S., Bastari, A., Piazza, F.: Nonlinear Speech Enhancement: An Overview, pp. 217–248. Springer, Berlin (2007)

7. Krizhevsky, A., Sutskever, I., Hinton, G.E.: Imagenet classification with deep convolutional neural networks. In: Advances in Neural Information Processing Systems, pp. 1097–1105 (2012)
8. LeCun, Y., Bottou, L., Bengio, Y., Haffner, P.: Gradient-based learning applied to document recognition. Proc. IEEE **86**(11), 2278–2324 (1998)
9. Loizou, P.C.: Speech Enhancement: Theory and Practice. CRC Press, Boca Raton, FL (2013)
10. Lopes, N., Ribeiro, B.: Towards adaptive learning with improved convergence of deep belief networks on graphics processing units. Pattern Recogn. **47**(1), 114–127 (2014)
11. McLoughlin, I., Song, Y.: Low frequency ultrasonic voice activity detection using convolutional neural networks. In: Proceedings of Interspeech. Dresden, Germany (2015)
12. Mohamed, A., Hinton, G., Penn, G.: Understanding how deep belief networks perform acoustic modelling. In: Proceedings of ICASSP, pp. 4273–4276. Kyoto, Japan (2012)
13. Morales-Cordovilla, J.A., Hagmuller, M., Pessentheiner, H., Kubin, G.: Distant speech recognition in reverberant noisy conditions employing a microphone array. In: Proceedings of EUSIPCO, pp. 2380–2384. Lisbona, Portugal (2014)
14. Price, R., Iso, K.I., Shinoda, K.: Wise teachers train better DNN acoustic models. Eurasip J. Audio Speech Music Process **2016**(1) (2016)
15. Principi, E., Squartini, S., Bonfigli, R., Ferroni, G., Piazza, F.: An integrated system for voice command recognition and emergency detection based on audio signals. Expert Syst. Appl. **42**(13), 5668–5683 (2015)
16. Rotili, R., Principi, E., Squartini, S., Schuller, B.: A real-time speech enhancement framework in noisy and reverberated acoustic scenarios. Cogn. Comput. **5**(4), 504–516 (2013)
17. Rumelhart, D.E., Hinton, G.E., Williams, R.J.: Learning representations by back-propagating errors. Nature **323**, 533–536 (1986)
18. Thomas, S., Ganapathy, S., Saon, G., Soltau, H.: Analyzing convolutional neural networks for speech activity detection in mismatched acoustic conditions. In: Proceedings of ICASSP, pp. 2519–2523. Florence, Italy (2014)
19. Ullrich, K., Schlüter, J., Grill, T.: Boundary detection in music structure analysis using convolutional neural networks. In: Proceedings of ISMIR, pp. 417–422. Taipei, Taiwan (2014)
20. Vacher, M., Caffiau, S., Portet, F., Meillon, B., Roux, C., Elias, E., Lecouteux, B., Chahuara, P.: Evaluation of a context-aware voice interface for ambient assisted living: qualitative user study vs. quantitative system evaluation. ACM Trans. Access. Comput. **7**(2), 5:1–5:36 (2015)
21. Vesperini, F., Vecchiotti, P., Principi, E., Squartini, S., Piazza, F.: Deep neural networks for multi-room voice activity detection: advancements and comparative evaluation. In: Proceedings of IJCNN, pp. 3391–3398. Vancouver, Canada (2016)
22. Zhang, X.L., Wang, D.: Boosting contextual information for deep neural network based voice activity detection. IEEE/ACM Trans. Audio Speech Lang. Process. **24**(2), 252–264 (2016)

Part IV
Special Session on Industrial Applications of Computational Intelligence Approaches

Chapter 17
A Hybrid Variable Selection Approach for NN-Based Classification in Industrial Context

Silvia Cateni and Valentina Colla

Abstract Variable selection is an important concept in data mining, which can improve both the performance of machine learning and the process knowledge by removing the irrelevant and redundant features. The paper presents a hybrid variable selection approach that merges a combination of filters with a wrapper in order to obtain an informative subset of variables in a reasonable time, improving the stability of the single approach of more than 36% in average, without decreasing the system performance. The proposed method is tested on datasets coming from the UCI repository and from industrial contexts.

17.1 Introduction

Variable selection is an important step of the development of Artificial Neural Network (ANN)-based models due to the negative effect that an inadequate selection can have on the performance of ANN during training. The goal of selecting the input variables is common to the implementation of all statistical models and depends on the relationships between input variables and output targets. In the case of parametric models, the difficulty of variable selection is lower due to the a-priori hypothesis of the functional form of the model, which usually derives from a physical interpretation of the process under consideration. When dealing with ANNs, variable selection is more difficult and critical, due to their extreme flexibility and generality. ANNs are supposed to be capable to identify redundant and noise variables during the training phase and that the trained network considers only the important input variables. On this basis, ANN are often developed without considering this aspect. The first obvious effect obtained by including a large number of input variables is the increasing of the size of the ANN; this increases the computational burden and

S. Cateni · V. Colla (✉)
Scuola Superiore Sant' Anna, TeCIP Institute, Via Alamanni 13D, 56010 Pisa, Italy
e-mail: colla@sssup.it

S. Cateni
e-mail: s.cateni@sssup.it

© Springer International Publishing AG 2018
A. Esposito et al. (eds.), *Multidisciplinary Approaches to Neural Computing*,
Smart Innovation, Systems and Technologies 69,
DOI 10.1007/978-3-319-56904-8_17

affects the training time. Moreover the noise introduced by redundant variables can hide or mask the real input-output relationship [21]. Another aspect regards the so-called *curse of dimensionality* [2] which states that, as the dimensionality of a model increase linearly, the total volume of the modelling process domain increase exponentially. ANN architectures, like for example MLP, are particularly sensitive to the course of dimensionality, due to the considerable increase of the number of connection weights with the number of input variables. Finally, in many applications, ANN is considered as a black box model but rule-extraction could be important for several purposes, such as defining input domains that generate determined ANN outputs, the discovery of novel relationships between inputs and output or the validation of the ANN behaviour that increases confidence in the ANN predictions. Based on the previous considerations, the identification of an optimal set of input variables clearly allows to create a more accurate, efficient, cost-effective and more easily interpretable ANN model. Here a hybrid variable selection approach is presented, which merges a combination of filters with a wrapper approach in order to obtain an informative subset of variables, by improving the stability of the single approach without decreasing the system performance. The method is applied to the design of binary classifiers, which have a high importance from the practical point of view, as many real world applications are approached as binary classification problems [3, 10, 17, 22].

The paper is organised as follows: Sect. 17.2 provides a description of the variable selection task; in Sect. 17.3 a description of the proposed method is provided; the obtained results are then shown in Sect. 17.4 and finally in Sect. 17.5 some concluding remarks are provided.

17.2 Variable Selection Approaches

Variable selection approaches are exploited in order to reduce the dimension of the feature space selecting a subset of input variables capable to describe the phenomenon under consideration [24]. In literature variable selection approaches are categorized into three classes: filter, wrapper and embedded methods. Filter approaches are considered as a pre-processing method, as they are independent on the classifier. The variables subset is generated by calculating the association between input and output of the system under consideration, also the variables are ranked considering their relevancy to the target by evaluating statistical tests. The main advantage of filter approaches lies in their low computational complexity, which makes them fast and suitable for complex and large dataset [9]. Many variable selection techniques use feature ranking as a principal selection algorithm due to its scalability, simplicity and satisfactory results. The ranking algorithm evaluates an index quantifying the strength of the link between an input and an output variable. Then, variables are ranked according to such index.

The *Fisher criterion* [13] is one of the most popular filter feature selection methods. The Fisher index $F(i)$ of the i-th variable is computed as:

$$F(i) = \left| \frac{\mu_1(i) - \mu_0(i)}{\sigma_1^2(i) + \sigma_0^2(i)} \right| \tag{17.1}$$

where $\mu_1(i)$, $\mu_0(i)$ and $\sigma_1(i)$, $\sigma_0(i)$ represent the mean value and the standard deviation of the i-th input variable computed on the samples belonging to class 1 and 0, respectively.

A second well known filter approach is the *t-test* which evaluates the importance of each variable using a popular statistical test [23]. The $T - test$ index is computed as:

$$T(i) = \frac{|\mu_1(i) - \mu_0(i)|}{\sqrt{\frac{\sigma_1^2(i)}{n_1} + \frac{\sigma_0^2(i)}{n_0}}} \tag{17.2}$$

where n_1 and n_0 are the number of instances in the unitary and null classes, respectively.

Another ranking filter approach is the one based on the computation of the *Relative Entropy* (E), also known as *Kullback-Leibler divergence* [19], that is used as a measure of the difference between two distributions. In the case of binary classifiers, the following index can be computed [26]:

$$E(i) = \frac{[\frac{\sigma_1^2(i)}{\sigma_0^2(i)} + \frac{\sigma_0^2(i)}{\sigma_1^2(i)} - 2]}{2} + \frac{[\mu_1(i) - \mu_0(i)]^2 \cdot [\frac{1}{\sigma_1^2(i)} + \frac{1}{\sigma_0^2(i)}]}{2} \tag{17.3}$$

A further simple example of a filter method is the popular *correlation-based approach*, which computes the correlation coefficient between each input variable and the target; variables are then ranked and a subset is extracted including the variables with the highest Pearson correlation coefficient [28]. It is commonly used to find the correlation between two continuous variables but it can also be adopted to compute the correlation between a feature and a binary target [29].

The correlation coefficient varies in the range $[-1, 1]$: $C = |1|$ corresponds to a perfect correlation while $C = 0$ means no correlation.

While filter approaches select the subset of variables independently from the classifier in a pre-processing phase, wrapper approaches consider the machine learning as a black box selecting subsets of variables on the basis on their predictive power [5, 18]. Wrapper methods are computationally more expensive than filter methods, but they are still quite simple and universal [6, 7]. A simple example of wrapper approaches is represented by the exhaustive search that becomes impracticable when the number of input variables is too large. In fact, if the dataset includes n variables, then 2^n possible subsets need to be evaluated. A popular wrapper approach, less expensive than the exhaustive search, is represented by the *Greedy Search Strategy* (GSS) that can be divided into two different directions: Sequential Forward

Selection (SFS) and Sequential Backward Selection (SBS). SFS starts with a n empty set of variables and the other variables are iteratively added until stopping criteria are reached [14]. On the other hand SBS starts with the whole variable set and then the less informative variables are removed one by one. Unlike filter and wrapper approaches, embedded approaches perform the variable selection during the learning algorithm; in fact the variables are selected in the training phase in order to reduce the computational cost and improve the efficiency [16].

17.3   Proposed Method

Lee in [20] introduced the idea to reduce the feature space by combining forward and backward elimination techniques before the implementation of an exhaustive search. In [29] a combination of a filter and a wrapper is presented and, finally, in [9], a reduction of the feature space is proposed exploiting filter feature selection methods before performing the exhaustive search. In this work, in addition to the combination of filters and wrappers methods, the stability of the variables selection is investigated. The stability notion was introduced in 1995 by Turney [27] and is a crucial aspect, especially when variable selection is aimed at knowledge discovery in addition to the improvement of the performance of a learning machine. In fact an acceptable variable selection algorithm should not only increase the classifier accuracy but also provide stable selection results when the training data sets is varied [4].

   The proposed method combines some traditional filter methods in order to work with a minimal dataset including only the useful features. Thus it does not need a priori conditions about the process under consideration, a crucial aspect dealing with real processes. In fact, firstly a combination of four filter variable selection methods is computed in order to reduce the feature space. Then a popular wrapper method (SFS or SBS) is performed in order to further reduce the number of variables in a reasonable time by also improving the stability of the approach. The criterion used to select features is the ANN performance. Let us consider a multivariate dataset including N instances, M input variables and a binary target to predict. For K times the dataset is shuffled and partitioned into two dataset: Training Set (75% of the available data) and Validation Set (25%). Four filter variable selection methods are applied on each of the K training datasets and each filter approach provides a score for each variable. Scores are firstly normalized in order to have a value in the range [0, 1] and then they are combined by selecting the average value of the four scores computed. The variables with a score greater than the mean score are selected and fed as input to the sequential selection methods. The SFS (or SBS) method is applied to the reduced dataset; the objective function is implemented with a feed-forward ANN that evaluates candidate the subsets and returns a measure of accuracy of the classification. Figure 17.1 represents the flow diagram of the developed approach. The adopted neural prediction model is a classical two layer perceptron [12] with l inputs, h neurons in the hidden layer with sigmoidal activation function, and one linear neuron in the output layer. h is computed using a popular empirical formula [15] which states that the

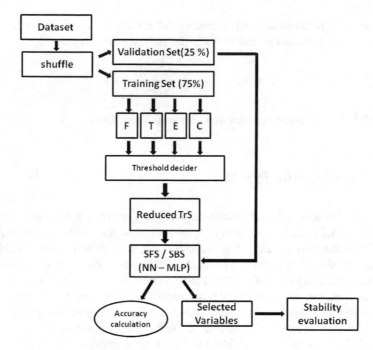

Fig. 17.1 Scheme of the proposed approach

number of patterns in the training set should be 5 times greater than the number of free parameters, namely: $N_P \geq 5(l \cdot h + 2 \cdot h + 1)$. Thus h is fixed as $h = int[0.7 \cdot N_p/(5 \cdot l + 10)]$.

Where $int(\cdot)$ represents the biggest integer value while 0.7 is an empirical factor. A threshold decider, fixed to 0.5, is then used to assign to the output of the net the unitary or null value. Finally the accuracy of the classifier is evaluated through the Average Accuracy [25], also known as Balance Classification Rate (BCR), that is suitable also for imbalance datasets [8] and is given by:

$$BCR = \frac{1}{2} \cdot \left[\frac{TP}{TP + FN} + \frac{TN}{TN + FP}\right] \qquad (17.4)$$

where True Positive TP is the percentage of correctly classified unitary samples, True Negative TN is the percentage of correctly classified null samples, False Positive FP is the percentage of null samples incorrectly classified and finally False Negative FN is the percentage of unitary incorrectly classified.

The *Tanimoto distance* [11] is used to quantify the stability of the variable selection approach, as it measures the similarity between two binary vectors. A subset of input variables can be represented by a binary vector, where each entry corresponds to one variable and its unitary value means that the variable is included in the subset. Given two binary vectors v_1 and v_2, their Tanimoto distance is given by:

$$T(v_1, v_2) = \frac{|v_1 \cdot v_2|}{|v_1| + |v_2| - |v_1 \cdot v_2|} \tag{17.5}$$

where $| \cdot |$ is the norm operator and $\cdot$ is the scalar product.

17.4 Experimental Results

The proposed hybrid variable selection approach has been applied to several datasets included in UCI learning repository [1] as well as to three datasets coming from industrial contexts, in particular the quality check in metal industry, where the binary classification should indicate if a product is defective or not. A description of the main characteristics of the exploited datasets is provided in Table 17.1.

Ten independent runs of both the standard SFS (or SBS) and the proposed hybrid method have been performed on each dataset in order to evaluate the stability of the variable selection solutions. Table 17.2 shows the obtained results performing GSS (SFS and SBS). In particular *BCR* refers to the mean accuracy of the ANN-based classifier, l is the length of the selected subset and finally T is the average Tanimoto distance between the ten selected subsets.

Table 17.2 shows that the proposed approach improves the stability by also reducing the number of selected input variables with respect to the traditional wrappers, without decreasing the accuracy. This hybrid method actually selects a few variables which are actually those ones which mainly affect the target, as shown by the increased stability improving knowledge of the process under consideration.

Table 17.1 Datasets description

Dataset	#Instances	#Features	#class 0	#class 1
ACA	690	14	383	307
BCW	699	9	458	241
Heart	270	13	120	150
MM	830	5	427	403
PID	768	8	500	268
Saheart	462	4	302	160
Ind-I	3756	6	3709	47
Ind-II	1915	10	1454	461
Ind-III	1235	26	517	718

Table 17.2 Classification results

Method	Dataset	BCR			1		T	
		ALL	GSS	Hyb	GSS	Hyb	GSS	Hyb
SFS	ACA	0.85	0.87	0.87	2	1.6	0.62	0.66
	BCW	0.98	0.98	0.98	4.1	2.9	0.52	0.93
	Heart	0.73	0.77	0.79	3.6	3.4	0.44	0.65
	MM	0.80	0.76	0.78	1.4	1.3	0.70	0.87
	PID	0.72	0.73	0.73	3.2	2.6	0.50	0.78
	Saheart	0.73	0.73	0.73	2.4	1	0.54	1
	Ind-I	0.80	0.81	0.81	6	2.8	0.53	0.88
	Ind-II	0.82	0.82	0.82	2.6	1.2	0.50	0.76
	Ind-III	0.71	0.77	0.80	4.3	2.7	0.33	0.83
SBS	ACA	0.85	0.84	0.84	12.5	4.3	0.75	0.81
	BCW	0.98	0.98	0.98	7.7	2.6	0.76	0.78
	Heart	0.73	0.81	0.82	11	5.2	0.74	0.76
	MM	0.80	0.79	0.79	4	3.2	0.65	0.66
	PID	0.72	0.74	0.74	6.8	2.7	0.80	0.84
	Saheart	0.73	0.74	0.74	3.2	1	0.71	1
	Ind-I	0.80	0.82	0.82	8.7	2.9	0.78	0.93
	Ind-II	0.82	0.82	0.82	3.7	2	0.45	0.82
	Ind-III	0.71	0.72	0.80	23.7	10.6	0.79	0.84

17.5   Conclusions

A hybrid algorithm (filter-wrapper) for variable selection is described. The main idea is the combination of four filter methods for variable selection and with a traditional sequential selection method applied to an already reduced in order to provide a more informative subset in a reasonable time. This method can be applied to all kind of datasets without any a priori assumption on the data and is suitable to large or imbalanced datasets. The proposed method has been successfully applied on several datasets coming from a public repository and three datasets coming from industrial field. Future work deals with the application of this approach to multi-class classifiers and for clustering or prediction purposes. Moreover the choice of two thresholds values will be automated in order to improve the accuracy of the classifier.

References

1. Asuncion, A., Newman, D.: Uci machine learning repository (2007). http://archive.ics.uci.edu/ml/datasets.html
2. Bellman, R.: Adaptive Control Processes: A Guided Tour. Princeton University Press (1961)
3. Cateni, S., Colla, V.: Improving the stability of wrapper variable selection applied to binary classification. Int. J. Comput. Inf. Syst. Ind. Manag. Appl. **8**, 214–225 (2016)

4. Cateni, S., Colla, V.: Improving the stability of sequential forward and backward variables selection. In: 15 th International Conference on Intelligent Systems design and applications ISDA 2015, pp. 374–379 (2016)
5. Cateni, S., Colla, V., Vannucci, M.: General purpose input variable extraction: a genetic algorithm based procedure give a gap. In: 9th International Conference on Intelligence Systems design and Applications ISDA'09, pp. 1278–1283 (2009)
6. Cateni, S., Colla, V., Vannucci, M.: Variable selection through genetic algorithms for classification purpose. In: Proceedings of the 10th IASTED International Conference on Artificial Intelligence and Applications, AIA 2010, pp. 6–11 (2010)
7. Cateni, S., Colla, V., Vannucci, M.: A genetic algorithm based approach for selecting input variables and setting relevant network parameters of som based classifier. Int. J. Simul. Syst. Sci. Technol. **12**(2), 30–37 (2011)
8. Cateni, S., Colla, V., Vannucci, M.: Novel resampling method for the classification of imbalanced datasets for industrial and other rreal-world problems. Int. Conf. Intell. Syst. Des. Appl. ISDA **2011**, 402–407 (2011)
9. Cateni, S., Colla, V., Vannucci, M.: A hybrid feature selection method for classification purposes. In: 8th European Modeling Symposium on Mathematical Modeling and Computer simulation EMS2014 1 Pisa (Italy), pp. 1–8 (2014)
10. Cateni, S., Colla, V., Vannucci, M.: A method for resampling imbalanced datadata in binary classification tasks for real-world problems. Neurocomputing **135**, 32–41 (2014)
11. Duda, R., Hart, P., Stork, D.: Pattern Classification. Wiley, New York (USA) (2001)
12. Fausett, L.: Foundamentals of Neural Networks. Prentice Hall (1994)
13. Golub, T., Slonim, D., Tamayo, P., Huard, C., Gaasenbeek, M., Mesirov, J., Coller, H., Loh, M., Downing, J., Caligiuri, M., lander, C.B.E.: Molecular classification of cancer: class discovery and class prediction by gene expression monitoring. Science **286**, 531–537 (1999)
14. Guyon, I., Elisseeff, A.: An introduction to variable and feature selection. Mach. Learn. **3**, 1157–1182 (2003)
15. Haykin, S.: Neural Networks: A Comprehensive Foundation. MacMillman Publishing (1994)
16. He, X., Cai, D., Niyogi, P.: Laplacian score for feature selection. In: Advances in Neural Information Processing Systems, pp. 507–514 (2005)
17. Koc, L., Carswell, A.D.: Network intrusion detection using a hnb binary classifier. In: 17th UKSIM-AMSS International Conference on Modelling and Simulation (2015)
18. Kohavi, R., John, G.: Wrappers for feature selection. Artif. Intell. **97**, 273–324 (1997)
19. Kullback, S., Leibler, R.: On information and sufficiency. Ann. Math. Stat. **22**, 79–86 (1951)
20. Lee, K.: Combining multiple feature selection methods. Ph.D. Thesis, The Mid-Atlantic Student Workshop on Programming Languages and Systems Pace University (2002)
21. May, R., Dandy, G., Maier, H.: Review of input variable selection methods for artificial neural networks. Artif. Neural Netw. Methodol. Adv. Biomed. Appl. (2011)
22. Nikooienejad, A., Wang, W., Johnson, V.E.: Bayesian variable selection for binary outcomes in high dimensional genomic studies using non-local priors. Bioinformatics **32**(2) (2016)
23. Rice, J.A.: Mathematical Statistics and Data Analysis. Third Edition (2006)
24. Sebban, M., Nock, R.: A hybrid filter/wrapper approach of feature selection using information theory. Pattern Recogn. **35**, 835–846 (2002)
25. Sokolova, M., Lapalme, G.: A systematic analysis of performance measures for classification tasks. Inf. Process. Manag. **45**, 427–437 (2009)
26. Theodoridis, S., Koutroumbas, K.: Pattern Recogn. (1999)
27. Turney, P.: Techncal note:bias and the quantification of stability. Mach. Learn. **20**, 23–33 (1995)
28. Yu, L., Liu, H.: Feature selection for high-dimensional data: a fast correlation basedfilter solution. In: Proceedings of the 20th International Conference on Machine Learning ICML, vol. 1, pp. 856–863 (2003)
29. Zhang, K., Li, Y., Scarf, P., Ball, A.: Feature selection for high-dimensional machinery fault diagnosis data using multiple models and radial basis function networks. Neurocomputing **74**, 2941–2952 (2011)

Chapter 18
Advanced Neural Networks Systems for Unbalanced Industrial Datasets

Marco Vannucci and Valentina Colla

Abstract Many industrial tasks are related to the problem of the classification of unbalanced datasets. In these cases rare patterns of interest for the particular applications have to be detected among a much larger amount of patterns. Since data unbalance strongly affects the performance of standard classifiers, several ad–hoc methods have been developed. In this work the main techniques for handling class unbalance are depicted and three methods developed by the authors and based on the use of neural networks are described and tested on industrial case studies.

18.1 Introduction

In the industrial field many practical tasks are related to the identification of the occurrence of unfrequent events within the manufacturing process. These particular situations may correspond, for instance, to machine malfunctions, sensors failures, umpreviewed variations in the input material or assembled parts, defects formation on the semi-finished products. These circumstances are usually quite rare, if compared to *normal* situations, corresponding to correct operation of a machine or a process. Nevertheless the correct identification of these sensible patterns is fundamental in the industrial practice: the early detection of machine failures can allow the users to promptly implement suitable countermeasures and maintenance procedures which avoid harmful and costly stops of the production chain. The detection of a defective final product can avoid its placing on the market, with consequent complaints from the customers and need for replacement [2]. If defects are detected on a semi-manufactured product, early discarding, downgrading or re-working of this product is possible, with a consequent saving of energy and primary raw material. It is thus evident that the identification of rare patters is often much more important

M. Vannucci (✉) · V. Colla
TeCIP Institute, Scuola Superiore Sant'Anna, Via G. Moruzzi, 1, 56124 Pisa, Italy
e-mail: mvannucci@sssup.it

V. Colla
e-mail: colla@sssup.it

© Springer International Publishing AG 2018
A. Esposito et al. (eds.), *Multidisciplinary Approaches to Neural Computing*,
Smart Innovation, Systems and Technologies 69,
DOI 10.1007/978-3-319-56904-8_18

181

than the identification of the patterns belonging to other categories corresponding to more "standard" process conditions. Therefore, in many of these industrial applications, the eventual generation of an acceptable rate of *false alarms* is preferable with respect to the missed detention of an even small rate of rare patterns.

The classification problems, which are characterized by a noticeable level of unbalance among the different classes, are commonly referred in literature as *classification of unbalanced (or uneven) datasets* and are mostly focused on binary classification tasks where the main aim consists on the correct identification of the minority class samples. Unfortunately, unfrequent patterns within large industrial dataset are hard to detect by means of standard methods. In facts, most classification techniques such as Neural Networks (NN) or Decision Trees (DT) assume that the training samples are evenly distributed among the different classes. These standard classifiers, when trained on unbalanced datasets, tend to ignore the less represented class, since their goal is the achievement of an *overall optimal performance* [8], which is not in line with most of the previously described industrial applications. In [7] it was empirically demonstrated that unbalance in the training dataset affects the performance of any kind of classifier and that the more the unbalance rate the worse the classifier performance when handling the minority class samples.

The degree of unbalance is not the only factor that negatively affects the classifiers performance. The dimension of training data plays a role as well: [20] shows that the lack of minority class examples prevents classifiers to correctly characterize them, find regularities and separate them from the frequent ones. Moreover in [21] it was highlighted that, with the availability of ever larger datasets, even with constant the unbalance rate, the performance of the classifiers improves since the informative content provided by the growing number of minority class samples allows their characterization. The complexity of the concept associated to rare samples is critical as well. In [9] it is shown that, in the case of simple datasets, where classes are separate, the effect of class imbalance is null, whilst, in the case of heavily overlapping classes and high complexity, the performance of classifiers degrades and classifiers are biased toward the majority class. Due to the critical nature of the problem and its strategic importance in the industrial field, several works can be found in literature related to the development of methods for the identification of rare patterns through different techniques. Despite this wide interests, among all the approaches no one seems to overcome the others and to be applicable successfully on any problem: it rather emerges that different families of approaches are suitable for different kinds of problems.

In this paper the focus is on a set of NN-based advanced methods based developed for handling particular problems coming from the industrial fields, where standard classifiers as well as general approaches for coping with class unbalance did not lead to satisfactory results.

The paper is organized as follows: in Sect. 18.2 an overview on the main approaches for designing classifier able to handle uneven datasets is presented. In Sect. 18.3 the NN-based methods are described in detail and their main achievements on industrial tasks are outlined. Finally in Sect. 18.4 some considerations and remarks are drawn together with the future perspective of the presented methods.

18.2   Classification of Unbalanced Datasets

Numerous literature works concern methods for the classification of unbalanced dataset [16]. These methods can be grouped into two sub–categories according to the way they try to mitigate the effect of class imbalance in the training dataset: the *internal* methods, which are based on algorithms expressly designed for this purpose, and the *external* methods, which operate on the training dataset itself through the so–called *resampling*, that directly reduces the unbalance degree of the training dataset so as it can be fed to standard classifiers.

Internal approaches involve, as a basis, different types of classifiers that are modified to favour the detection of minority class samples. Support Vector Machines (SVMs) have been modified in [1] for such purpose, while in [11] the outcomes of multiple independent SVMs are combined achieving interesting results. In [14] a special kind of radial basis function network employing Rectangular Basis Functions (RecBF) in the hidden layer has been investigated. This particular activation function was demonstrated to allow a higher precision in the detection of the boundaries of the regions associated to different classes.

Several methods are based on the concept of the recognition of the minority class rather than on the differentiation of the minority and majority classes. These approaches, commonly referred as One–Class–Learning (OCL), exploit the concept of similarity with respect to the rare samples. A successful example of OCL method using a modified SVM architecture, the so–called v-SVM, is proposed in [13]. A particular kind of internal methods employs an altered cost matrix that penalizes the misclassification of rare samples during the training in order to promote their detection. This approach can be coupled to several standard classification methods such as NNs or DTs [12]. The main drawback of this approach lies in the empirical determination of the optimal cost matrix on the basis of the application and classes distributions. Ensemble methods which combine the output of a set of so-called *weak learners* in order to improve the classification performance of the whole system are used also. Boosting techniques in particular have been applied: in [10], during the learning process, weak learners that achieve good performance in the detection of rare patterns are progressively added to the ensemble until the desired performance is reached on the training set.

External methods reduce the unbalance ratio of the training dataset in order to avoid the classifier to bias toward the majority class. Resampling can be achieved in two ways: by removing frequent samples from the original dataset (*under–sampling*) or by adding samples belonging to the minority class (*over–sampling*). The main risk related to under–sampling lies in the potential loss of information due to the elimination of samples containing useful information. On the other hand, over–sampling can introduce misleading knowledge into the training dataset with detrimental effect when classifying patterns outside of the training phase.

A further critical point concerns the unbalance ratio to be achieved by resampling methods. Optimal ratio is unknown and there is actually no method to compute it; it is rather dependant upon applications and original datasets features and, in practice, is empirically estimated. This issues have been tackled by using advanced resampling techniques that suitably select for removal the majority class samples lying close to boundary regions of the input space and replicate minority class samples located close to the boundary with the majority class with the effect of spreading the regions that the classifier will associate to the minority class [6]. The SMOTE algorithm [5], differently form other over–sampling methods, does not duplicate existing minority samples but synthetically creates and places them where they *likely* could be (i.e. close to existing minority samples or between them). Other advanced approaches combining under- and over-sampling are presented in [3, 4, 17].

18.3 Advanced Neural Networks Based Systems

In this section three approaches aiming at the classification of unbalanced datasets are presented. These methods have been developed by the authors for the solution of particular industrial problems and are based on the advanced use of NNs. They belong both to the families of internal and external approaches. All the methods are described and main achievements on two industrial case studies are reported. In the following it is assumed that the minority class is associated to the output value 1 while the majority class to 0 value.

18.3.1 The Thresholded Neural Network

The Thresholded Artificial Neural Network (TANN) [18] combines a standard two–layers Feed–Forward NN (FFNN) to a threshold operator to counteract the biasing of standard classifiers toward the majority class by the effect of class imbalance. The structure of the FFNN is arbitrary apart from the output layer charcterized by a logarithmic sigmoid function. The output of the network, which lies in the range [0;1], is processed by a threshold operator that associates 1 (rare sample) to values higher than the threshold t and 0 otherwise. The value of t determines the sensitivity of the TANN to rate patterns: the lower t the more TANN is encouraged to classify an arbitrary pattern as belonging to the minority class. The TANN aims at improving the response to rare patterns by means of the tuning t in order to maximize the performance of the classifier on the basis a merit function that takes into account the overall accuracy (*Corr*), the rate of minority class samples detected (*Det*) as well as the rate of false positives (i.e. false alarms) (*FA*). The training of the TANN can be summarized into two subsequent phases: the first one devoted to the usual NN training while the second one to the optimal selection of t among a set of candidate thresholds spanning in [0;1] according to the following merit function (the higher the better):

$$E(t) = \frac{\gamma Det(t) - FA(t)}{Corr(t) + \mu FA(t)} \qquad (18.1)$$

γ and μ are two empirical parameters. This merit function formalizes the requirements of the obtained learner in terms of general performance and detection of rare events.

18.3.2 The LASCUS Method

LASCUS (Labelled SOM Clustering for Unbalanced Sets) [15] is an advanced method for the identification of rare patterns within unbalanced classification problems. LASCUS is designed to grant a high detection rate of minority class patterns despite the generation of an acceptable rate of false alarms, where they are not as critical as missed detection of rare patterns. Machine malfunctionings, for instance, often belong to this category of problems. LASCUS partitions the input space according to training data distribution and assigns the so–formed regions to the majority or minority class according to the density of rare patterns lying in each region. In order to promote the detection of rare patterns, the areas of the input space where the concentration of such patterns is *high enough* are assigned to minority class. The key ideas of LASCUS concern the data partitioning and the determination of the rare events concentration rate that determines the assignment of a particular partition to the rare class. Data partition is achieved through a Self Organizing Map (SOM), that creates a set of clusters maintaining the original data distribution. The density of minority samples within each cluster is determined by assigning all the samples to the cluster according to their distance to its centroid. The determination of the critical rare events concentration is obtained by calculating a set of performance measures including false alarms, minority samples detection rates and overall accuracy for the set of rare events concentrations reported by all the clusters. These indexes are fed to a fuzzy inference system which implements a human driven criterion for the LASCUS performance evaluation associated to each critical concentration value. The higher rated concentration is selected.

18.3.3 Dynamic Resampling for NN Training

The *Dynamic resampling* (DYNRES) is a concept introduced in [19], belongs to the family of external approaches and exploits the characteristics of NNs training procedures. This method mitigates the effect of class unbalance by feeding the different phases of the NN training with different resampled *versions* of the original dataset: in each phase only a part of majority patterns is used according to a resampling rate to be defined by the user. The training process is divided into *blocks of epochs* that exploit different subsets of the original dataset for the NN training. Each subset, which is different throughout the blocks, includes all the minority samples and a subset of frequent samples that are probabilistically selected according to two criteria: probability is higher for previously less frequently selected samples; probability of selecting an arbitrary pattern p is proportional to the classification performance achieved by the FFNN during the training limited to the blocks of epochs including p. This particular resampling technique allows both a balanced training and the retention of the informative content of all the majority class samples, of which none is neglected throughout the whole training. Moreover this procedure tends to select with higher probability those samples whose exploitation is most convenient in terms of classification performance.

18.3.4 Experimental Tests

The approaches introduced in Sec. 18.3 have been tested on two industrial case studies affected by the problem of class unbalance. For both of them the detection of the situations associated to the minority classes is fundamental due to the nature of the applications. Minority patterns represent industrial critical situations that can affect the production processes or the quality of the products.

The *occlusion* problem concerns the detection of nozzles occlusions during the continuous casting of steel. This event (1.2% of observations) dramatically slows down productivity and its detection is a key issue of the problem. The available dataset contains measures related to process parameters and steel characteristics. The *Metal Sheet Quality* (MSQ) problem accounts the automatic grading of metal sheets quality on the basis of the information provided by a set of sensors that inspect sheets surface. The grading depends on the type, number and shape of the reported defects. Defective products (10% of observations) have to be spotted in order not to be put into the market.

The results achieved by the tested methods are shown in Table 18.1 in terms of overall accuracy (ACC), rare patterns detection (DET) and false alarms (FA) rates and compared to other approaches that include the combination of FFNN based classifiers with standard oversampling, SMOTE–oversampling and undersampling.

Table 18.1 highlights the advantages derived from the use of approaches developed for handling unbalanced datasets. In facts for both the applications the results achieved by standard approaches (DT, FFNN) are poor and draw attention to the biased nature of these classifiers. The proposed NN–based approaches obtain a clear improvement in terms of performance with respect to standard methods and even to classical approaches for handling class unbalance: the rate of detected minority samples is—for all methods and case studies—much higher, keeping acceptable the rate of FA. More in detail, in the MSQ application FA rate is comparable to the one obtained by standard approaches (or even better). In the occlusion problem FA is higher but, in the particular application, it does not represent a problem since it only requires the activation of some simple and not costly countermeasures. Among the proposed advanced NN–based techniques LASCUS is the one that correctly recognizes the higher number of rare patterns and raises the higher number of false alarms. This behaviour is not surprising, since such method is designed to this purpose. TANN performs slightly better than DYNRES, but it is worth to note that DYNRES

Table 18.1 Performance of tested approaches on the case studies. The percentual value reported in the method column is the nominal resampling unbalance ratio

Database	Method	ACC %	DET %	FA %
Occlusion	DT	98	17	2
	FFNN	98	17	2
	FFNN + Overs. 25%	90	40	9
	FFNN + SMOTE. 50%	86	53	13
	FFNN + Unders. 25%	88	45	11
	v-SVM	97	0	0
	TANN	77	71	16
	LASCUS	80	77	20
	DYNRES 40%	82	65	18
MSQ	DT	83	68	7
	FFNN	72	29	4
	FFNN + Overs. 25%	88	70	4
	FFNN + SMOTE. 50%	89	68	4
	FFNN + Unders. 25%	89	69	4
	v-SVM	87	65	1.5
	TANN	83	77	3
	LASCUS	84	86	4
	DYNRES 25%	89	75	5

performance depends on the resampling rate selected by the user: although different nominal resampling rates have been tested in this work, an exhaustive test or an optimization of such rate could improve the performance.

18.4   Conclusions and Future Works

In this work the problem of the classification of unbalanced dataset in the industrial framework has been handled, by putting into evidence the particular criticalities related to this problem from an industrial point of view. Three approaches developed by the authors for this purpose and based on artificial NNs have been described and tested on two industrial case studies. The performance achieved by these methods—compared with other existing approaches—highlights not only the need for *ah*–hoc methods for tackling the issues connected to class unbalance but also the suitability of NNs to act as a basis for advanced methods for unbalanced classification, by exploiting the well known NNs robustness and generalization capability. In the future, since none of the methods or family of methods seems to outperform the other ones, hybrid approaches that merge internal and external techniques will be investigated.

References

1. Akbani, R., Kwek, S., Japkowicz, N.: 15th European Conference on Machine Learning ECML 2004, Pisa, Italy, Sept. 20–24, pp. 39–50. Springer, Berlin (2004)
2. Borselli, A., Colla, V., Vannucci, M., Veroli, M.: A fuzzy inference system applied to defect detection in flat steel production. In: 2010 IEEE International Conference on Fuzzy Systems (FUZZ), pp. 1–6 (2010)
3. Cateni, S., Colla, V., Vannucci, M.: Novel resampling method for the classification of imbalanced datasets for industrial and other real-world problems. Int. Conf. Intell. Syst. Des. Appl. ISDA **2011**, 402–407 (2011)
4. Cateni, S., Colla, V., Vannucci, M.: A method for resampling imbalanced datasets in binary classification tasks for real-world problems. Neurocomputing **135**, 32–41 (2014)
5. Chawla, N.V., Bowyer, K.W., Hall, L.O., Kegelmeyer, W.P.: Smote: Synthetic minority over-sampling technique. J. Artif. Int. Res. **16**(1), 321–357 (2002)
6. Chawla, N.: C4.5 and imbalanced data sets: investigating the effect of sampling method, probabilistic estimate, and decision tree structure. In: Proceedings of ICML03 Works on Class Imbalances (2003)
7. Estabrooks, A., Jo, T., Japkowicz, N.: A multiple resampling method for learning from imbalanced datasets. Comp. Intell. **20**(1), 18–36 (2004)
8. He, H., Garcia, E.A.: Learning from imbalanced data. IEEE Trans. Knowl. Data Eng. **21**(9), 1263–1284 (2009)
9. Japkowicz, N., Stephen, S.: The class imbalance problem: a systematic study. Intell. Data Anal. **6**(5), 429–449 (2002)
10. Leskovec, J., Shawne-Taylor, J.: Linear programming boosting for uneven datasets. In: 20th International Conference on Machine Learning (ICML'03), pp. 456–463. AAAI Press, event Dates: 21–24 August (2003)
11. Li, P., Chan, K., Fang, W.: Hybrid kernel machine ensemble for imbalanced data sets. In: 18th International Conference on Pattern Recognition (ICPR'06), vol. 1, pp. 1108–1111 (2006)
12. Ling, C., Yang, Q., Wang, J., Zhang, S.: Decision trees with minimal costs. In: Proceedings of the 21-st International Conference on Machine Learning ICML '04, p. 69. ACM, New York, NY, USA (2004)
13. Schölkopf, B., Smola, A.J., Williamson, R.C., Bartlett, P.L.: New support vector algorithms. Neural Comput. **12**(5), 1207–1245 (2000)
14. Soler, V., Prim, M.: 17th International Conference on Artificial Neural Networks—ICANN 2007, vol. I, pp. 511–519. Springer, Berlin (2007)
15. Vannucci, M., Colla, V.: Novel classification method for sensitive problems and uneven datasets based on neural networks and fuzzy logic. Appl. Soft Comput. J. **11**(2), 2383–2390 (2011)
16. Vannucci, M., Colla, V., Cateni, S., Sgarbi, M.: Artificial intelligence techniques for unbalanced datasets in real world classification tasks, chap. In: Computational Modeling and Simulation of Intellect: Current State and Future Perspectives, pp. 551–565. IGI Global (2011)
17. Vannucci, M., Colla, V., Nastasi, G., Matarese, N.: Detection of rare events within industrial datasets by means of data resampling and specific algorithms. Int. J. Simul. Syst. Sci. Technol. **11**(3), 1–11 (2010)
18. Vannucci, M., Colla, V., Sgarbi, M., Toscanelli, O.: Thresholded neural networks for sensitive industrial classification tasks. Lecture Notes in Computer Science (including subseries Lecture Notes in Artificial Intelligence and Lecture Notes in Bioinformatics) 5517 LNCS(PART 1), pp. 1320–1327 (2009)

19. Vannucci, M., Colla, V., Vannocci, M., Reyneri, L.: Dynamic resampling method for classification of sensitive problems and uneven datasets. In: Communications in Computer and Information Science 298 CCIS (PART 2), pp. 78–87 (2012)
20. Weiss, G.M.: Mining with rarity: a unifying framework. SIGKDD Explor. Newsl. 6(1), 7–19 (2004)
21. Weiss, G.M., Provost, F.: Learning when training data are costly: the effect of class distribution on tree induction. J. Artif. Int. Res. 19(1), 315–354 (2003)

Chapter 19
Quantum-Inspired Evolutionary Multiobjective Optimization for a Dynamic Production Scheduling Approach

Maurizio Fiasché, Diego E. Liberati, Stefano Gualandi and Marco Taisch

Abstract The Production Scheduling is an important phase in a manufacturing system, where the aim is to improve the productivity of one or more factories. Finding an optimal solution to scheduling problems means to approach complex combinatorial optimization problems, and not all of them are solvable in a mathematical way, in fact a lot of them are part of the class of NP-hard combinatorial problems. In this paper a joint mixed approach based on a joint use of Evolutionary Algorithms and their quantum version is proposed. The context is ideally located inside two factories, partners and use cases of the *white'R* FP7 FOF MNP Project, with high manual activity for the production of optoelectronics products, switching with the use of the new robotic (re)configurable island, the white'R, to highly automated production. This is the first paper approaching the problem of the dynamic production scheduling for these types of production systems proposing a cooperative solving method. Results show this mixed method provide better answers and is faster in convergence than others.

Keywords Evolutionary algorithms · PSO · QEA · QiEA · QPSO · Quantum EA · Scheduling

M. Fiasché (✉) · M. Taisch
Department of Management, Economics and Industrial Engineering,
Politecnico di Milano, Milan, Italy
e-mail: maurizio.fiasche@polimi.it

D.E. Liberati
IEIIT, National Research Council of Italy, Rome, Italy

S. Gualandi
Antoptima, SA, Lugano, Switzerland

19.1 Introduction

In general, the *Production Scheduling* is the activity of assigning a starting time
point to a collection of activities that must be executed by a collection of machines in
order to obtain the production of a given quantity of goods [1]. More formally, given
a collection of Job $J = \{J_1,...,J_n\}$ and a collection of machine $M = \{M_1,...M_m\}$, we
have to find a **schedule**, that is a mapping of jobs to machines and processing times
that satisfies some constraints and it is optimal with respect to some criteria, such as,
for instance, to minimize the makespan, to parallelize tasks and operations, etc.

The context of this paper is the white'R FP7 European Project, which develops a
highly automated, self-contained, white room based, multi-robotic island designed
for the motion, assembly, and dis-assembly of optoelectronic products. The physical outcome of the project is the production of two demonstrators where the
white'R island will be configured differently to be used in two existing shop floors
of two end-users (hereafter called EU1 and EU2), dealing respectively with laser
diodes (EU1) processing and customized solar energy systems (EU2). However this
new technological tool brought along the need to design and implement a conceptual management framework, for sustaining the robotic island integration into an
existing shop floor [2] and for investigating models and techniques able to solve the
decisional problems of planning and scheduling in this particular case study,
extending them also for other re-configurable production systems. In this context,
we need to be able to schedule the production of given quantities of the devices that
could be produced within a White'R island (hereafter called simply white'R), such
as the multi-emitter diode lasers assembled by EU1, and the string of solar cells
produced by EU2. For both type of devices, in order to be produce a single unit of
product it is necessary to execute a sequence of so-called Mechatronic Tasks using
different workstations. Each mechatronic task must respects a set of precedence
relations with other tasks, can require the assignment of one or more workstation,
and it has a nominal duration that changes along the life cycle of the island.

A key feature of the production scheduling for the White'R is the possibility to
account for the degradation of the system, that is, the ability to consider that each
Mechatronic task can work in different nominal conditions, to which correspond
different time duration to complete a task. Therefore, in choosing a scheduling
problem to formulate and to solve the production scheduling for the White'R
project, we need to be as flexible as possible with respect with these requirements.
However the solving techniques presented and implemented are general and usable
also for other scheduling problems. In this work a joint use of Particle Swarm
Optimization (PSO) technique and the quantum inspired PSO (QPSO) version is
proposed analyzing results comparing them also with the use of single methods.
The rest of the paper is structured like follow: in Sect. 19.2 the scheduling problem
modeling is presented, in Sect. 19.3 the optimization issues are approached,
describing and introducing classical, quantum and cooperative approaches, in
Sect. 19.4 experimental setup and results are presented and in Sect. 19.5 conclusions are inferred with possible future directions.

19.2   Resource Constrained Project Scheduling Problem (RCPSP)

The literature on scheduling models for manufacturing [3, 4] is very rich of alternatives, and almost every single manufacturing application domain has developed a very specialized model. However, in the context of White'R we have decided to formulate the Production Scheduling problem as a Resource Constrained Project Scheduling Problem (RCPSP), that is:

> Given a set of *resources* (i.e., the robotic arm and the workstations of the islands) of limited availability, and a set of *activities* (i.e., the mechatronic tasks) of known *duration* and *resources* requests, and linked by *precedence* and *non overlapping relations* the RCPSP consists of finding a schedule of minimal duration (i.e., minimum makespan) with the assignment of a start time to each activity such that the precedence relations and the resource availabilities are respected [5].

In White'R, examples of resources are: (i) the robotic arm, which is the most important unary resource of the island, (ii) the cutter, (iii) the wire bonding machines, and all other workstations that have been introduced in the previous paragraphs. Basically, in this context, all the resources are unary: they can be used by a single task at the time. For an example of possible resources/workstations in the case of the production of multi-emitter diode lasers, refer to the next Fig. 19.1, which shows a preliminary screenshot of the Mechatronic task scheduler developed and delivered during the project.

The activities to be scheduled, i.e., the Mechatronic tasks, are produced by the planning system presented in [2, 6] and however given as input to the production scheduler, and they must be described in terms of data detailed in Table 19.1 (Fig. 19.2).

For a small example of a sequence of mechatronic tasks for the assembly of multi-emitter laser, we refer to Fig. 19.1, which again shows a screenshot from the Mechatronic Task Scheduler under development. The screen shot shows on the left

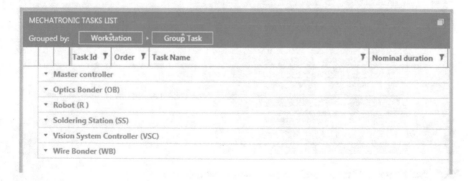

Fig. 19.1 Collection of resources available for the mechatronic tasks in the production of multi-emitter diode lasers (OPI use case)

Table 19.1 List of input data to schedule the mechatronic tasks

	Data	Comment
1.	Task ID	Unique ID for each single tasks
2.	Nominal duration	With consistent unit of time (e.g., msec)
3.	Setup time	With consistent unit of time (e.g., msec)
4.	Resource consumption	A list of resource that must be acquired to complete the activity related to the task
5.	Precedence relations between tasks	Represented as binary relations between task IDs
6.	Non overlapping relations between tasks	Represented as binary relations between task IDs

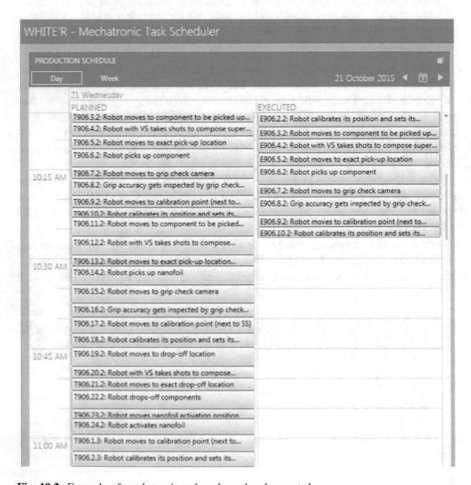

Fig. 19.2 Example of mechatronic tasks: planned and executed

side, under the "PLANNED" column, a short sequence of tasks. For each task is given a unique ID, such as, T906.3.2. The first letter 'T' denotes a planned task, the next digits '906.2' identifies the given task, while the last digit '2' represent the repetitions of the same type of task within the given sequence. On the right side, under the "EXECUTED" column, the same task is marked by the 'E' character, which marked the task as completed (with a small time delay), that is E906.3.2. Please, note that in this figure the time scale is purely qualitative.

In formulating the production-scheduling problem as a RCPSP, we need in addition to introduce two dummy activities that represent the *start time* and the *end time*. In addition, we consider the possibility to introduce the concept of group tasks, which correspond to a collection of mechatronic tasks that are tighten together for representing precedence or non-overlapping relations between group of tasks.

19.2.1 Formal Description of RCPSP

In the RCPSP, we are given a set J of non-preemptable jobs (i.e. =, mechatronic tasks) and a set M of resources (i.e., workstations). Each resource has a unary capacity. Every job j has a processing time p_j and resource demands r_{jk} of each resource.

The starting time S_j of a job is constrained by its predecessors that are given by a precedence graph $D = (V,A)$ with $V \subseteq J$. An arc $(i, j) \in A$ represents a precedence relationship, i.e., job i must be finished before job j starts. The goal is to schedule all jobs with respect to resource and precedence constraints, such that the latest completion time of all jobs, that is, the so-called end-job, is minimized [7].

The RCPSP can be modeled easily as a constraint program using the global cumulative constraint [8], which enforces that at each point in time, the cumulated demand of the set of jobs running at that point does not exceed the given capacities. Given a vector S of start time variables S_j for each job j, the RCPSP can be modeled as follows:

$$\begin{array}{lll} \min & max_{j \subset J}\{S_j + p_j\} & \\ \text{subject to} & S_i + p_i \leq S_j & \forall(i,j) \in A \\ & cumulative(S,p,r_k) & \forall k \in M \end{array} \quad (19.1)$$

This basic model will be developed and solved as Constraint Programming problem.

19.2.2 Multimodal RCPS

In order to consider the degradation of the system that has an impact of the actual duration of each activity, we can use also a Multimodal RCSP formulation of the

Table 19.2 Tabular data for each task $J = \{J_1,...,J_n\}$

	Task Mode	Data
1.	Mode $h = 1$ (e.g. *normal*)	• nominal duration • nominal setup time
2.	Mode $h = 2$ (e.g. *warning*)	• nominal duration • nominal setup time
3.	Mode $h = 3$ (e.g. *degraded*)	• nominal duration • nominal setup time

production-scheduling problem. In the multimodal RCSP, we have that each activity can be executed in a different Mode, that corresponds to a different degradation status: the basic examples is to have two modes, *normal* and *degraded*, with the second mode requiring more execution time than the first one.

In general, with respect to the formulation of the previous paragraph, to formulate a Multimodal RCPSP we need to provide for each activity, i.e., for each mechatronic tasks, and for each mode the following data:

(1) The nominal duration for each mode
(2) The nominal setup time for each mode

With respect to model (19.1) of the previous section, the Multimodal RCPSP has an additional dimension for each processing time, that is, p_{jh} where the index h is used to index the degradation state in which the mechatronic task corresponding to job S_j is working.

Indeed, the production task scheduler will expect as input to account for degraded conditions, the following tabular data for each task $J = \{J_1, \ldots, J_n\}$ (Table 19.2):

19.3 Optimization Challenges

For solving this type of problem a mathematical approach is possible in a simple production environment, but with the increment of variable, tasks, operations, especially for the scheduling phase in the framework presented in [2, 6] thought for more general application of white'R or other complex machineries, a heuristic approach can need for a right solving approach. We used the PSO, the Quantum PSO and a joint use of them for obtaining: fast convergence of particles, more solution than using only the QPSO algorithm and good accuracy on the synthetic database used for testing the production environment.

19.3.1 *Quantum Inspired Particle Swarm Optimization*

Particle Swarm Optimization (PSO) is a population based optimization method presented in 1995 by Eberhart and Kennedy [9]. Individuals in PSO work together, for solving a certain problem being interested in their own performance and in those of other particles of the swarm. Each particle has its own fitness value computed in the optimization process. The best fitness value achieved until then in this way is stored, this is the personal or individual best (*pbest*). The overall best fitness value obtained by any particle in the population is called global best (*gbest*), the solution is always stored. Each particle accelerate at a new position by calculating the velocity of that position where the value of *pbest* and *gbest* would influence the direction of the particle in the next iteration. In Eq. (19.2) the velocity update formula is presented and in Eq. (19.3) the new particle position computation is shown.

$$v = w * v n_{t-1} + c_1 * rand()(gbest_n - x_n) + c_2 * rand()(pbest_n - x_n) \qquad (19.2)$$

$$x_n = x n_{t-1} + v n \qquad (19.3)$$

where
  $c_1 > 0$ and $c_2 > 0$ are constant variables called cognitive and social parameters, and
  $w > 0$ is a constant variable called the inertia parameter.
  Moreover to create a swarm of i = 1,...,N particles, for all points in time, each particle i has

1. A current position x_i,
2. A record of the direction it follows to get to that position v_i,
3. A record of its own best previous position $pbest = (pbest1,...,pbestD)$,
4. A record of the best previous position of any member in its group

$$gbest = (gbest1, \ldots, gbestD).$$

The random number can assume values between 0 and 1. c_1 and c_2 control the particle acceleration towards personal best or global best. As previously reported a QEA has been presented in 2002 [10], it is inspired by the concept of quantum computing. In fact in accordance with the classical computing concept, information can be represented in bits, where each bit must hold either 0 or 1. Instead, in quantum computing, information is represented by a qubit whose value could be 0, 1, or a superposition of both. Superposition allows the possibility for a qubit to stay in both states, 0 and 1 simultaneously based on its probability. The quantum state is represented in the Hilbert space and is defined by: $|\Psi> = \alpha|0> + \beta|1>$, where α and β are complex numbers, and represent the probability at which the corresponding state is likely to appear [11]. Probability fundamentals require:

$$|\alpha|^2 + |\beta|^2 = 1,$$

where the value $|\alpha|^2$ represents the probability that a qubit is in the OFF (0) state and the value $|\beta|^2$ represents the probability that a qubit is in the ON (1) state. It is possible to define an individual with several qubits with the following general notation:

$$\begin{bmatrix} \alpha_1 & \alpha_2 & \ldots & \alpha_m \\ \beta_1 & \beta_2 & \ldots & \beta_m \end{bmatrix}$$

The Quantum inspired PSO (QPSO) was first discussed in [12] and there are several variants of QPSO [13, 14]. The main idea of QPSO is to use a standard PSO function to update the position represented in a qubit for that particle. In order to update the probability of a qubit during the run of the classical PSO, the quantum angle, θ, can be used. Quantum angle θ is here represented as $\begin{bmatrix} \cos(\theta) \\ \sin(\theta) \end{bmatrix}$.

The algorithm described before is characterized by a number of iterations where the quantum particle swarm related parameters are less than that of the general particle algorithm, so the computational speed is much higher than the general PSO and also the implementation is easier.

19.3.2 PSO and QPSO Joint Approach

For both QPSO and PSO algorithms, getting trapped into the local optima, is easy near the end of iteration, becoming impossible, in this case, to find the global optimal solution when the scheduling is complicated. The simulated annealing uses probabilistic concept to prevent the local optima, so, here, the simulated annealing is used inside each population to avoid the local optima [13].

A hybrid joint approach is presented in this section. It adopts several populations of independent evolution, and the QPSO presented in the previous section is used for optimizing some of them while the others are optimized by the GA technique. All sub-populations are connected together through the transport operators during the iterative process. In this way, during the first steps of the evolution, each population can get close fast to the optimal solution; in the last phases, while some populations are trapped into the local optima, it can carry the best individual of the entire population into this group by transport operators increasing the populations differentiation and escaping by the local optima. With this approach, an evolving mechanism of stimulation each other is present during the entire evolution process [15, 16].

19.4   Experimentations and Results

Data of 10 workstations and 10 work-pieces are adopted for testing this approach and make a comparison with the other single techniques. Therefore, other run tests have been done with other objective functions. Each population is composed of 40

Table 19.3 A comparison on a example test case

	Optimal solution	Minimum time
QPSO	8	85
PSO	10	100
PSO + QPSO	6	87

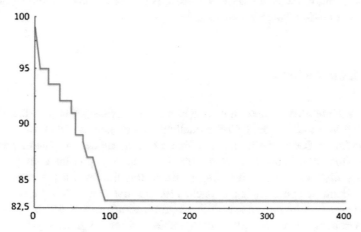

Fig. 19.3 The convergence function

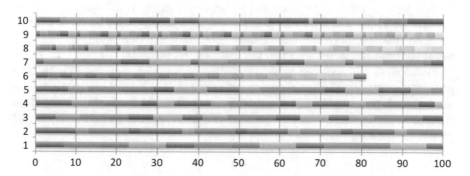

Fig. 19.4 Gantt chart of the optimal scheduling

individuals; the quantum particle swarm populations and PSO populations are both composed of 2; the maximum number of iterations is 400; simulation has been executed 10 times. In Table 19.3 the simulation results is shown. The convergence curve of the joint approach of a simulation at a time t is shown in Fig. 19.3. The Gantt Chart of the optimal scheduling is shown in Fig. 19.4.

The solution obtained with the use of the joint optimization algorithm is shown in Table 19.3 and It results being better than that obtained by QPSO. While the QPSO seems to be faster than others in convergence, QPSO seems to be too easy to be trapped into the local optima so as not to obtain the best solution. This could be an important issue when solving large-scale scheduling problem. The solution quality for some complex scheduling problems is comparatively poor, but it increases to great extension with the enhancement of the number of populations and of iteration. The situation has been repeated in a lot of our test cases. In these experimental setup the parallelization of machineries use has been taken in account, for the working mode of the activity when it is longer of the duration cut off (10 in this case), a mode change is considered.

19.5 Conclusions

This paper presents a new mixed approach of PSO-QPSO for the dynamic scheduling in a multimodal RCPS modelling of a (re)configurable robotic system, in particular of the white'R, but not limited to it. In fact a general production planning and scheduling framework has been already presented in [2, 6], like previously discussed. The scheduling phase of this framework presented in detail under different perspectives is approached for the first time in this paper. Moreover the challenge is solving the scheduling optimization process for the white'R but also generalizing it for complex configurable and reconfigurable production systems. Thus in the first part of the paper we presented the problem modelling and in the second part we proposed a solving method based on the joint use of PSO and QPSO. Results have been discussed in some examples confirming our mixed approach is more suited for complex scheduling systems when we have a lot of tasks on different machines/workstations, obtaining a faster algorithm and with better performance in terms to find the best optimal solution.

Future work will focus on better comparison of solving techniques [17, 18] also proposing a general QEA for numerical optimization [19–21] and comparing it with the approach presented in this paper. The proposed method will also be tested on different types of dataset such as general benchmark dataset as well as other real production dataset.

References

1. Graham, R.L., Lawler, E.L., Lenstra, J.K., Rinnooy Kan, A.H.G.: Optimization and approximation in deterministic sequencing and scheduling: a survey. Ann. Discr. Math. **4**, 287–326 (1979)
2. Sisca, F.G., Fiasché, M., Taisch, M.: A novel hybrid modelling for aggregate production planning in a reconfigurable assembly unit for optoelectronics. In: Arik, S. et al. (eds.) ICONIP 2015, Part II, LNCS 9490, pp. 571–582. Springer International Publishing, Switzerland (2015). doi:10.1007/978-3-319-26535-3_65S
3. Pinedo, M.: Scheduling: Theory, Algorithms, and Systems. Springer, New York (2008)
4. Pinedo, M.: Planning and Scheduling in Manufacturing and Services. Springer (2005)

5. Artigues, C., Demassey, S., Neron, E.: Resource Constrained Project Scheduling—Models, Algorithms, Extensions and Applications. Wiley, New York (2008)
6. Fiasché, M., Ripamonti, G., Sisca, F.G., Taisch, M., Tavola, G.: A Novel Hybrid Fuzzy Multi-Objective Linear Programming Method of Aggregate Production Planning. Springer Smart Innovation, Systems and Technologies, Advances in Neural Networks, pp. 489–501 (2016). doi:10.1007/978-3-319-33747-0_49
7. Berthold, T., Heinz, S., Lübbecke, M.E., Möhring, R.H., Schulz, J.: A Constraint Integer Programming Approach for Resource-Constrained Project Scheduling. Lecture Notes in Computer Science, Springer, vol. 6140, pp. 313–317 (2010)
8. IBM Contraint Programming Optimizer, Part of the IBM Optimization Studio. http://www-01.ibm.com/software/commerce/optimization/cplex-cp-optimizer/
9. Hamed, H.N.A., Kasabov, N., Shamsuddin, S.M.: Integrated feature selection and parameter optimization for evolving spiking neural networks using quantum inspired particle swarm optimization. SoCPaR 2009—Soft Computing and Pattern Recognition, pp. 695–698 (2009)
10. Han, K.H., Kim, J.H.: Quantum-inspired evolutionary algorithm for a class of combinatorial optimization. IEEE Trans. Evol. Comput. 580–593 (2002)
11. Hamed, H.N.A., Kasabov, N., Michlovský, Z., Shamsuddin, S.M.: String pattern recognition using evolving spiking neural networks and quantum inspired particle swarm optimization. Volume 5864 LNCS, Issue PART 2, pp. 611–619 (2009)
12. Sun, J., Feng, B., Xu, W.B.: Particle swarm optimization with particles having quantum behavior. Proc. Congr. Evol. Comput. 1, 325–331 (2004)
13. Lu, T.-C., Juang, J.-C.: Quantum-inspired space search algorithm (QSSA) for global numerical optimization. Appl. Math. Comput. 218, 2516–2532 (2011)
14. Quanke, P., Wenhong, W., Qun, P., Zhu, J.: Particle swarm optimization algorithm for job shop scheduling problems. Mech. Sci. Technol. 25(6), 675–679 (2006)
15. He, J.-J., Ye, C.-M., Xu, F.-Y., Ye, L., Huang, H.: Solve job-shop scheduling problem based on cooperative optimization. In: *Proceedings of the International Conference on E-Business and E-Government, ICEE 2010*, pp. 2599–2602 (2010)
16. Feifei, L., Kun, Y., Xiyu, L.: Multi-particle swarm co-evolution algorithm. Comput. Eng. Appl. 43(22), 44–46 (2007)
17. Nastasi, G., Colla, V., Cateni, S., Campigli, S.: Implementation and comparison of algorithms for multi-objective optimization based on genetic algorithms applied to the management of an automated warehouse. J. Intell. Manuf. 1–13 (2016)
18. Colla, V., Nastasi, G., Cateni, S., Vannucci, M., Vannocci, M.: Genetic algorithms applied to discrete distribution fitting. In: *Proceedings—UKSim-AMSS 7th European Modelling Symposium on Computer Modelling and Simulation, EMS 2013*, pp. 30–35 (2013)
19. Defoin-Platel, M., Schliebs, S., Kasabov, N.: Quantum-inspired evolutionary algorithm: a multimodel eda. IEEE Trans. Evol. Comput. (in print, 2009)
20. Fiasché, M.: A quantum-inspired evolutionary algorithm for optimization numerical problems. In: ICONIP 2012, Part III, LNCS 7665 (PART 3), pp. 686–693 (2012). doi:10.1007/978-3-642-34487-9_83
21. Hamed, H.N.A., Kasabov, N., Shamsuddin, S.M.: Quantum-inspired particle swarm optimization for feature selection and parameter optimization in evolving spiking neural networks for classification tasks. In: Kita, E. (ed.) Evolutionary Algorithms. InTech (2012). doi:10.5772/10545. ISBN: 978-953-307-171-8

Chapter 20
A Neural Network-Based Approach for Steam Turbine Monitoring

Stefano Dettori, Valentina Colla, Giuseppe Salerno
and Annamaria Signorini

Abstract This paper presents a Neural Network (NN) approach for steam turbines modelling. NN models can predict generated power as well as different steam features that cannot be directly monitored through sensors, such as pressures and temperatures at drums outlet and steam quality. The investigated models have been trained and validated on a dataset created through the internal sizing design tool and tested by exploiting field data coming from a real-world power plant, in which a High Pressure and a Low Pressure turbines are installed. The proposed approach is applied to identify the variation of the characteristics from data measurable on the operating field, by means of suitable monitoring and control algorithms that are implemented directly on the PLC.

Keywords Neural networks · Steam turbine modelling · Monitoring

20.1 Introduction

In the field of complex industrial system an ever increasing demand is observed for accurate system modelling for simulation, monitoring and control purposes. An exemplar case can be found in the field of monitoring and control of steam turbines, in the context in which power generation undergoes variable operating conditions, characterized by a variable steam production profile as a function of daily

S. Dettori · V. Colla (✉)
Scuola Superiore Sant' Anna, TeCIP Institute, Via Alamanni 13D, 56010 Pisa, Italy
e-mail: colla@sssup.it

S. Dettori
e-mail: s.dettori@sssup.it

G. Salerno · A. Signorini
General Electric Oil & Gas, Firenze, Italy
e-mail: giuseppe.salerno@bhge.com

A. Signorini
e-mail: annamaria.signorini@bhge.com

© Springer International Publishing AG 2018
A. Esposito et al. (eds.), *Multidisciplinary Approaches to Neural Computing*,
Smart Innovation, Systems and Technologies 69,
DOI 10.1007/978-3-319-56904-8_20

203

cycle such as in Concentrated Solar power plants, or where the steam turbines are started depending on the energy cost during the day. Steam turbines were originally designed for energy generation with stable steam production and rare start-up and shut down cycles, thus control systems have been designed starting from simplified models which are not appropriate far from nominal operating points. Under variable operating conditions, steam turbines are often subjected to continuous thermal stresses and windage effects and thus to premature aging and to the relative lowering of the efficiencies. In this cases a strong point of an accurate modelling is the possibility to follow phenomena that cannot be directly monitored through sensors. A physic-based approach needs an accurate modeling, especially as regards the relationship between the thermodynamic efficiencies of the blades, or more for large-scale models, of the machine drums. Non-linear equations for efficiencies, mass flow and enthalpy, that reproduce the real behavior of the machine, have to be accurately tuned. An intermediate modelling approach is represented by the use of particular semi-empirical relations [1–3] or by the approximation of fundamental thermodynamic equations [4]. The work described in [5], proposes two different approaches for the modelling of different parts of a power plant: in particular, the steam turbine modelling is based on energy and mass balance and exploits a Neural Network (NN). A black box approach shows no or weak link with the represented system of phenomenon, but allows to model the non-linear relationships between steam features and the generated mechanical power or thermodynamic characteristics of the steam in the turbine drums.

The proposed modelling approach is based on a feedforward NN (FFNN), which identifies the variation of the characteristics from data measurable on the operating field, by means of suitable monitoring and control algorithms that are implemented directly on the PLC and meet its stringent requirements in terms of memory usage and computational speed.

The paper is organized as follows: Sect. 20.2 provides a description of modeling issues related to the implementation of the thermodynamic equations to be tuned in different conditions of steam state and rotational speed. Section 20.3 describes the application of NNs; Sect. 20.4 describes the strategy for defining the dataset used to train and build the models, the data set used for model validation and presents results of NN modelling approach, when compared to field data. Finally, Sect. 20.5 provides some concluding remarks and hints for future work.

20.2   Background on the Modelling Issues

A modelling approach suitable for a final implementation in the PLC language has to balance two fundamental and often counteracting characteristics: accuracy and simplicity. In order to achieve good accuracy, a steam turbine thermodynamic model must estimate the fundamental parameters of each drum (mass flow, pressure, temperature, enthalpy and power) for a given set of boundary conditions: inlet steam condition, inlet steam valve stroke, mass extractions, extraction pressures and condenser pressure. A good turbomachinery model has to keep into account different factors:

- the heavy non-linearity behaviors of efficiencies, which depend on the rotational speed and ratios between drum pressures;
- the mass flow, estimated by means of the Stodola equation [6];
- the windage effect, which typically occurs on the last blades of the turbine and which reduces their generated power, by introducing friction losses;
- the kinetic energy contributions, which depends on steam velocity and geometrical features of the blades.

The thermodynamic models [6] estimate the steam flow along the turbine blades, with relationships that need iterative algorithms to match the boundary conditions of pressure or steam mass flow, therefore they are not suitable to the implementation of PLC platform. The simplest models are based on an extension of the concept of Willans' Line concept (e.g. [7, 8]), with the limitation that are valid between 40–100% of the load, require the measure of steam mass vapor flow and a map of isentropic efficiency in function of power and inlet pressure or steam mass flow, with results errors on power estimate around 10–15% with an average error not specified, typically useful for the design of industrial steam systems, but not the monitoring of machines.

An further interesting feature to estimate is the steam quality at turbine outlet defined as is the percentage of vapor mass in a liquid-vapor mixture:

$$x_{out} = \frac{m_v}{m_l + m_v} \tag{20.1}$$

where m_v and m_l are respectively the vapor mass and the liquid mass. Generally, the last stages of low pressure steam turbines (and sometimes also the High Pressure turbines) operates with a condensing wet steam flow, which can cause problems such as blade corrosion and efficiency reduction. In some cases, it is important to monitor or control this feature with the aim of optimizing the power production systems [9]. Steam quality is very difficult to measure, therefore in most industrial plants it is not monitored at all.

Due, on one hand, to the difficulty of developing a simple but accurate mathematical model which can consider all the above mentioned parameters and, on the other hand, to the availability of numerical data from both field experiments and simulations, data-driven modelling approaches appear suitable and NNs are flexible and generic enough for a successful application.

20.3   The Neural Network-Based Approach

The NN-based model taking into account all the factors listed in Sect. 20.2 should be integrated with a steam turbine mechanical model in order to estimate not only the actual generated power, power losses or speed, but also operating condition of the machine. Different NNs have been developed, with different aims: prediction of the power on each steam turbine drum or the overall turbine power; prediction of the pressures and temperatures at exit of each drum and of the steam quality at last stage of the turbine.

The architecture of the hidden layer (number of neurons and number of hidden layers), has been defined by means of an optimization study, performed with the aim of minimizing the error of the NN committed on the validation set (the experimental optimization is reported in Sect. 20.4.2). The transfer function of hidden layers and output layer are respectively, hyperbolic tangent and a linear function. The training of the neural network was performed using Neural Network Toolbox (MATLAB 8.0), by means of Levenberg-Marquardt back-propagation [10] and early stopping algorithm. In particular the maximum number of epochs has been set to 1000. The target function to be minimized was the mean squared error between the prediction and the desired output.

The following NNs have been trained:

- A NN for overall power estimation, net_1;
- A NN to estimate the power of each drum D_i, net_2;
- A NN to estimate pressures and temperatures at outlet of each drum D_i, net_3;
- A NN to estimate the steam quality at turbine outlet, net_4.

The training set of the NN was defined as a couple input/output with input of each NN, $I_{net_i} = \begin{bmatrix} P_{in} & T_{in} & Q_{e1} & \cdots & Q_{en} & P_{out} \end{bmatrix}^T$, composed by inlet steam pressure P_{in} and temperature T_{in}, mass flow of n bleedings or extractions $\begin{bmatrix} Q_{e1} & \cdots & Q_{en} \end{bmatrix}$ and outlet pressure P_{out}. The output for each NN is defined as follows:

- for the net_1, the overall power, $O_{net_1} = W_{overall}$
- for the net_2, the power W_{Di} of each drum D_i, with $i = 1, \ldots, m$ drums,
  $O_{net_2} = \begin{bmatrix} W_{D1} & \cdots & W_{Dm} \end{bmatrix}^T$
- for the net_3, the steam pressures P_i and Temperatures T_i at the exit of each i-th drum of the turbine, $O_{net_3} = \begin{bmatrix} P_1 & T_1 & \cdots & P_m & T_m \end{bmatrix}^T$
- for the net_4, the steam quality at exit of the turbine x_{out}, $O_{net_4} = x_{out}$

20.4 Training and Test Results

20.4.1 Dataset Selection Strategy

Two different steam turbines have been modelled by means of the NN-based approach: a High Pressure (HP) steam turbine with impulse stage and 3 reaction drums and a Low Pressure (LP) steam turbine, with 4 reaction drums and a condenser. The power size of the HP turbine is about 18 MW, with inlet rated steam pressure of 106 BarA and temperature of 378 °C. The power size of the LP turbine

Table 20.1   HP and LP turbine training dataset

Input	Description	N. operating points	Ranges
HP turbine			
P_{in}	Inlet steam pressure	19	15–105 (BarA)
T_{in}	Inlet steam temperature	5	330–390 (°C)
Q_1	Bleeding mass flow	4	0.001–4.5 (Kg/s)
P_{out}	Exit steam pressure	9	4–20 (BarA)
LP turbine			
Q_{in}	Inlet mass flow	10	5–50 (kg/s)
T_{in}	Inlet steam temperature	5	330–390 (°C)
Q_1	1st bleeding mass flow	3	0–3 (Kg/s)
Q_2	2nd bleeding mass flow	3	0–3 (Kg/s)
Q_3	3rd bleeding mass flow	3	0–3 (Kg/s)
Q_4	4th bleeding mass flow	3	0–3 (Kg/s)
P_{out}	Condenser pressure	3	0.04–0.08 (BarA)

is about 40 MW, with inlet rated steam pressure of 18.3 BarA and temperature of 378 °C. The HP and LP turbines have, respectively, 1 and 4 bleedings.

A dataset of operating conditions has been created by mean the GE internal sizing tool, which contains the turbine geometrical and mechanical data, with a wide range of operating conditions. The dataset has been divided in two parts: a training dataset, 80% of the original dataset randomly sampled and a validation dataset, for the early stopping algorithm, the remaining 20% of the original dataset. The training/validation dataset of the HP and LP turbines consist, respectively, of 2420 and 8445 different conditions shown in Table 20.1.

A second dataset for the testing of the trained NNs has been created by means field data measured during one full day. In particular, inlet steam pressure, temperature, outlet pressure have been measured from site during one day activity and used as boundary condition for GE internal sizing tool. The outputs of this tool have been used to evaluate the accuracy of the models. The particularity of this database is that as regard to the LP turbine, the condenser pressure in some operating hours of the day was particularly outside of the nominal ranges.

20.4.2   Results

In order to evaluate the accuracy of the NN models, Mean Absolute Error (Mean$|e|$), Max Absolute Error (Max$|e|$), Max Percent Absolute Error (Mean$|\%e|$) and Standard Deviation (STD) have been computed.

As mentioned in Sect. 20.3, the optimal architecture of each NN (number of neurons and number of hidden layers) has been defined to the aim of minimizing the error on the validation set. In a FFNN with n_i inputs, n_u outputs, n_h hidden layers and N neurons in each hidden layers, the number of weights n_W is given by the following well-known formula:

$$n_W = (n_h - 1)N^2 + (n_i + n_h + n_u)N + n_u \qquad (20.2)$$

The optimization has been conducted by varying the number of weights n_W and number of hidden layers n_h. Each NN architecture has been trained 40 times and the optimal NN architecture has been selected to the aim of achieving a good compromise between the average error on the validation dataset, the minimization of maximum % error of the NN and a good standard deviation over 40 trainings. The optimal architecture of each NN is shown in Table 20.2. Finally each optimal architecture, and in particular the best network over 40 trainings, has been tested by means of the field dataset. The results are shown in Tables 20.3 and 20.4.

Table 20.2 Optimal architecture of neural networks

Turbine	HP				LP			
Net	net_1	net_2	net_3	net_4	net_1	net_2	net_3	net_4
n_h	2	2	2	2	2	2	2	2
n_n	10 10	19 19	19 19	8 8	16 16	16 16	16 16	7 7

Table 20.3 Test results for HP steam turbine neural networks

NN	Section	Output	Mean$|e|$	Max$|e|$	Max$|\%e|$	STD
net_1	Overall	$W_{Overall}$(kW)	7.8	44	0.4	9.3
net_2	Impulse	W_{Imp} (kW)	2.1	6.1	0.9	2.2
	D1	W_{D1} (kW)	2.2	11	0.7	2.7
	D2	W_{D2} (kW)	1.8	9.2	0.3	2.1
	D3	W_{D3} (kW)	5.3	23	1.2	5.3
	Overall	$W_{Overall}$(kW)	8.3	44	0.5	10
net_3	Impulse	P_{Imp} (BarA)	0.02	0.2	0.8	0.03
		T_{Imp} (°C)	0.05	0.3	0.1	0.04
	D1	P_{D1} (BarA)	0.04	0.1	1	0.04
		T_{D1} (°C)	0.1	0.6	0.2	0.1
	D2	P_{D2} (BarA)	0.03	0.08	1.3	0.02
		T_{D2} (°C)	0.08	0.7	0.3	0.1
	D3	T_{D3} (°C)	0.1	1.4	0.8	0.2
net_4	Last stage	x_{out}	2e-4	1e-3	0.1	3e-4

Table 20.4   Test results for LP steam turbine neural networks

NN	Section	Output	Mean$\|e\|$	Max$\|e\|$	Max$\|\%e\|$	STD
net_1	Overall	$W_{Overall}$ (kW)	63	1250	9.3	71
net_2	D1	W_{D1} (kW)	30	139	5.8	11
	D2	W_{D2} (kW)	33	144	4.0	11
	D3	W_{D3} (kW)	52	166	5.1	15
	D4	W_{D4} (kW)	19	810	21	57
	Cond	W_{Cond} (kW)	20	509	293	39
	Overall	$W_{Overall}$ (kW)	106	1021	7.6	89
net_3	D1	P_{D1} (BarA)	0.1	0.2	6.2	0.06
		T_{D1} (°C)	6.0	15	5.7	2.8
	D2	P_{D2} (BarA)	0.04	0.1	5.1	0.03
		T_{D2} (°C)	5.6	13	7.4	2.3
	D3	P_{D3} (BarA)	0.02	0.06	8.8	0.02
		T_{D3} (°C)	2.7	10	10	1.7
	D4	P_{D4} (BarA)	0.02	0.1	91	0.02
		T_{D4} (BarA)	0.5	16	35	1.2
	Cond	T_{Cond} (°C)	0.1	1	2.5	0.2
net_4	Last stage	x_{out}	3e-3	0.01	1.2	1e-3

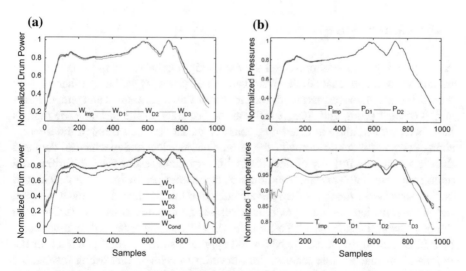

Fig. 20.1   **a** Test results on HP (*top figure*) and LP (*figure below*) turbine drum power estimation. **b** Test results on HP Pressures (*top figure*) and LP Pressures (*figure below*) at outlet of each drum. Daily profile (*continuous line*) and esteem (*dashed line*)

Figures 20.1a and b show, respectively, the daily profile of the power generation of each HP and LP turbine drum and the estimate of each drum power obtained through the NN-based models, the Pressures and Temperatures of HP turbine. For confidentiality constraints, all the variables have been normalized with respect to their maximum value in the day and the figures on the estimation of HP and LP outlet steam quality are not reported.

The test results are very interesting and encouraging: in particular the NN models net_1 and net_2 for the prediction of overall power and Drums power are capable to predict the generated power with very low errors and a good accuracy for both HP and LP turbines. As far as the results on LP Turbine are concerned, the maximum errors are due to the condenser pressure outside the nominal ranges of operating conditions. Results on net_3 show very low errors for the HP Turbine, whereas for the LP Turbine, temperatures estimates show a low accuracy, while the pressure estimate is quite accurate. In this case also for the LP turbine the NN show bigger errors during the operating hours of the day, when the condenser pressure is particularly far from of the nominal ranges. The NN models for the estimate of steam quality (net_4) show very good results for the HP Turbine, and good results for the LP Turbine.

To sum up, all HP Turbine models show very good results, whereas LP Turbine models show good results during the nominal operating conditions and poor results when condenser pressure is too far from the sampled dataset, by highlighting a lack of generalization. In order to solve this lack of precision and accuracy, the space of the condenser pressure samples should cover more extensive ranges and more dense sampling.

20.5   Conclusion

A NN-based approach is presented for modelling a generic steam turbine, and, in particular, for the estimate of the overall generated power as well as of other relevant variables that cannot be measured, such as drums power, pressures and temperatures at exit of each drum, and steam quality at the last stage of the turbine. The models have been trained and validated on a massive dataset created through the internal sizing design tool, which contains the turbine geometrical and mechanical data and then tested on one full day of field data with very good results for HP Turbine and good results for LP Turbine. The obtained results are promising and the achieved performance show the feasibility and advantage of the NN-based approach in steam turbine monitoring, including extremely off design conditions for HP Turbine. On the other hand, in the case of the LP turbine, which is characterized by more complex input-output relations, the NN-based approach shows a good potential for the prediction of inaccessible features, but, in order to achieve interesting results, the training dataset has to be designed with a denser sampling covering more extensive ranges of the input variables.

The future work will cover the above-highlighted gaps and will explore other Neural paradigms and more complex network structures, including hybrid approaches.

References

1. Medina-Flores, J.M., Picn-Nez, M.: Modelling the power production of single and multiple extraction steam turbines. Chem. Eng. Sci. **65**(9), 2811–2820 (2010)
2. Chaibakhsh, A., Ghaffari, A.: Steam turbine model. Simul. Model. Pract. Theory **16**(9), 1145–1162 (2008)
3. Varbanov, P.S., Doyle, S., Smith, R.: Modelling and optimization of utility systems. Chem. Eng. Res. Des. **82**(5), 561–578 (2004)
4. Ray, A.: Dynamic modelling of power plant turbines for controller design. Appl. Math. Model. **4**(2), 109–112 (1980)
5. Lu, S., Hogg, B.W.: Dynamic nonlinear modelling of power plant by physical principles and neural networks. Int. J. Electr. Power Energy Syst. **22**(1), 67–78 (2000)
6. Larry, D.S., Hall, C.: Fluid Mechanics and Thermodynamics of Turbomachinery. Butterworth-Heinemann (2013)
7. Mavromatis, S.P., Kokossis, A.C.: Conceptual optimisation of utility networks for operational variationsI. Targets and level optimisation. Chem. Eng. Sci. **53**(8), 1585–1608 (1998)
8. Aguilar, O., Perry, S.J., Kim, J.K., Smith, R.: Design and optimization of flexible utility systems subject to variable conditions: Part 1: modelling framework. Chem. Eng. Res. Des. **85**(8), 1136–1148 (2007)
9. Ganjehkaviri, A., Jaafar, M.M., Hosseini, S.E.: Optimization and the effect of steam turbine outlet quality on the output power of a combined cycle power plant. Energy Convers. Manag. **89**, 231–243 (2015)
10. Hagan, M.T., Menhaj, M.B.: Training feedforward networks with the Marquardt algorithm. IEEE Trans. Neural Netw. **5**(6), 989–993 (1994)

Chapter 21
A Predictive Model of Artificial Neural Network for Fuel Consumption in Engine Control System

Khurshid Aliev, Sanam Narejo, Eros Pasero
and Jamshid Inoyatkhodjaev

Abstract This paper presents analyses and test results of engine management system's operational architecture with an artificial neural network (ANN). The research involved several steps of investigation: theory, a stand test of the engine, training of ANN with test data, generated from the proposed engine control system to predict the future values of fuel consumption before calculating the engine speed. In our paper, we study a small size 1.5 L gasoline engine without direct fuel injection (injection in intake manifold). The purpose of this study is to simplify engine and vehicle integration processes, decrease exhaust gas volume, decrease fuel consumption by optimizing cam timing and spark timing, and improve engine mechatronic functioning. The method followed in this work is applicable to small/medium size gasoline/diesel engines. The results show that the developed model achieved good accuracy on predicting the future demand of fuel consumption for engine control unit (ECU). It yields with the error rate of 1.12e-6 measured as Mean Square Error (MSE) on unseen samples.

Keywords Artificial neural network · Predictive models · Engine control systems

K. Aliev (✉) · S. Narejo · E. Pasero
Department of Electronics & Telecommunications, Politecnico di Torino,
Turin, Italy
e-mail: khurshid.aliev@polito.it

S. Narejo
e-mail: sanam.narejo@polito.it

E. Pasero
e-mail: eros.pasero@polito.it

J. Inoyatkhodjaev
Department of Applied Science, Turin Polytechnic University in Tashkent,
Tashkent, Uzbekistan
e-mail: inoyatkhodjaev@gmail.com

© Springer International Publishing AG 2018   213
A. Esposito et al. (eds.), *Multidisciplinary Approaches to Neural Computing*,
Smart Innovation, Systems and Technologies 69,
DOI 10.1007/978-3-319-56904-8_21

21.1   Introduction

ANN have received much attention in recent years due to the number of functions which can be applied to engines such as modelling, controller design, on-board testing and diagnostics. ANN is an information processing paradigm that is inspired by biological nervous systems, functions, and mathematical and physical methods of information processing. The latter is composed of many simple processing units working in parallel to each other which are connected insome way and depend on the status of the dynamic response of the external input information computer system. It is a simple simulation of the brain with specific smart features and rapid processing abilities to perform [1].

Neural networks generally contains three layers: input layer, hidden and output layer. Each of them consists of nodes or neurons. Moreover, ANN can solve classification problems and function approximation. Applications of ANN model start operate correctly if the design of data satisfies all the dynamics of the system. Furthermore, structures and combinations of networks considerably affect the performance level of system. This article will describe data acquiring procedures, construction of the experiments and neural network description. Collection of data needs to be done precisely and the size of data is another important factors of ANN. The data needs to be determined if transient behavior or steady-state operations provide sufficient features for training and validation. The more features the training data covers, the better the network is trained for the generalization of engine behavior. Designing of experiments has a significant impact on the model performance and especially for engine systems [2].

21.2   Research Background

Neural networks, and fuzzy systems can be significant in the usage of advanced control strategies. The authors of [3] used IC (Internal combustion) engines with highly nonlinear characteristics containing variable time constant terms and delays.

The capability of smart systems performances, for instance neural networks and fuzzy methods: if we relate these nonlinear properties, ANN becomes excellent tool in the modeling of engines [4]. The authors of [5] discuss a technique where static neural networks (SNN), time delayed neural networks (TDNN) and dynamic neural networks are used for modeling an spark-ignition (SI) engine.

Similarly, in our investigation, computational intelligence is used at 2 stages of our work. At first, the management of ECU is directed by creating an application of fuzzyneural network (FNN). Afterwards, once the developed model of ECU passes the hot test procedure, we created another model of ANN for prediction. The objective for designing this model was, to predict the future demand of fuel consumption in engine by taking previous time interval and speed (rotation per minute) as an input.

21.3   Experimental Setup

The experimental setup of our work consists of three steps. In the first step we develop a model for engine control system of vehicle motor. The controlling and management unit is based on neuro-fuzzy system. Afterwards, the hot test procedure is implemented to test the developed ECU which is the second step in our procedure. The next step is to construct a predictive model, which is capable to estimate the unknown values of fuel consumption for future time intervals. The steps are further discussed in details.

21.3.1   Engine Control System

Neural networks can only come into play if the problem is expressed by a sufficient amount of observed examples. These observations are used to train the black box (as element of ECU). In this case no prior knowledge about the problem needs to be given. In our case to develop engine management (controlling) algorithm, we use the attributes of FNNs. The learning procedure is constrained to ensure the semantic properties of the underlying fuzzy system. A neuro-fuzzy system approximates n-dimensional unknown functions which and partly represented by training examples. A neuro-fuzzy system is represented as a special three-layered feedforward neural network as shown in Fig. 21.1.

The first layer corresponds to the input variables—acquired data from sensors. The second layer symbolizes the fuzzy rules—processes in the engine control unit.

Fig. 21.1   The architecture of a neuro-fuzzy system: the use of neural networks in an engine management system

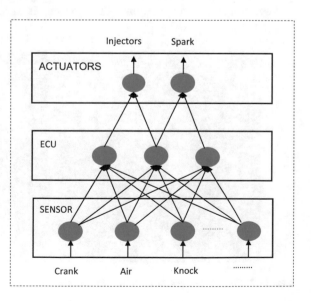

The third layer represents the output variables—output signals from the control unit to actuators. The fuzzy sets are converted as (fuzzy) connection weights.

21.3.2 Engine Hot Test Process

The hot test process consists of two main offline tests which includes production and durability hot tests. A production hot test (idle running engine) was used for the engine start verification prior to shipment to avoid costly final assembly plant warranty issues. These stations are often used as a confidence test for the quality process, and also act as a low cost electrical test. Often operator subjective, an noise vibration harshness (NVH) test looking for odd engine noise may also be used in this test, as well as for fuel and oil leakage. On the other hand, a durability hot test which (puts a well-controlled load on the engine during running) is used in large diesel markets, where horsepower verification and fuel consumption, emission and calibration can only be done while running the engine.

In our case, the engine hot test process included the checking of the main engine parameters and engine performance data acquisition. The engine was rigged up on a trolley outside the test container as shown in Fig. 21.2. Subsequently, the trolley was connected automatically through the connection plate to the test stand. After manual release by the worker, the trolley is automatically pulled to the test position and fixed.

On the test stand and on the trolley, connection plates with quick fit connectors, guarantee a fast docking. If the engine is properly connected to the test stand, the test stand program will start the test cycle. During the hot test the following parameters test the performance of power: torque with either oil, water or fuel leakages under warm conditions. The next step is to record engine control module

Fig. 21.2 Engines hot test machine (container type)

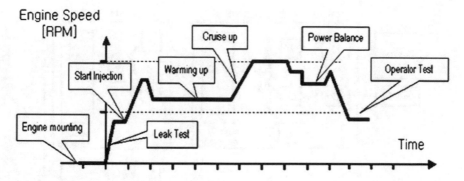

Fig. 21.3 Hot test sequences

(ECM) trouble codes, noise, electrical and mechanical functions, fuel consumption and emission. The test consist of different sequences and steps illustrated in Fig. 21.3.

After the automatic test run, the observed results were transferred to the test field server via an ethernet interface. The outcomes were reported in an Excel file. The measured results of other test runs were saved simultaneously on the server. The last observations run was saved and automatically opened in the post-processing of personal computer (PC) [6].

21.3.3   Prediction for Fuel Consumption

The engine hot test machine generates 2 types of output data: graphical—curves of separate engine characteristics and numerical data. The numerical data are used for ANN training process with detail breakdown which gives outputs from 24 engine parameters recorded frequently after every 0.47 s.

As much information we get from the training process as "smarter" our controlling system performs. In engine control unit system, amount of fuel being used is very crucial. Fuel consumption is the weight flow rate of fuel required to produce a unit of power or thrust. That is the reason to know in advance the consumption of fuel. Fuel consumption in the designed engine management control system is predicted one time step earlier with ANN model.

The two attributes speed and time are taken from the dataset generated by ECU, to train the ANN model for prediction of fuel consumption. The data is scaled in the range of [0, 1] for the artificial neural network to learn the underlying nonlinear structure and dependencies present in the dataset. Further to this, in contemplation to predict accurate fuel consumption values over step ahead, as it is discussed earlier that the choice of ANN used here is nonlinear autoregressive exogenous

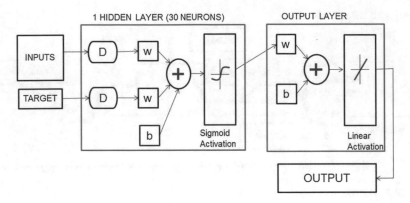

Fig. 21.4 Architecture of model NARX ANN

model (NARX), which is a recurrent neural network. The expression of that model is given in Eq. (21.1).

$$y(t) = f(u(t-nu), \ldots, u(t-1), u(t), y(t-ny) \ldots, y(t-1)) + \in \qquad (21.1)$$

which identifies that in order to predict the future values of y, the past values of same parameters and the current and the past values of external variable u are far most crucial. It is the realization in literature that the recurrent neural networks with sufficient number of hidden neurons having nonlinear activation function behaves in the manner of nonlinear autoregressive moving average (ARMA) method for time series forecasting [7, 8]. Inputs to our adopted model comprises of time (sec) and engine speed (rpm).

Output of model is a future value of fuel consumption. Figure 21.4 depicts the architecture is based on input layer, 1 hidden layer containing 30 hidden nonlinear neurons with sigmoid activation function, and 1 output layer containing one linear neuron. The selection of hidden neurons is based on Monte-Carlo simulation [9, 10].

21.4 Results and Discussion

The NARX model is trained on 40,000 samples of data set. In order to avoid the over fitting of model on data and to improve generalization, the dataset is divided in 70% for training, 15% for validation check and 15% for testing the model performance. The rest of dataset samples are kept aside to evaluate the efficiency of network on unseen data. The performance of the network is measured by Mean Square Error (MSE) as given in Eq. (21.2).

$$MSE = \frac{1}{N} \sum_{i=1}^{N} [y(t) - \hat{y}(t)]^2 \qquad (21.2)$$

where $\hat{y}$ is a vector of n predictions, and y is the vector of observed values corresponding to the inputs to the function which generated the predictions. Practices of model can be observed from Fig. 21.5a and b which shows the measures related with the training set of model.

The model is trained using Levenberg-Marquardt training Algorithm, training stops as the error is not reduced further on validation set by choosing the early stopping criteria. However the model completes 333 epochs of iteration to minimize the error on predictions. The overall network performance measure is 3.5603e-05 MSE.

Another validation parameter R is used, involves analyzing the goodness of fit of the regression. Monitoring whether the regression residuals are random, and checking whether the model's predictive performance deteriorates substantially when applied to data that were not used in model estimation. The graph of the regression is illustrated in Fig. 21.6. The solid line represents the best fit linear regression line between outputs and targets. The R value is an indication of the relationship between the outputs and targets. If R = 1, this indicates that there is an exact linear relationship between outputs and targets. On the other hand, if R is close to zero, then there is no linear relationship between outputs and targets. Figure 21.7 depicts the behavior of actual fuel consumption and predicted fuel consumption with time. The deviation of predicted samples from the actual ones can be properly seen in Fig. 21.8a and b.

Figure 21.8b shows 20 samples prediction of fuel consumption using one step ahead prediction, the average difference between the actual and the predicted samples is around 1.1242e-06, which is nearly accurate. This model helps to know

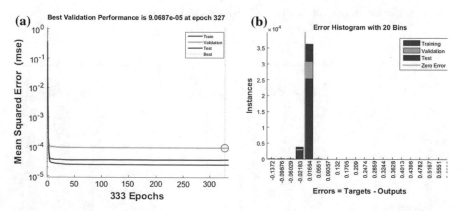

Fig. 21.5 a Performance of model on training, validation and test sets **b** error histogram of training, validation and testing

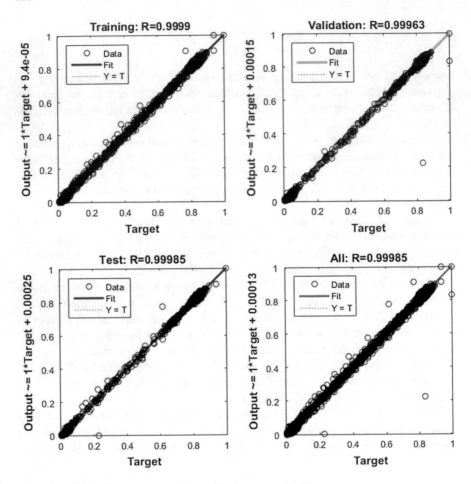

Fig. 21.6 Regression plots of training, validation, test and all in one

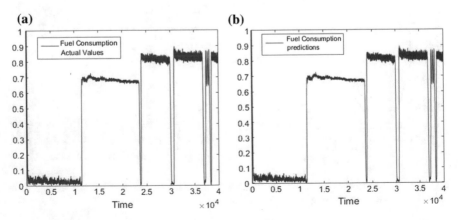

Fig. 21.7 **a** Actual fuel consumption samples scaled in [0.1] **b** outputs of step ahead prediction from model

(a)   (b)

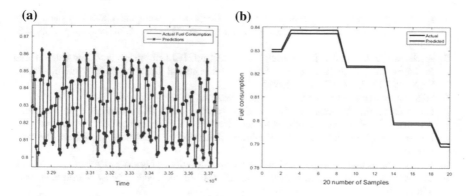

Fig. 21.8 **a** Actual and predicted values **b** 20 samples of actual and predicted FCs

in early time stamp what will be fuel consumption, even the speed is unavailable but focusing the rotation at previous time and the previous fuel consumptions one is able to infer the approximate fuel consumption for future.

21.5 Conclusion

In this paper we presented neural network approach to predict the fuel consumption of the engine based on NARX-models. The input to NARX is extracted from the resultant parameters of engine hot test. The hot test was conducted on ECU which is again based on neuro-fuzzy system. The ECU is further using ANN architecture which can manage engine controlling system precisely and outputs gave good results of learning and prediction. By predicting the fuel consumption earlier, we can manage more accurately air/fuel ratio, exhaust managing by exhaust gas recirculation (EGR) system and cam timing (if it is variable). As a result, it improves the main engine characteristic, effectiveness, fuel consumption, performance and pollutant volume.

References

1. Jiang, F., Zhenhua, L.: Electronic control unit design of artificial neural network-based natural gas engine. Adv. Mater. Res. **998–999**, 617–620 (2014). ISSN: 1662-8985
2. Deng, J., Stobart, R., Maass, B.: Artificial neural networks—Industrial and control engineering application. Loughborough University, UK (2011)
3. Howlett, R.J., de Zoysa, M.M., Walters, S.D., Howson, P.A.: Neural network techniques for monitoring and control of internal combustion. In: International Symposium on Intelligent Industrial Automation, Genova, Italy (1999)

4. Jones, R.P., Cherry, A.S., Farrall, S.D.: Application of intelligent control in automotive vehicles. In: IEEE International Conference on Control, Coventry, UK, vol. 389, pp. 159–164 (1994)
5. Ayeb, M., Lichtenthaler, D., Winsel, T., Theuerkauf, H.J.: SI engine modeling using neural networks. In: Proceedings of the 1998 SAE International Congress & Exposition. Detroit, USA, vol. 1357, pp. 107–115 (1998)
6. "GM Global Hot test procedure" standards process
7. Diaconescu, E.: Prediction of chaotic time series with NARX recurrent dynamic neural networks. In: Proceedings 9th WSEAS International Conference on Automation and Information. World Scientific and Engineering Academy and Society, Bucharest, Romania (2008)
8. Haykin, S.: Neural networks, 2nd edn. Pearson Education, Singapore (1999)
9. Dorffner, Georg: Neural networks for time series processing. Neural Netw. World 6(4), 447–468 (1996)
10. Pasero, E.G., Moniaci, W.: Artificial neural networks for meteorological Nowcast. In: International Symposium on Computational Intelligence for Measurement Systems and Application CIMSA (2004)

Chapter 22
SOM-Based Analysis to Relate Non-uniformities in Magnetic Measurements to Hot Strip Mill Process Conditions

Gianluca Nastasi, Claudio Mocci, Valentina Colla, Frenk Van Den Berg and Willem Beugeling

Abstract The paper describes the application of a Self-Organising Map for the analysis and the interpretation of measurements taken by a Non-Destructive Testing system named IMPOC® and related to the hardness of steel coils after the hot rolling mill. This work addresses the problem of understanding whether distinct process conditions may lead to non-uniform mechanical properties along the coil. The proposed approach allows to point out, for each specific steel grade, some process conditions that are more frequently associated to disuniformities.

22.1 Introduction

In steel manufacturing novel Non-Destructive Testing (NDT) techniques [12, 14] are introduced complementary to destructive mechanical testing, as they provide information over the full product length in a much faster way. In the assessment of mechanical properties, a frequently used strategy exploits the IMPOC® system [9, 11, 13]. Such system is composed by two identical heads, mounted face-to-face, at a distance of 50 mm. When the steel strip passes through the heads, it is magnetized and the residual field is recorded every two meters. The so-called IMPOC value is defined as the maximum intensity of such magnetic field and is used to quantify

G. Nastasi · C. Mocci · V. Colla (✉)
Scuola Superiore Sant' Anna, TeCIP Institute, ICT-COISP Center, Pisa, Italy
e-mail: colla@sssup.it

G. Nastasi
e-mail: g.nastasi@sssup.it

C. Mocci
e-mail: c.mocci@sssup.it

F. Van Den Berg · W. Beugeling
Tata Steel, IJmuiden, The Netherlands
e-mail: frenk.van-den-berg@tatasteel.com

W. Beugeling
e-mail: willem.beugeling@tatasteel.com

© Springer International Publishing AG 2018
A. Esposito et al. (eds.), *Multidisciplinary Approaches to Neural Computing*,
Smart Innovation, Systems and Technologies 69,
DOI 10.1007/978-3-319-56904-8_22

223

strength of the material. This device is thus capable to provide a measurement of an important mechanical property of the semi-finished product along its length, which can be very high, up to 3000 m. This is a very important feature, as the customer provides specifications for the steel product, which must be respected in any portion of the product itself. For this reason, considering the relevant length of the produced steel strip, a discrete sampling and testing of portions of the strip itself is not adequate to ensure compliance with respect to the customer requirements. However, the interpretation of the measurement provided by the IMPOC® [10] to the final aim of improving the process and providing higher uniformity of the mechanical properties along the product is not straightforward and relevant research efforts are spent by the researchers involved in the steel field in this direction. In particular the project entitled "Product Uniformity Control" (PUC) has started in 2013 under the leadership of Tata Steel, involves many major steel producers as well as a number of research institutions and it addresses the optimisation of the product uniformity through the adoption of online NDT systems. In order to establish a relationship between microstructural (non-)uniformity and online measured (non-)uniformity, yet only partly understood, PUC follows a concerted approach integrating experimental research, physical modelling and data mining. The work described in this paper is developed within this ongoing project.

The IMPOC® system, which is considered in the work described in this paper, is placed after the Hot-Rolling Mill (HRM) and the present work addresses the problem of understanding whether distinct HRM process conditions may lead to non-uniform strength values along the coil taking into account IMPOC values. To this aim, an algorithm is presented to capture process conditions that affect strength uniformity along the strip. To this aim, a Self-Organizing Map (SOM) is exploited. This Neural Network (NN) mainly considers the position, shape and topology of the input data, allows the visualization of complex input spaces [8] and is widely applied in data processing applications [15, 16]. This approach was developed and validated on a database provided by Tata Steel, one of the leading steel industries, which includes HRM process parameters and corresponding IMPOC® measurement for about 6700 coils.

The paper is organised as follows: Sect. 22.2 provides some basic background on the industrial application; Sect. 22.3 provides a description of the needed pre-processing steps on the available data in order to make it suitable for the proposed application; Sect. 22.4 describes the analyses that can be performed by exploiting the SOM-based data representation. Finally in Sect. 22.5 some concluding remarks and hints for future work are provided.

22.2 The Industrial Problem

The IMPOC® system which provided the measurements that have been processed in the work described in this paper is placed after a Hot-Rolling Mill (HRM). A HRM is the process which aims at reducing the thickness of the semi-finished steel product

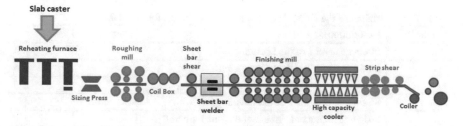

Fig. 22.1 Scheme of the processing of the steel in the hot rolling mill

(slab) after it has been reheated in the reheating furnace to produce a longer and flat semi-finished product called Hot Rolled Coil (HRC). Figure 22.1 schematically depicts the relevant processing steps undergone by the casted slab.

The HRM is composed by different subprocesses, among which the roughing mill, which is responsible for the first gross thickness reduction, and the finishing mills, which provides the HRC with the desired final thickness. The HRC is finally cooled and coiled to be stored. The HRC will be further processed in the Cold Rolling Mill (CRM), which produce the final coil, that is sold to manufacturers producing a wide variety of goods. The HRM is therefore an intermediate but fundamental stage for the determination of the final mechanical properties of the products and a critical task is to understand whether particular HRM process conditions could lead to non-uniform coils in terms of IMPOC values. The task is even more complex considering that any steel plant produces several kind of steels, that are grouped by grade, and each grade group has its own characteristic chemical composition and target mechanical properties.

22.3 Data Pre-processing and Selection

The present work addresses the problem of understanding whether distinct HRM process conditions may lead to non-uniformities taking into account IMPOC values. Two datasets have been exploited, which were provided by Tata Steel. The first dataset contains the HRM process parameters of almost 7000 coils. The second dataset contains IMPOC measurements, sampled every 2 s in a position immediately following the HRM. Both datasets needed some pre-processing in order to reduce the number of variables to consider as well as to remove unreliable data.

The HRM dataset contains around 90 variables, but most of them are not meaningful for the analysis or are strongly linear correlated. The number of input variables of the original HRM dataset has been reduced by jointly exploiting process knowledge

Table 22.1 Variables of the HRM dataset considered as inputs for the SOM

Symbol	Description	Unit	Range
S_{f1}	Strain at stand 1 of the finishing mill		
S_{f7}	Strain at stand 7 of the finishing mill		
ρ_{f1}	Strain rate at stand 1 of the finishing mill		
ρ_{f7}	Strain rate at stand 7 of the finishing mill		
T_{f1}	Strip temperature at stand 1 of the finishing mill		
T_{f7}	Strip temperature at stand 7 of the finishing mill		
T_{v11}	Strip temperature at the roughing mill		

and variable selection procedures [3–5]. Hence just a small subset of meaningful and independent HRM process parameters were taken into account: the final set of considered input variables is listed in Table 22.1.

Several preprocessing steps are necessary to cleanse the data provided by the IMPOC®. In particular, in order to remove unrealistic data, a simple rule has been applied, which is an extension of the classic Grubbs' test [7]. Two thresholds $U = \mu + 5\sigma$ and $L = \mu - 5\sigma$ are computed on the basis of the mean value μ and the standard deviation σ of the IMPOC signal computed on the whole dataset. Measurements lower that U or higher than L are considered unrealistic and are thus removed from the dataset.

A fixed number of samples have been discarded from the heads and the tails of coils due to the possibility for the IMPOC signal to be affected by the presence of weldings between consecutive coils processed in continuous lines.

As the focus of the present analysis is on the coils that show not homogeneous IMPOC measurement, such characteristic must be codified in terms of features of the measurement associated to that coil. Presently a simple rule has been implemented, which derives from the standard Statistical Process Control [6]: for each coil the mean value μ_c of the IMPOC signal is calculated and for each steel grade g, the standard deviation σ_g of the IMPOC signal is computed. The coils which shows two consecutive measurements of IMPOC value which differ from μ_c for an amount exceeding $3\sigma_g$ is considered anomalous and labelled as non-uniform coil sections. Figure 22.2 shows an exemplar anomalous sequence of IMPOC values.

The whole cleansed dataset containing both uniform and non-uniform samples is then fed to the SOM for its training.

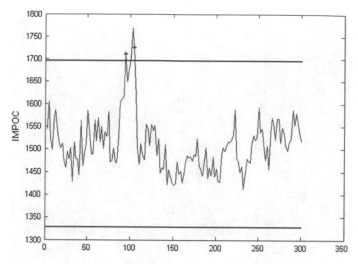

Fig. 22.2 Exemplar sequence of IMPOC values associated with a non-uniform coil

22.4   SOM-Based Analysis

The SOM is a particular kind of NN, which mainly looks at the shape, position and topology of the input patterns and allows the 2D visualization of complex input spaces [8]. The structure of a SOM is composed of one or two-dimensional array of identical neurons. The input pattern is broadcast in parallel to all those neurons and for each input pattern the most responsive neuron is located.

In the present case, due to the huge amount of input patterns stored in the HRM dataset, a 16×16 (256) neurons SOM has been defined and trained for the analysis. The SOM has been fed with the selected process variables and two different approaches were pursued: the analysis of specific HRM process conditions and the detection of anomalous behavior of those conditions.

After the training phase, each neuron of the SOM represents a restricted range of HRM process conditions. The process conditions also depend on the steel grade as highlighted by the clusters shown in Fig. 22.3. In this figure, the numbers written inside the hexagons representing the neurons, correspond to the most active steel grade; in other words, most of the input patterns activating a neuron labelled with the number n are related to coils belonging to steel grade n.

The SOM-based representation of data features has been exploited in order to identify which process conditions most likely may lead to a non-uniformity occurrence. To this purpose, the same network has been fed only with non-uniform coils, by thus highlighting the HRM parameter sets that appear to be most frequently associated with disuniformities, by visualizing which are the neurons activated by only non-uniform coils. Such analysis can be performed separately for each steel grade. For instance, Fig. 22.4 is obtained by feeding the NN only with non-uniform coils

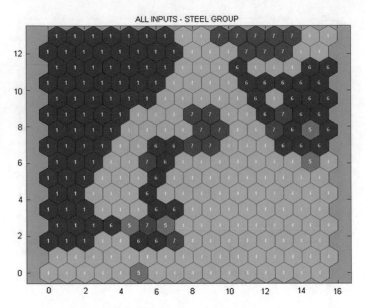

Fig. 22.3 Exemplar trained SOM

belonging to steel grade 1: the most active neuron (highlighted by a green circle) is the number 49 with 71 activations, which means that the corresponding range of HRM process parameters are most frequently observed in conjunction with non-uniform coils.

As many IMPOC measurements are taken along the coil length, a further analysis can be performed aimed at pointing out possible reasons for the non-uniformities, by considering the evolution to the HRM process parameters which may lead to non-uniformity, such as, for instance a sudden significant temperature variation. To this aim, the SOM plot has been customized in order to graphically show the "trajectory" of the process parameters for a single coil along its length in order to visualize possible strong values variations. Figure 22.5a, b show the two exemplary trajectories of a uniform and a non-uniform coil, respectively. These preliminary analyses seem to highlight that disuniformity are quite frequently associated to large variations of the HRM process parameters.

Afterwards, each coil has been divided into three sections: head (first 250 m), tail (last 250 m) and middle (the remaining part). An analysis of position of non-uniformities highlights that their majority (91.78%) occur at the head or the tail of the coil. This can be attributed to the fact that for the first and the last 150 m of a coil the strip is not under tension in the run out table, where the cooling process occurs. Hence, for these sections, the cooling process is less controlled.

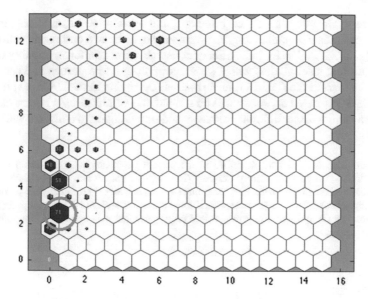

Fig. 22.4 Trained SOM fed only with non-uniform coils belonging to steel grade 1: the sizes of the hexagons are proportional to the activations of each neuron

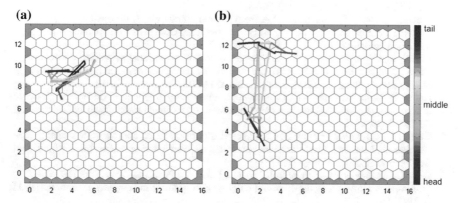

Fig. 22.5 Exemplar trajectories **a** uniform coil; **b** non-uniform coil

22.5 Conclusions

A SOM-based approach to the analysis and interpretation of the data coming from the IMPOC®, a Non-Destructive Testing system providing a continuous magnetic measurement related to the mechanical strength of the steel strip has been presented. Although this work is still ongoing, the first results provide encouraging indications on the potential of these neural networks to support the analysis of the process conditions that can be associated to a lack of uniformity in the material properties along the product length.

Future work will concern, on one hand, the application of more sophisticated techniques for data preprocessing, such as more sophisticated outliers detection techniques [1, 2], as well as for indication of coils that are not uniform, possibly targeted to the different steel grades, which show quite different properties and behaviours during the thermo-mechanical transformations occurring in the HRM. On the other hand, the SOM-based approach will be refined and validated on a bigger database, with data coming from different producers and/or belonging to coils in a wider range of steel grades.

Acknowledgements The work described in the present paper has been developed within the project entitled "Product Uniformity Control" (Ref. PUC, Contract No. RFSR-CT-2013-00031) that has received funding from the Research Fund for Coal and Steel of the European Union, which is gratefully acknowledged. The sole responsibility of the issues treated in the present paper lies with the authors; the Commission is not responsible for any use that may be made of the information contained therein.

References

1. Cateni, S., Colla, V., Nastasi, G.: A multivariate fuzzy system applied for outliers detection. J. Intell. Fuzzy Syst. **24**(4), 889–903 (2013)
2. Cateni, S., Colla, V., Vannucci, M.: A fuzzy system for combining different outliers detection methods. In: Proceedings of the IASTED International Conference on Artificial Intelligence and Applications, AIA 2009, pp. 87–93 (2009)
3. Cateni, S., Colla, V., Vannucci, M.: General purpose input variable extraction: a genetic algorithm based procedure GIVE a GAP. In: 9th International Conference on Intelligence Systems Design and Applications ISDA'09, pp. 1307–1311 (2009)
4. Cateni, S., Colla, V., Vannucci, M.: Variable selection through genetic algorithms for classification purpose. In: IASTED International Conference on Artificial Intelligence and Applications AIA2010, pp. 6–11 (2010)
5. Cateni, S., Colla, V., Vannucci, M.: A genetic algorithm based approach for selecting input variables and setting relevant network parameters of som based classifier. Int. J. Simul. Syst. Sci. Technol. **12**(2), 30–37 (2011)
6. Doty, L.: Statistical Process Control, 2nd edn. Industrial Press Inc., New York (1996)
7. Grubbs, F.E.: Procedures for detecting outlying observations in samples. Technometrics **11**, 1–21 (1969)
8. Kohonen, T.: Self-organized formation of topologically correct feature maps. Biol. Cybern. **43**(1), 59–69 (1982)
9. Meilland, P., Lombard, P., Reboud, C., Skarlatos, A., Svaton, T., Martinez-De-Guerenu, A., Labbé, S., Stolzenberg, M.: B-h curves approximations for modelling outputs of non-destructive electromagnetic instruments (2016)
10. Nastasi, G., Mocci, C., Colla, V., Van-Den-Berg, F., Mulder, R., Beugeling, W., Kebe, T.: Chill marks effect detection algorithm for plant IMPOC data. In: Proceedings of the 19th World Conference on Non-destructive Testing WCNDT (2016)
11. Scheppe, D.: IMPOC—increasing production yield by online measurement of material properties. SEAISI Q. (South East Asia Iron and Steel Institute) **83**(3), 41–46 (2009)
12. Sgarbi, M., Colla, V., Cateni, S., Higson, S.: Pre-processing of data coming from a laser-emat system for non-destructive test of steel slabs. ISA Trans. **51**(1), 181–188 (2012)
13. Skarlatos, A., Reboud, C., Svaton, T., de Guerenu, A.M., Kebe, T., Van-Den-Berg, F.: Modelling the IMPOC response for different steel strips. In: Proceedings of the 19th World Conference on Non-destructive Testing WCNDT (2016)

14. Van-Den-Berg, F., Kok, P., Yang, H., Aarnts, M., Vink, J.J., Beugeling, W., Meilland, P., Kebe, T., Stolzemberg, M., Krix, D., Peyton, A., Zhu, W., Martinez-De-Guerenu, A., Gutierrez, I., Jorge-Badiola, D., Gurruchaga, K., Lundin, P., Volker, A., Mota, M., Monster, J., Wirdelius, H., Mocci, C., Nastasi, G., Colla, V., Davis, C., Zhou, L., Schmidt, R., Labbé, S., Reboud, C., Skarlatos, A., Svaton, T., Leconte, V., Lombard, P.: In-line characterisation of microstructure and mechanical properties in the manufacturing of steel strip for the purpose of product uniformity control. In: Proceedings of the19th World Conference on Non-Destructive Testing WCNDT (2016)
15. Vannucci, M., Colla, V.: Novel classification method for sensitive problems and uneven datasets based on neural networks and fuzzy logic. Appl. Soft Comput. J. $11(2)$, 2383–2390 (2011)
16. Vannucci, M., Colla, V., Cateni, S.: An hybrid ensemble method based on data clustering and weak learners reliabilities estimated through neural networks. In: Lecture Notes in Computer Science (including subseries Lecture Notes in Artificial Intelligence and Lecture Notes in Bioinformatics), vol. 9095, pp. 400–411. Springer (2015)

Chapter 23
Vision-Based Mapping and Micro-localization of Industrial Components in the Fields of Laser Technology

C. Theoharatos, D. Kastaniotis, D. Besiris and N. Fragoulis

Abstract This paper proposes a methodology to visual assist a robotic arm to accurately micro-localize different optoelectronic components by means of object detection and recognition techniques. The various image processing tasks performed for the effective guidance of the robotic arm are analyzed under the scope of implementing a specific production scenario with localization accuracy in the scale of a few microns, proposed by a laser diode manufacturer. In order to elaborate the necessary procedures for achieving the required functionality, the required algorithms engaged in every step are presented. The analysis of possible implementations of the vision algorithms for achieving the required image recognition tasks take into account the possible content of the camera data, the accuracy of the results based on predefined specifications, as well as the computational complexity of the available algorithmic solutions.

Keywords Visual servoing · Robot visual assistance · Pattern recognition · Object localization · Image processing

C. Theoharatos (✉) · D. Kastaniotis · D. Besiris · N. Fragoulis
Computer Vision Systems, IRIDA Labs S.A., Patras Innovation Hub, Kato-Ano
Kastritsiou 4, Magoula, 26504 Patras, Greece
e-mail: htheohar@iridalabs.gr

D. Kastaniotis
e-mail: dkastaniotis@iridalabs.gr

D. Besiris
e-mail: dbes@iridalabs.gr

N. Fragoulis
e-mail: nfrag@iridalabs.gr

© Springer International Publishing AG 2018
A. Esposito et al. (eds.), *Multidisciplinary Approaches to Neural Computing*,
Smart Innovation, Systems and Technologies 69,
DOI 10.1007/978-3-319-56904-8_23

23.1   Introduction

During the past years, vision-based guidance of robotic arms has received increased attention due to the trend for building more advanced and automated mechatronic systems. The settlement of visual sensors in the mechatronic operational loop and the appropriate processing of visual information, can provide accurate guidance of the robotic arm, control the positioning of the end-effector for object manipulation operations, as well as enhance system's precision, flexibility and autonomy [1, 2]. Visual servo control (or visual servoing), that is the task of using visual information to control the pose of the robot's end-effector relative to a target object or a set of target features, has found a wide variety of applications in the robotics industry [3–5].

This work describes and analyses the solutions proposed by our team to address the challenge of highly accurate robot micro-localization for laser diode manufacturing [6]. Accurate object localization involves two major challenges (1) the way to detect and recognize object models that lie within the field-of-view (FoV) of the camera and (2) the means to jointly infer object foreground/background regions and precisely localize the entire object regions. In our work, robot visual assistance utilizes a set of two cameras. The first one is positioned at a static position on top of a loading tray, aiming to detect, recognize and coarsely localize (i.e. map) the different optoelectronic components lying within the storage tray. The second one is mounted on the robot arm and aims to precisely micro-localize the components for accurate picking and placement after guiding the robot arm on top of the components, based on the information provided in the initial mapping step. In order to provide constant lightning conditions, a robust illumination system has been designed consisting of proper LED bars and LED rings attached to the respective cameras.

The various image processing and recognition tasks performed for the effective guidance of the robotic arm are analyzed under the scope of implementing a specific production scenario proposed by a laser diode manufacturer. In order to elaborate the necessary procedures for achieving the required functionality, the necessary algorithms engaged in every step are presented. The analysis of possible implementations of the vision algorithms for achieving the required image recognition tasks take into account the possible content of the camera data, the accuracy of the results based on predefined requirements, as well as the computational complexity of the available algorithmic solutions. In addition, the design and consolidation of the vision sensing system that provide input to the various software components responsible for the visual checks is presented. The design ensures proper visual information at all processing blocks. This is a fairly complicated task, since the placement of the cameras as well the different types of cameras are also related with the production phase, the design of the robot and the planning of the procedures.

Specifically regarding object detection and localization, an intelligent, yet simple, technique is implemented based on a gradual multi-thresholding technique operating on the intensity channel. This method produces different segmentation masks at different intensity levels, which are next merged efficiently to provide the object region under study. In all our experiments, component localization is achieved with sub-pixel accuracy. This is because the center of the component and also its roll angle is calculated as a mathematical property (statistical moments) which accounts for all the pixels and thus end-up to sub-pixel result. In addition, recognition is performed on the detected components based on specific shape analysis criteria. Extensive evaluation of this approach ended-up to impressive results with adequate robustness under illumination changes.

23.2 Background and Overview Material

23.2.1 Object Detection and Localization

Vision-based procedures for assisting the effectiveness and accuracy of the robot lies in the field of object detection/recognition and precise localization. Object localization, in particular, is an essential task in machine vision for discriminating different objects within a scenery and analyzing their spatial relations.

In classical object recognition and localization, a sliding window is commonly used across the image to search the objects under investigation, at multiple scales. This approach, even being time-consuming, has been widely used in various application for the detection of rigid or articulated objects such as faces, pedestrians, moving vehicles, etc. [7, 8]. In order to speed up the search and improve localization efficiency, object parts can be first identified through a restricted number of sliding windows, which can be then ensembled into entire objects [9]. This can be realized by identifying clusters of promising local features and search can follow over windows around the neighborhood of these clusters. Feature selection is one of the most vital steps in this line of research. Several keypoint detectors are made available over the years such as Harris [10] and Kanade-Tomasi features [11]. Recently, most of these descriptor were inspired by the pioneer work of SIFT (Scale Invariant Features Transform) [12], providing suitability in object detection and recognition. Quite robust and faster alternatives have been also proposed such as SURF (Speeded Up Robust Features) features [13]. Robust feature extraction is then followed by a proper classification scheme, such as an SVM classifier, providing a complete image detection and recognition framework. These techniques lie in the field of Bag-of-Features (BoF) approaches in which local features are first extracted from an image to learn a visual vocabulary, followed by a feature quantization step to convert feature vectors to "codewords" and, finally, object recognition is performed by learning a classifier to assign the BoF representations of images to different classes [14]. At the end, object localization can be realized by taking the maximum of the classification score as indication of the presence of the identified object in the image. However, these methods are not robust, since they are susceptible to noise and due to the fact that the components to be detected are pure in texture and size.

The use of global features have been also proposed in the literature to predict the expected position and scale of an object, which combined with local features can increase the detection and recognition accuracy [15]. Moreover, branch-and-bound schemes have presented good object detection and localization results [16]. These methods maximize a class of classifier functions over a set of different subimages, returning in this way the same object locations that an exhaustive sliding window approach would do, but converging to global optimal solutions in sublinear times. Other approaches reported in the literature include graph partitioning and spectral graph clustering for detecting regions of a given object model and provide a segmentation mask using this configuration [17], or probabilistic methods that combine scaled cues to detect and segment object regions that are then merged into segmentation masks [18].

All previous techniques usually fail in the case of inadequate local statistics e.g. if the objects are very small or lightning conditions are not stable. In addition, the use of heuristics to reduce computational complexity often occurs to false alarms and misdetection problems. To tackle these restrictions, which specifically occur in the industrial machine vision community, a precise object detection and localization technique is presented here. Our approach is partly inspired from the recent work Marszalek and Schmid [19], which directly generate and cluster shape masks thus producing more informative results for object class localization. In our work, micro-localization accuracy is achieved by merging shape masks generated from multiple thresholding of the grayscale images. The technique involves a gradual multi-thresholding operation on the intensity channel, producing different segmentation masks at different intensity levels, which are next merged efficiently to provide the object region under study. In all our experiments, a thorough refinement of shape corners using sub-pixel accuracy is provided to increase localization accuracy and achieve the localization requirements within a few micrometers. The proposed scheme is explained in Sect. 23.3, after a short description of the optoelectronic components that need to be detected and localized by the vision system, as well as the camera system set-up utilized for obtaining the needed micro-localization accuracy.

23.2.2 Optoelectronic Objects Under Study

Vision based mapping and localization is mainly concentrated on two very key optoelectronic objects: Chip-on-Carrier (CoC) and NanoFoil (NF). A CoC is a p-n junction with the p-side down mounted on an optimized submount providing very low thermal resistance, and is utilized in single- or multi-emitter diodes. A NF is a reactive multi-layer foil that provides instantaneous localized heat up to 1500 °C in for a variety of applications, totally controllable and affordable material. Both these two components have to be soldered on an emitter's package, after been detected

(a)   **(b)**

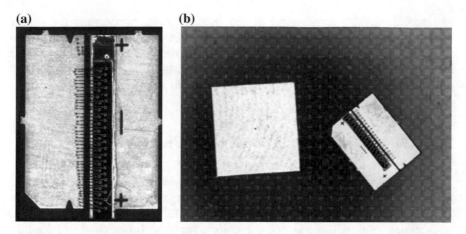

Fig. 23.1 **a** A Chip-on-Carrier and **b** a NanoFoil along with a CoC within a gel-pack

on appropriate gel-packs and waffle packs that are placed on a storage tray. Figure 23.1a illustrate the CoC component, while Fig. 23.1b a CoC and a NanoFoil in a gel-pack. The pick and place operations are realized by the robot with the support of special end-effector tips. The correct picking and the placing process is assisted by a vision system composed by a camera mounted on the vision structure combined with an on board camera directly mounted on the robot's end effector.

23.2.3   Vision System Set-Up

The hardware infrastructure of the vision system consists of a common PC as vision system controller, two cameras with appropriate lenses and proper illumination for keeping constant lightning conditions. Camera selection was made based on the housing form factor as well as the underlying image sensor, whose pixel size fulfill the very tight process requirement of ± 5 µm specified i.e. for the positioning of chip on carrier chip in multi emitter assembly use case.

More specifically, the vision system is composed by the "Storage" and the "Robot" cameras. The "Storage" camera is a 12 Mps CCD global shutter camera (GCP4241C) for capturing images of the components placed on the storage tray, aiming to map the entire area under inspection. The camera is vertically mounted on a vision dedicated structure keeping it at a fixed distance of 100 cm from the storage area's surface. The chosen lens is Ricoh Lens FL-BC2518-9 M, allowing the inspection of the storage area of 400 × 500 mm. In order to properly illuminate the storage area, two LED bars are positioned on each side of the storage tray with an angle of 45° in order to avoid surface reflections and provide proper illumination to the inspected area. The "Robot" camera is an 11Mps CMOS rolling shutter small size camera (GC3851CP) mounted on the robot arm, aiming to calibrate its position and provide precise localization of the optoelectronic components for proper picking-up. The camera is equipped with a Kowa Lens LM50JC10 M, allowing the inspection of an area with dimension 11.7×16.5 mm, accompanied by a LED ring (FLDR-i90A) for providing constant illumination. In both cases, camera's properties such as focus, ISO and zoom are fixed to specific values.

23.3 Blob Detection and Localization Framework

The proposed object detection and localization framework is presented here. At first, the color image is converted to intensity image in order to achieve a color invariant representation. This transformation is followed by a multiple level thresholding approach to separate the objects from the background and identify objects' surface. Object surface detection is performed by scanning the intensity image at different intensity levels using a predefined number of steps. In every step a binary mask is provided, which is also filtered using morphological operations and, at the end, the information extracted from all masks is fused to detect the objects' surface. Since the surface characteristics of the CoC and NanoFoil are known a priori, the method detects the surface boundary using the binary mask estimated previously. If specific requirements are met, the object's boundary is accepted. From this boundary initially the diagonal is estimated as the two points of the diagonal have the maximum Euclidean distance. Then the farthest point with respect to those two points is also estimated. The procedure of blob detection through image binarization is shown in Fig. 23.2 (top) for an image captured using the "Storage" camera and in Fig. 23.2 (bottom) for an image captured using the

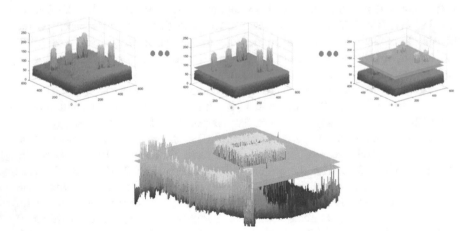

Fig. 23.2 Blob/object detection based on image binarization for an image captured using the Storage (*top diagram*) and Robot (*bottom diagram*) camera

"Robot camera, in which intervals between different thresholds are presented as flat surfaces with different colors.

While the boundary provides as a highly accurate estimation of the corner coordinates, it is important to benefit from subpixel accuracy refinement. Sub-pixel accurate corner locator is based on the observation that every vector from the center q to a point p located within a neighborhood of p is orthogonal to the image gradient at p subject to image and measurement noise. Consider the expression:

$$e_i = DI_{pi}^T(q - p_i) \tag{23.1}$$

where DI_{pi} is an image gradient at one of the points p_i in a neighborhood of q. The value of q is to be found so that e_i is minimized. A system of equations may be set up with e_i set to zero:

$$\sum_i \left(DI_{pi} DI_{pi}^T \right) - \sum_i \left(DI_{pi} DI_{pi}^T p_i \right) \tag{23.2}$$

where the gradients are summed within a neighborhood of q. Calling the first gradient term G and the second gradient term b gives:

$$q = G^{-1}b \tag{23.3}$$

The algorithm sets the midpoint of the neighborhood window at this new center q and then iterates until the center stays within a set threshold. Finally, the refined corners are returned. After this refinement step, the detection and the rotation angle of the CoC and NanoFoil is achieved at the required accuracy.

Next, the central point and rotation angle of the components need to be estimated. Regarding central point, the middle point of the object's diagonal is used in order to compute the X-Y coordinates of the component. The X-Y values are provided in micrometers, as the displacement of the central point of the component from the center of the captured image. Concerning the rotation angle θ, this is estimated based on the direction and angle of the two longest detected boundaries of the component. Angle convention is provided in the range of [−180:180].

23.4   Experimental Results

23.4.1   Vision-Based Mapping

The goal here is to be able to map the area inspected by the "Storage" camera, that is detect, recognize and coarsely localize the positions (in terms of X-Y-θ coordinates) of the CoCs and NanoFoils lying within respective gel- and waffle-packs on a loading tray. Initial detection and recognition approach was based mainly on the color distribution function of each component, followed by some additional image processing steps such as morphological filtering, blob analysis for region merging and a pattern classification engine. In an effort to make our algorithm more robust, we investigated new methods for detecting objects, including Harris features and SURF features amongst others. These methods, however, were not robust since they were found susceptible to noise and due to the fact that the components to be detected are pure in texture and size.

The proposed mapping scheme is compared to SIFT and SURF keypoint extraction, utilized in three steps. In the first step, e.g. SURF features within a particular scale are detected, described as a vector encoding the local information around keypoints. Each vector is then classified using a machine learning model (i.e. logistic regression) learnt offline, using several annotated examples. Classification accuracy is presented on CoC and NanoFoil components, as well as on some Calibration Markers that have been placed on the loading tray so as the robot to be able to register its position with high accuracy in terms of translation, rotation and height, prior the picking-up of components. Figure 23.3 illustrates the mapping result obtain using the proposed scheme on an image captured using the Storage

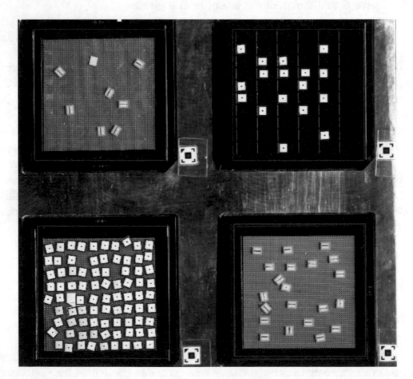

Fig. 23.3 Vision-based mapping for the detection, recognition and coarse localization of CoC (*green rectangle*), NanoFoil (*blue rectangle*) and calibration marks (*red-dot*)

Table 23.1 Confusion matrix of CoC, NanoFoil (NF), Calibration Marks (CM) and Background (BGR) classification of the proposed approach, compared to SURF-based classification

		Classified as CoC	Classified as NF	Classified as CM	Classified as BGR
SURF	CoC	0.80	0.09	0.03	0.08
	NF	0.04	0.83	0.01	0.12
	CM	0.02	0.01	0.91	0.06
	BGR	0.12	0.08	0.02	0.78
Proposed	CoC	**0.96**	0.02	0.00	0.02
	NF	0.02	**0.95**	0.00	0.03
	CM	0.00	0.00	**1.00**	0.00
	BGR	0.00	0.01	0.00	**0.99**

camera. Detected components are shown using different colors. In addition, the experimental results for the SURF descriptors and the proposed approach are presented in Table 23.1.

23.4.2 Component Micro-localization

The goal here is to be able to precisely localize the position and orientation of the CoC and NanoFoil component prior proper picking by the robot's end-effector. Several methodologies were initially investigated in this task, none of which being able to accurately detect the components' corners with a precision of a few micron ($\pm$ 5 μm) required by the laser diode manufacturer. Table 23.2 provides the results of the proposed approach described in Sect. 23.3, compared to approaches that presented the most valid localization outputs such as the FAST descriptor [20], which evaluates the intensity of a small region using a decision tree trained offline, Harris corner detector [9] and Hough transform [21]. In addition, execution times are provided for all techniques as well as the number of times that the detector failed to precisely localize the CoC over 1000 tests, given as a percentage of failure. The localization results for each one of the three components under study are also illustrated in Fig. 23.4.

Table 23.2 Localization accuracy results of different detectors, along with execution times

Evaluation criterion	Execution time (s)	Accuracy (μm)	Percentage of failure (%)
Fast detector	0.22	18.57 ± 3.06	9.9
Harris detector	2.37	22.56 ± 4.94	12.2
Hough transform	8.02	28.20 ± 14.10	32.5
Proposed	1.67	2.35 ± 0.07	6.2

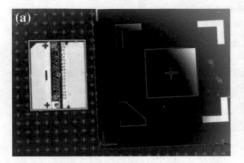

Fig. 23.4 Detection, recognition and micro-localization of **a** a CoC next to a calibration mark and **b** a CoC next to a NanoFoil

23.5 Conclusions

This study proposes a methodology to visual assist a robotic arm for accurately micro-localizing different optoelectronic components by means of object detection and recognition techniques. The various image processing tasks performed for the effective guidance of the robotic arm are analyzed under the scope of implementing a specific production scenario with localization accuracy in the scale of a few microns, proposed by a laser diode manufacturer [22]. The proposed method provides robustness to object detection and localization in the difficult scenarios of industrial machine vision, which is characterized by inadequate object local statistics and strict computational requirements. The presented results are compared with other well-known existing techniques, providing significantly better results in terms of effectiveness and efficiency.

Acknowledgements This work was supported by the 7th Framework Programme (FP7), FoF. NMP.2013-2: Innovative Re-Use of Modular Equipment Based on Integrated Factory Design, in the context of the WhiteR project under Grant GA609228. All vision related hardware components were provided by Framos GmbH.

References

1. Kouskouridas, R., Amanatiadis, A., Gasteratos, A.: Guiding a robotic gripper by visual feedback for object manipulation tasks. In: IEEE International Conference on Mechatronics, pp. 433–438. IEEE Press, Istanbul, Turkey (2011)
2. Gacsádi, A., Tiponuţ, V., Szolgay, P.: Image-based visual servo control of a robotic arm by using cellular neural networks. In 15th International Workshop on Robotics in Alpe-Adria-Danube Region, Balatonfüred, Hungary (2006)
3. Hutchinson, S., Hager, G.D., Corke, P.I.: A tutorial on visual servo control. IEEE Trans. Robot. Autom. **12**, 651–670 (1996)

4. Papanikolopoulos, N.P., Khosla, P.K., Kanade, T.: Visual tracking of a moving target by a camera mounted on a robot: a combination of control and vision. IEEE Trans. Robot. Autom. **9**, 14–35 (1993)
5. Oh, P.Y., Allen, K.: Visual servoing by partitioning degrees of freedom. IEEE Trans. Robot. Autom. **17**, 1–17 (2001)
6. Sisca, F.G., Fiasché, M., Taisch, M.: A novel hybrid modelling for aggregate production planning in a reconfigurable assembly unit for optoelectronics. In: Arik, S., et. al. (eds.) ICONIP 2015, Part II, LNCS, vol. 9490, pp. 571–582. Springer International Publishing, Switzerland (2015)
7. Viola, P., Jones, M.: Robust real-time object detection. Int. J. Comput. Vis. **57**, 137–154 (2004)
8. Dalal, N., Triggs, B.: Histograms of oriented gradients for human detection. In: 2005 IEEE Conference on Computer Vision and Pattern Recognition, San Diego, CA, USA, pp. 886–893 (2005)
9. Mohan, A., Papageorgiou, C., Poggio, T.: Example-based object detection in images by components. IEEE Trans. Pattern Anal. Mach. Intell. **23**, 349–361 (2001)
10. Harris, C., Stephens, M.: A combined corner and edge detector. In: 4th Alvey Vision Conference, pp. 147–151 (1988)
11. Lucas, B.D., Kanade, T.: An iterative image registration technique with an application to stereo vision. In: 7th International Joint Conference on Artificial Intelligence, San Francisco, CA, USA, pp. 674–679 (1981)
12. Lowe, D.G.: Distinctive image features from scale-invariant keypoints. Int. J. Comput. Vis. **60**, 91–110 (2004)
13. Bay, H., Ess, A., Tuytelaars, T., Van Gool, L.: SURF: speeded up robust features. Comput. Vis. Image Underst. **110**, 346–359 (2008)
14. Lazebnik, S., Schmid, C., Ponce, J.: Beyond bags of features: spatial pyramid matching for recognizing natural scene categories. In: 2006 IEEE Conference on Computer Vision and Pattern Recognition, New York, NY, USA, pp. 2169–2178 (2006)
15. Murphy, K., Torralba, A., Eaton, D., Freeman, W.: Object detection and localization using local and global features. In: Ponce, J., Hebert, M., Schmid, C., Zisserman, A. (eds.) Towards Category-Level Object Recognition, LNCS, vol. 4170, pp. 382–400. Springer, Heidelberg (2006)
16. Lampert, C.H., Blaschko, M.B., Hofmann, T.: Beyond sliding windows: object localization by efficient subwindow search. In: IEEE Conference on Computer Vision and Pattern Recognition, Los Alamitos, CA, USA, pp. 1–8 (2008)
17. von Luxburg, U.: A Tutorial on Spectral Clustering. Technical report, Max-Planck-Institut für biologische Kybernetik (2007)
18. Kumar, P.M., Torr, P.H.S., Zisserman, A.: OBJ CUT. In: IEEE Conference on Computer Vision and Pattern Recognition, pp. 18–25 (2005)
19. Marszalek, M., Schmid, C.: Accurate object localization with shape masks. In: IEEE Conference on Computer Vision and Pattern Recognition, pp. 1–8 (2007)
20. Rosten, E., Drummond, T.: Machine learning for high-speed corner detection. In: 9th European Conference on Computer Vision, pp. 430–443. Springer, Berlin, Heidelberg (2006)
21. Ballard, D., Brown, C.: Computer Vision. Ch. 4. Prentice-Hall (1982)
22. Fiasché, M., Ripamonti, G., Sisca, F.G., Taisch, M., Valente, A.: management integration framework in a shop-floor employing self-contained assembly unit for optoelectronic products. In: 1st IEEE International Forum on Research and Technologies for Society and Industry Leveraging a better tomorrow (RTSI), pp. 569–578 (2015)

Part V
Special Session on Social and Biometric Data for Applications in Human-Machine Interactions: Models and Algorithms

Chapter 24
Artificial Neural Network Analysis and ERP in Intimate Partner Violence

Sara Invitto, Arianna Mignozzi, Giulia Piraino, Gianbattista Rocco,
Irio De Feudis, Antonio Brunetti and Vitoantonio Bevilacqua

Abstract The aim of this work is to analyze, through artificial neural network
models, cortical pattern of women with Intimate Partner Violence (IPV) to investigate representative models of sensitization or habituation to the emotional stimulus in IPV. We investigate the ability of high emotional impact images, during a
recognition task, analyzing the electroencephalogram data and event related
potentials. Neural network analysis highlights an impairment in IPV group in
cortical arousal, during the emotional recognition task. The alteration of this
capacity has obvious repercussions on people's lives, because it involves chronic
difficulties in interpersonal relationships.

Keywords Intimate partner violence · ERP · Artificial neural network ·
Student t-test features reduction · Binary classification

24.1 Introduction

Intimate partner violence (IPV) is an important problem of public and social health
and involves men and women all over the world regardless of culture, religion and
demographic characteristics. This term is often used interchangeably with the
phrase domestic violence. The biological perspective also has provided information
about the factors that might constitute IPV indicators [1, 2]; skin conductance,

S. Invitto (✉) · A. Mignozzi
Department of Biological and Environmental Sciences and Technologies,
University of Salento, Lecce, Italy
e-mail: sara.invitto@unisalento.it

G. Rocco · I. De Feudis · A. Brunetti · V. Bevilacqua (✉)
Department of Electrical and Information Engineering,
Polytechnic of Bari, Bari, Italy
e-mail: vitoantonio.bevilacqua@poliba.it

G. Piraino
Istituto Santa Chiara, Lecce, Italy

© Springer International Publishing AG 2018   247
A. Esposito et al. (eds.), *Multidisciplinary Approaches to Neural Computing*,
Smart Innovation, Systems and Technologies 69,
DOI 10.1007/978-3-319-56904-8_24

as well as the levels of testosterone, and salivary immunoglobulin in the aggressors represent the main markers of a high reactivity to stressful situations, and correlate to impulsive and aggressive behaviors. Also in this perspective, these indicators may be useful within prevention programs of the phenomenon, but even both in terms of analysis of the risk of recurrence of the phenomenon and in terms of a greater understanding of its underlying causes.

In addition, the biological perspective has highlighted not only the correlation between the location of the violence and the consequent impairment of cognitive functions of the women involved, but also made it possible to highlight how these alterations refer in post-traumatic stress disorder (PTSD) [3]. In fact, numerous data show that 90% of lesions caused by IPV are around the area of head, face and neck of the victims; such injuries often lead to both structural and functional traumatic brain injury (TBI), of multiple severity. Small TBI involves mental status abnormalities, neurological and neuropsychological deficits. Women with IPV histories show alterations in the regulation of both emotions and pain sensitivity, and in management control of fear; such alterations correspond with functional deficits, which correlate with structural alterations and represent the main characteristics of PTSD [3]. It is possible to catalog the focal characteristics of PTSD symptoms in three main categories: intrusive, avoidance and hyper-excitability symptoms.

In intrusive symptoms, the patient relives in persistent and persecutory form the traumatic experience, through recurring thoughts and feelings, which recur vividly in the form of flashbacks. In the symptoms of avoidance, the patient tends to avoid stimuli and situations, which can evoke memories of the traumatic experience; finally, the symptoms of hyper-excitability involve manifestation of high levels of arousal and physiological agitation [4].

Other secondary symptoms include an impoverishment of autobiographical memory for positive events [5], an impaired working memory functioning [6] and an alteration of attentional capacity. For this reason, studies that have analyzed the structural abnormalities of the brain resulting from PTSD have focused on the hippocampus, a gray matter structure of the limbic system involved in declarative memory, working memory and episodic memory [7]. The importance of PTSD, and its correlation with the domestic violence, is revealed through the consequences of such phenomena, corresponding to an alteration of the victims' routine. These consequences are manifested although a late and acute onset, consequently to stressful events or threatening and catastrophic situations, causing serious emotional suffering protracted in time.

Because of the importance of the implications, PTSD was recognized as mental disorder by the American Psychiatric Association in 1980, and subsequently introduced in DSM-III. This disorder was then conceptualized as the inability to integrate the traumatic experience, which generates fear, stress and helplessness action [8]. In recent years a growing number of studies have tried to improve the understanding of the relationship between IPV and PTSD, starting from cognitive impairments mentioned above; for this reason, it has been particularly used a survey technique capable of accurately detecting cognitive alterations, from a spatial and temporal point of view: electroencephalography (EEG). This technique permits to analyze the Event Related Potential (ERP), dwelling mainly on the N200 and P300 components, which are both associated with attentional and mnemonic skills but refer to different processes [9].

The N200 component appears to be related to the presence of stimuli with a localization waiting; it seems to reflect a stimulus discrimination process, and not a simple detection of the same. The process of discrimination that correlates with the elicitation of the component is also manifested in a greater level of arousal, which is the corresponding portion to the physiological found in women with histories of IPV and resulting PTSD [10].

About the P300 component, a meta-analysis of studies examining the response of the P3a elicited by destroyers related to trauma found a higher amplitude of the component in the subjects with PTSD compared with control subjects [11].

24.2   Method

24.2.1   Subjects

The research sample was constituted by a total of 28 women. The experimental group was composed of 14 women (mean age = 39; s.d. = 11), recruited in a center against women violence, while the control group was composed of 14 women (mean = 33; s.d. = 7) without PTDS or IPV nor depressive/anxiety symptomatology. All subjects provided written, informed consent to participation and publication of data that could identify them under a code. Preliminary results were presented in SPR Congress [12]. The work was carried out in accordance with the Code of Ethics of the World Medical Association (Helsinki Declaration) [13] for experiments involving humans.

24.2.2   Experiments

This study included the EEG recording of subjects, through BrainAmp device, with the Brain Vision Recorder Software (© 2010 Brain Products GmbH), during the execution of a Go-NoGo Task with Emotional Visual Images; results were subjected to analysis using the Brain Vision Analyzer Software (© 2010 Brain Products GmbH). Before the execution of the task, all subjects filled informed consent according to the Helsinki declaration, anamnestic data, the Beck Depression Inventory-II, the Beck Anxiety Inventory—Coradeschi and a PTSD questionnaire.

The Beck Anxiety Inventory is one of self-report instrument consisting of 21 items, which allows to valuate an assessment of the chronicity and severity of anxiety disorder in adults, including those symptoms only minimally overlapped with those of depressive nature. In this experiment, we administered the Italian version of Coradeschi. The Beck Depression Inventory is a self-assessment instrument consisting of 21 multiple-choice items, which allows to measure the severity of depression in adults and adolescents aged from 13 to 80 years. The PTSD questionnaire was drafted by invoking diagnostic and temporal criteria of the DSM-IV-TR, which is also a self-report instrument consisting of 19 items, where each one is rated on a scale according to the criterion of the duration and frequency.

After filled the questionnaires, the first step consisted in preparing the subject to the EEG recording, which was through the international system 10–20. In our study, the subjects performed a Go-NoGo task, consisting in continuous repetition of two types of stimuli, one of which was presented repeatedly and the other presented only sporadically [14]. Each subject had to press a button every time the image emotionally significant appeared. For this task, we used E-prime 2.0, an application of Psychology software tools, Inc. The stimuli presented were 60: 12 target, called T2, had negative valence and 48 non-target, referred to as T1, had positive or neutral value. The task took 20 min and consisted of the random image play-back, with an inter-stimulus duration of 1000 ms.

24.3   Data Analysis and Results

24.3.1   EEG Dataset

Dataset included the features coming from the ERP wave (Figs. 24.1 and 24.2), considering amplitude and latency of the N200, P300 and N500 components. As already mentioned, the collected data concerned 28 individuals divided into 2 classes: 14 with PTSD and 14 in the control group. For each observation, we had 15 attributes (F1, F2, …, F15), corresponding to 15 electrodes. The correspondence between features and EEG channels is shown in Table 24.1.

24.3.2   Data Analysis

The first step of data analysis consisted in the recognition of a reduced number of features useful for discriminating between the mentioned two classes. For this purpose, different subsets of features were found using the two-tailed Student's t-distribution and then validated through an Artificial Neural Network (ANN). After the selection of the statistically significant features, we examined whether and how these subsets could be considered typical of the investigated phenomenon.

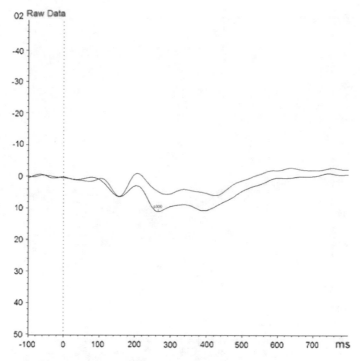

Fig. 24.1 Matching ERPs: *Red line* IPV group; *Black line* Control group on occipital right channel (O_2 electrode, feature F15)

We compared the samples of the PTSD class to the control class ones. For example, considering the feature F1 of N200 latency component: first we verified that the samples of both classes were approximately normally distributed using the *D'Agostino&Pearson omnibus normality test*, then we employed the *t*-Student test between the samples of the two classes. The *t*-test showed that F1 of N200 latency was significant with p-value = 0.0148 < 0.05 (significance level).

The same method was employed on all the features of every EEG wave component. Formerly, for each data collection (both the amplitude and the latency of N200, P300, N500) we firstly computed, for each index i, the average of the two classes for the i-th feature (see Eq. 24.1)

$$\mu_{i,c} = \frac{1}{14} \sum_{j=1}^{14} x_{ij,c}, \quad i = 1, 2, \ldots, 15; \quad c = 1, 2. \tag{24.1}$$

where j is the index of the sample and c is the class, $x_{ij,c}$ is the value of the j-th sample of the i-th feature in the c-th class.

Then we used the Student's *t*-distribution to determine the significant difference between the two mean values [15].

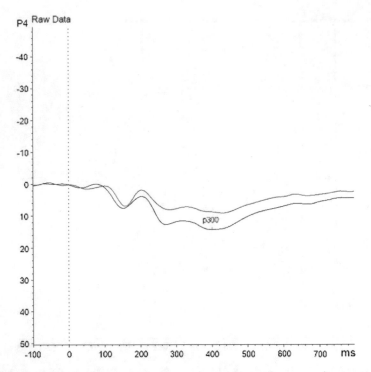

Fig. 24.2 Matching ERPs: *Red line* IPV group; *Black line* Control group on parietal right channel (P4 electrode, feature F13)

Table 24.1 Mapping of features and EEG channels

Feat	F1	F2	F3	F4	F5	F6	F7	F8	F9	F10	F11	F12	F13	F14	F15
EEG	Fp1	Fp2	Fz	Cz	Pz	F3	F4	F7	F8	C3	C4	P3	P4	O1	O2

The inferential statistical analysis revealed that all the datasets showed a certain number of significant features, except the N200 amplitude components one. Then, we built a subset for every ERP wave component made up of significant features. The significant subsets of features and the corresponding p-values are shown in Table 24.2.

24.3.3 Binary Neural Classifier

We designed, trained, validated and tested ten neural network binary classifiers based on the reduced number of features (PTSD vs. Control).

In Table 24.3 there is a representation of a confusion matrix, which shows the number of false positives (FP), false negatives (FN), true positives (TP), and true

Table 24.2 Subsets of significant features with the corresponding p-value, for each ERP wave component

ERP	Significant features	p-value < 0.05	ERP	Significant features	p-value < 0.05
N200 latency	F1	0.0148	P300 amplitude	F4	0.0309
	F2	0.0131		F5	0.0036
	F8	0.0099		F13	0.0083
	F9	0.0060		F15	0.0314
	F13	0.0290			
P300 latency	F4	0.0028	N500 amplitude	F4	0.0020
	F5	0.0058		F5	0.0010
	F10	0.0420		F10	0.0293
	F11	0.0286		F11	0.0262
	F13	0.0013		F12	0.0059
	F14	0.0074		F13	0.0011
N500 latency	F5	0.0493			
	F11	0.0065			

Table 24.3 A confusion matrix allows visualization of the classification algorithm performances

		Expert advices		Total
		PTSD	Control	
Prediction	PTSD	TP	FP	Class 1 (TP + FP)
	Control	FN	TN	Class 2 (FN + TN)
Total		PTSD (TP + FN)	Control (FP + TN)	

negatives (TN) resulting from the classification process. Sensitivity (Eq. 24.2) measures the proportion of positive individuals who are correctly identified as positive to the total number of positive individuals. Specificity (Eq. 24.3) measures the proportion of negative individuals who are correctly identified as negative to the total number of negative individuals. Accuracy (Eq. 24.4) is the global concordance of the true positive and negative results in the whole set of subjects [16–18].

$$sensitivity = \frac{TP}{TP + FN} \tag{24.2}$$

$$specificity = \frac{TN}{TN + FP} \tag{24.3}$$

$$accuracy = \frac{TP + TN}{TP + FP + FN + TN} \tag{24.4}$$

All the feed forward Multilayer Perceptron [19, 20] topologies had one hidden layer. For all neurons was used the log-sigmoid activation function.

The input data for the neural network were the five ERP wave components (amplitude and latency of N200, P300 and N500, independently). In training, we used the following settings: random initialization of synaptic weights with values between ±1, learning rate of 0.3 and momentum of 0.2, method of cross-validation with a training time of 500 epochs [20].

24.3.4 Results

The comparison of the results in classification obtained for each ERP wave component in Table 24.2, using both the full data set and the dataset restricted to the statistically significant features, allowed us to observe that only N200 latency and P300 amplitude produced the best classification performances employing the corresponding restricted datasets, as shown in Table 24.4. In all the other cases (P300 latency, N500 latency and amplitude), the performances were worse than using the full datasets composed of 15 features.

On average, the PTSD group showed, for each significant feature, higher values of N200 latency and lower values of P300 amplitude than the control one.

The corresponding electrode locations to the statistically significantfeatures of the ERP wave componentsrepresent the Regions of Interest (ROIs) of the specific phenomenon. In the case of N200 latency, the ROI wascomposed by F1, F2, F8, F9, and F13; in the case of P300 amplitude, the ROI covers the fully connected areacomposed byF4, F5, F13, and F15. In order to improve the classification performances reducing the number of features to be analysed, we looked for a smaller set of features. Final results show that the significant features for discriminating individuals in the two classes could be reducedto F2, F8 and F9 for the N200 latency (Table 24.4).

Table 24.4 Performances of the N200 latency and P300 amplitude, using the full dataset input and the dataset restricted to the statistically significant features

ERP	Dataset	Number of neurons in hidden layer	Avg accuracy (%)	Avg AUC	Avg sensitivity	Avg specificity
N200 latency	[F1, F2, ..., F15]	8	70.83	0.75	0.55	0.80
	[F1, F2, F8, F9, F13]	3	**81.67**	**0.85**	**0.95**	**0.75**
	[F2, F8, F9]	2	**85.00**	**0.86**	**0.95**	**0.80**
P300 amplitude	[F1, F2, ..., F15]	8	68.33	0.57	0.50	0.82
	[F4, F5, F13, F15]	3	**75.00**	**0.59**	**0.55**	**0.87**

24.4   Discussion

Results highlight how the presence of IPV and subsequent development of PTSD might impair the ability of people to recognize emotional stimuli. The alteration of this capacity has obvious repercussions on people's lives, because it involves daily difficulties in interpersonal relationships.

In light of the findings, it is clear that individuals with histories of IPV and suffering from PTSD exhibit a higher sensitivity threshold than stimuli involving violence in general or specifically to women. Consider the constant stress that these women receive or have received, it is likely that for them is necessary a more specific input to reset the threshold of sensitivity, for this type of stimulus, at a lower level; it is relevant to consider that the specific input depends on the traumatic experience of the person.

Considering the experience of these women, it is possible to suppose that a higher sensitivity threshold may represent the physiological part of an avoidance behavior; thismakes the woman unable to give the right interpretation to an emotional stimulus.

Although the P300 seems to be the elective component for the study of PTSD, specifically for the attention processes, the results of several studies reported difference regarding the methodological and the sampling. In addition, it is important to emphasize that the P300 can be influenced by factors such as medications, the chronicity of the disorder, the absence or presence of correlation between stimulus and trauma, or even if the emotional stimuli themselves may, or not, generate bias in the task [21].

These considerations deserve attention because other studies have shown different results with those of our experiment [22]. Future studies may help to clarify the contribution of P300 component and quantify the interference with this by the factors mentioned above.

Some studies suggest that possible attentional difficulties may be associated with the trauma itself rather than to PTSD [23–25].

In other words, the difficulties consistent with the development of PTSD in women with IPV become known in many forms. These difficulties are in equal measure intrusive in people's daily life; it therefore seems necessary to continue toward an understanding of the phenomenon to take care about all the variables related to the problem.

Acknowledgements 'Università del Salento — '5 for Thousand Research Fund'.

References

1. Romero-Martinez, A., Lila, M., Conchell, R., Gonzalez-Bono, E., Maya-Albiol, L.: Immunoglobulin a response to acute stress in intimate partner violence perpetrators: the role of anger expression-out and testosterone. Biol. Psychol. **96**, 66–71 (2014)

2. Romero-Martinez, A., Lila, M., Williams, R., Gonzalez-Bono, E., Moya-Albiol, L.: Skin conductance rises in preparation and recovery to psychosocial stress and its relationship with impulsivity and testosterone in intimate partner violence perpetrators. Int. J. Psychophysiol. 329–333 (2013)

3. Wong, J.Y.-H., Fong, D.Y.T., Lai, V., Tiwari, A.: Bridging intimate partner violence and the human brain: a literature review. Trauma Violence Abuse 15, 22–33 (2014)

4. Almli, L., Fani, N., Ressler, K., Smith, A.: Genetic approaches to understanding post-traumatic stress disorder. Int. J. Neuropsychopharmacol. 355–370 (2014)

5. Harvey, G., Bryant, R., Dang, S.: Autobiographical memory in acute stress disorder. J. Consul. Clin. Psychol. 500–506 (1998)

6. Vasterling, J., Duke, L., Brailey, K., Constans, J., Allain Jr., A., Sutker, P.: Attention, learning, and memory performance and intellectual resources in Vietnam veterans; PTSD and no disorder comparations. Neuropsychology 16, 5–14 (2002)

7. Wheeler, M., Buckner, R.: Functional-anatomic correlates of remembering and Knowing. Neuroimage 1337–1349 (2004)

8. Nemeroff, C., Sherin, J.: Post traumatic stress disorder: the neurobiological impact of psychological trauma. Dialogues Clin. Neurosci. 13(3), 263–278 (2011)

9. Luck, S.: An Introduction to the Event-Related Potential Technique, pp. 1–357. The MIT Press (2005)

10. Vogel, E.K., Luck, S.J.: The visual N1 component as an index of a discrimination process. Psychophysiology 37, 190–203 (2000)

11. Karl, A., Malta, L., Maercker, A.: Meta-analytic review of event-related potential studies in post-traumatic stress disorder. Biol. Psychol. 71, 123–147 (2006)

12. Invitto, S., Mignozzi, A., Quarta, M., Sammarco, S., Nicolardi, G., de Tommaso, M.; Intimate partner violence and emotional face recognition. In: Psychophysiology-SPR 54th Annual Meeting Society of Psychophysiological Research, at Atlanta, 2014, Wiley Online Library

13. Kemperman, C.J.F.: Helsinki declaration. The Lancet (1982)

14. Invitto, S., Faggiano, C., Sammarco, S., De Luca, V., De Paolis, L.: Haptic, virtual interaction and motor imagery: entertainment tools and psychophysiological testing. Sensors 16(3), 394 (2016)

15. Ahsen, M.E., Singh, N.K., Boren, T., Vidyasagar, M., White, M.A.: A new feature selection algorithm for two-class classification problems and application to endometrial cancer. In: 51st IEEE Annual Conference on Decision and Control (CDC), Maui, Hawaii, USA, pp. 2976–2982. IEEE (2012)

16. Menolascina, F., Tommasi, S., Paradiso, A., Cortellino, M., Bevilacqua, V., Mastronardi, G.: Novel data mining techniques in aCGH based breast cancer subtypes profiling: the biological perspective. In IEEE Symposium on Computational Intelligence and Bioinformatics and Computational Biology, 2007, CIBCB'07, pp. 9–16. IEEE (2007)

17. Scolaro, G.R., De Azevedo, F.M.: Classification of epileptiform events in raw EEG signals using neural classifier. In: 3rd IEEE International Conference on Computer Science and Information Technology (ICCSIT), vol. 5, pp. 368–372. IEEE (2010)

18. Bevilacqua, V., Mastronardi, G., Menolascina, F., Pannarale, P., Pedone, A.: A novel multi-objective genetic algorithm approach to artificial neural network topology optimisation: the breast cancer classification problem. In: International Joint Conference on Neural Networks (IJCNN), Vancouver, BC, Canada, pp. 1958–1965. IEEE (2006)

19. Bevilacqua, V., Cassano, F., Mininno, E., Iacca, G.: Optimizing feed-forward neural network topology by multi-objective evolutionary algorithms: a comparative study on biomedical datasets. In: Advances in Artificial Life, Evolutionary Computation and Systems Chemistry, pp. 53–64. Springer International Publishing (2015)

20. Sovierzoski, M.A., Argoud, F.I.M., de Azevedo, F.M.: Evaluation of ANN classifiers during supervised training with roc analysis and cross validation. In: International Conference on BMEI, vol. 1, pp. 274–278. IEEE (2008)

21. Kimble, M., Kaloupek, D., Kaufman, M.: Stimulus novelty differentially affects attentional allocation in PTSD. Biol. Psychiatry 880–890 (2000)

22. Javanbakht, A., Liberzon, I., Amirsadri, A., Gjini, K., Boutros, N.: Event-relates potential studies of post-traumatic stress disorder: a critical review and synthesis. Biol. Mood Anxiety Disord. 1–5 (2011)
23. Karl, A., Schaefer, M., Malta, L., Dorfel, D., Rohlender, N., Werner, A.: A meta-analysis of structural brain abnormalities in PTSD. Neurosci. Biobehav. Rev. 1004–1031 (2006)
24. Kimble, M., Frueh, B., Marks, L.: Dose the modified Stroop effect exist in PTSD? Evidence from dissertation abstract and the peer reviewed literature. J. Anxiety Disord. 23(5), 650–655 (2009)
25. Kimble, M., Fleming, K., Bandy, C., Kim, J., Zambetti, A.: Eye tracking and visual attention to threating stimuli in veterans of the Iraq war. J. Anxiety Disord. 24, 293–299 (2010)

Chapter 25
Investigating the Brain Connectivity Evolution in AD and MCI Patients Through the EEG Signals' Wavelet Coherence

Cosimo Ieracitano, Nadia Mammone, Fabio La Foresta
and Francesco C. Morabito

Abstract Mild cognitive impairment (MCI) is a neurological disorder that degenerates into Alzheimer's disease (AD) in 8–15% of cases. The MCI to AD conversion is due to a loss of connectivity between different areas of the brain. In this paper, a wavelet coherence approach is proposed for investigating how the brain connectivity evolves among cortical regions with the disease progression. We studied Electroencephalograph (EEG) recordings acquired from eight patients affected by MCI at time T0 and we also studied their follow up at time T1 (three months later): three of them converted to AD, five remained MCI. The EEGs were analyzed over delta, theta, alpha 1, alpha 2, beta 1 and beta 2 sub-bands. Differently from MCI stable subjects, MCI patients who converted to AD, showed a strong reduction of cortical connectivity in theta, alpha(s) and beta(s) sub-bands. Delta band showed high coherence values in each pair of electrodes in every patient.

Keywords Mild cognitive impairment · Alzheimer disease · Electroencephalography · Wavelet coherence

25.1 Introduction

The concept of mild cognitive impairment (MCI) has evolved over the last 2 decades to represent an intermediate state between the cognitive changes of aging and dementia. It describes a cognitive decline of mind that does not interfere particularly with day-to-day life and ordinary activities. It is a common condition in

C. Ieracitano (✉) · F. La Foresta · F.C. Morabito
DICEAM, Mediterranea University of Reggio Calabria, Reggio Calabria, Italy
e-mail: cosimo.ieracitano@unirc.it

N. Mammone
IRCCS Centro Neurolesi Bonino-Pulejo, Messina, Italy

© Springer International Publishing AG 2018
A. Esposito et al. (eds.), *Multidisciplinary Approaches to Neural Computing*,
Smart Innovation, Systems and Technologies 69,
DOI 10.1007/978-3-319-56904-8_25

259

15–20% of individuals aged 60 years and over. Although subjects with MCI could remain stable or return to be normal over time, a lot of them progress to dementia of Alzheimer's disease (AD) within 5 years. The annual rate of MCI progression to Alzheimer's disease is between 8 and 15% [1]. Therefore, longitudinal studies of neural changes of the brain due to the development of AD are emerging. However, a few follow up studies can be found in literature [2, 3], as most of papers focus on the AD versus MCI classification. Alzheimer's disease is a degenerative neurological disorder that causes irreversible loss of neurons, loss of intellectual abilities, including memory, reasoning, intellectual deficits and behavior disturbance. These abnormalities are thought to be the result of the cerebral death of neurons which causes functional disconnections between different cortical areas [4]. Of course, early diagnosis of AD could help improve life quality, but unfortunately there is no valid criteria or symptom able to make a diagnosis before the manifestation of illness. Currently there is not a single clinical test for the exact diagnosis of AD but EEG signal analysis has been becoming a promising and powerful tool for early detect the dementia of Alzheimer's disease [5]. EEG is a non-invasive method and measures the background electrical activity of the brain, generated by neurons of the cerebral cortex. EEG signal is usually decomposed in its sub-bands: delta (0–4 Hz), theta (4–8 Hz), alpha 1 (8–10 Hz), alpha 2 (10–12 Hz), beta 1 (12–18 Hz), beta 2 (18–30 Hz). Each sub-band relates to different functional and physiological area of the brain [6]. The hallmarks of EEG abnormalities in AD patients are slowing of the rhythms and a decline in functional signal connectivity between different brain regions [4]. The most common way to quantify functional connectivity is quantifying the coherence between EEG signals. Nowadays, there are still challenging issues: actually, we do not know how the EEG abnormalities evolve together with the disease progression. Researches have been investigating how the brain connectivity of MCI and AD patients changes over time, at T0 and after a predetermined range of time called "follow-up" (T1, T2 etc.), in order to determine possible biomarkers able to early detect the Alzheimer's disease. For this aim, coherence studies of the neural dynamic of the brain through EEG analysis have been emerging. Conventional coherence, based on Fourier analysis, analyzes just spectral components, losing time information. This approach is suitable for stationary time series. However, many time series are nonstationary, meaning that their frequency content changes over time. Because of the nonstationary nature of EEG signal a coherence in the time-frequency plane is needed. Hence, to overcome this issue a time-frequency analysis is evaluated: the Short-time Fourier transform (STFT) might be used, but it is limited by the time filtering, the fixed sliding window does not make it suitable for different signal frequencies [6]. The wavelet theory is the best method because it fields excellent time-frequency resolution. The wavelet coherence methodology, is a recent mathematical tool enables to provide the synchronization and the direction of information between two signals. Grinsted et al. [7], Torrente and Campo [8], discussed the cross wavelet transform and wavelet coherence for analyzing the time frequency behavior of two signals in a geophysical application. Lachaux et al. [9] applied wavelet coherence theory in EEG analysis. They introduced a wavelet-based method to follow the time-course of coherence between brain signals not affected by any disorder [9]. Sankari et al. presented a wavelet coherence study of electroencephalograph recordings acquired from AD and HC (healthy controls) patients. Wavelet coherence was estimated over the four sub-bands (delta, theta, alpha, and beta) using Morlet wavelet [5, 10]. Sankari and Adeli presented a probabilistic neural network (PNN) model for the classification of AD and HC using the one way analysis of variance (ANOVA) statistical test [11]. In [12], Klein et al. analyzed multichannel EEG recordings of 26 patients acquired in an experiment on associative learning, and compared the results of Fourier coherence and wavelet coherence, concluding that wavelet coherence detects features that were not accessible with Fourier analysis.

In this paper, we investigate how the brain connectivity changes in patients with MCI that converted to AD, and in patients with MCI that did not converted to AD. To our best knowledge, we propose the first longitudinal study on MCI patients through the wavelet coherence analysis. The paper is organized as follows: Sect. 25.2 will introduce the wavelet coherence, how the patients were selected and how the EEG was recorded and preprocessed. Section 25.3 will report the results and Sect. 25.4 will draw the conclusions.

25.2 Methodology

The brain connectivity can be investigated by coherence analysis of EEG recordings. In fact, the coherence can be used as marker to evaluate how the spectral activity of two EEG electrodes are similar in different scalp locations: it can be seen as a measure of temporal synchronization (or de-synchronization) between a couple of EEG recordings. Conventional Coherence [13] is suitable for stationary signals characterized by a non-varying FFT spectrum over time. It is well known, that the EEG is a non-stationary signal, so a time-varying spectral coherence, like wavelet coherence, is necessary to investigate the changes in cortical connectivity [12]. The wavelet coherence optimizes also the brain connectivity estimation in the EEG sub-bands by selecting properly the frequency range.

25.2.1 Wavelet Coherence Estimation

The time-frequency analysis is based on the continuous wavelet transform (CWT) theory. The CWT of a time series x is defined as:

$$CWT(a,b) = \int_{-\infty}^{+\infty} x(t)\Psi_{a,b}^{*}(t)dt \qquad (25.1)$$

Fig. 25.1 Time-frequency representation of the wavelet coherence of a couple of electrodes (ch1–ch19) in the mth window. The coloration goes from *dark blue* (0-low coherence) to *yellow* (1-high coherence)

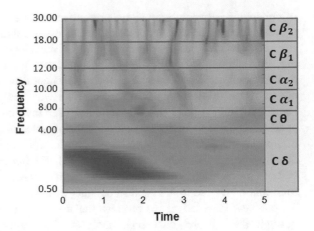

where Ψ is the mother wavelet, a the scaling parameter, b the shifting parameter. Since each scale refers to a frequency value, CWT is a function of time and frequency. There are several wavelets [14], in this study the Morlet wavelet is chosen. It has been proved as a good choice since it provides a good balance between time and frequency localization. In order to investigate the relationship between two signals in the time-scale domain, the wavelet cross spectrum $C_{xy}(a, b)$, is defined as:

$$C_{xy}(a,b) = C_x^*(a,b)C_y(a,b) \tag{25.2}$$

where $C_x(a, b)$ and $C_y(a, b)$ denote the CWTs at scale a and position b. The asterisk sign * is used to denote the complex conjugate operator. Areas in the time-frequency plane where two signals show common power or consistent phase behavior indicate a relationship between the signals.

The wavelet coherence (Fig. 25.1) of two EEG channels x and y is defined as:

$$WC(a,b) = \frac{\left|S\left(C_x^*(a,b)C_y(a,b)\right)\right|^2}{S\left(|C_x(a,b)|^2\right)S\left(|C_y(a,b)|^2\right)} \tag{25.3}$$

where S denotes a smoothing operator in time and scale. For real-valued time series, the wavelet coherence is real-valued if a real-valued analyzing wavelet is used, and complex-valued if a complex-valued analyzing wavelet is used.

25.2.2 EEG Data Processing

The steps of the algorithm are the following (Fig. 25.2): (1) recording and storing on a computer the n-channels EEG (n = 19 electrodes); (2) partitioning the EEG into m non-overlapping windows and processing it epoch by epoch (3) for each window, estimation of the wavelet coherence (C_{xy}) for all pairs of electrodes. C_{xy} is a k × p matrix with k samples and p frequency values; (4) after choosing the frequency range from 0.5 to 30 Hz, the C_{xy} matrix is decomposed in six sub-matrices corresponding to delta, theta, alpha 1, alpha 2, beta 1 and beta 2. In this way the wavelet coherence has been evaluated in the six EEG sub-bands: delta (0–4 Hz), theta (4–8 Hz), alpha 1 (8–10 Hz), alpha 2 (8–12 Hz), beta 1 (12–18 Hz) and beta 2 (18–30 Hz). Each C_{xy} sub-matrix is averaged over frequency:

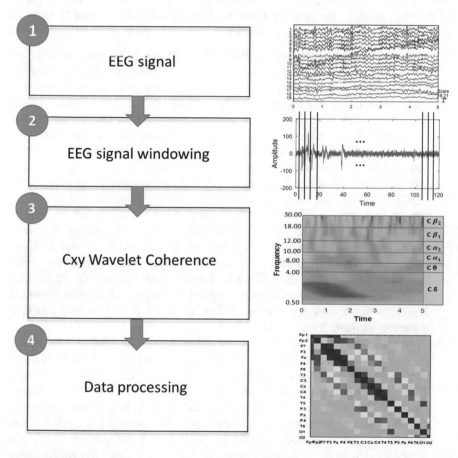

Fig. 25.2 The flowchart of the procedure. (*1*) n-channel EEG recording; (*2*) the EEG is splitted into m non-overlapping windows and processed epoch by epoch; (*3*) for each window, the wavelet coherence (C_{xy}) is evaluated for every pairs of electrodes and decomposed in the six EEG sub-bands (*4*); the m n × n coherence matrices are averaged over the time, obtaining the average coherence matrix C

$$A_{xy} = \frac{1}{U-L} \int_L^U C_{xy}(f)df \tag{25.4}$$

(U = upper frequency band L = lower frequency band) and over time

$$\widetilde{C_{xy}} = \frac{1}{N} \sum_1^N A_{xy} \tag{25.5}$$

in order to get a single average value of coherence. For each EEG window, a n × n matrix of average values of coherence has been evaluated, coming up with m coherence matrices. Finally, the m coherence matrices are averaged also over the time, in order to calculate, for each sub-band, the average coherence matrix C. We will now illustrate every step of the procedure.

(1) *Study population*: Eight MCI patients (4 males and 4 females), at different clinical stages, were recruited at the IRCCS Centro Neurolesi Bonino-Pulejo of Messina (Italy) and enrolled within an ongoing cooperation agreement that included a clinical protocol approved by the local Ethical Committee of IRCCS Centro Neurolesi Bonino-Pulejo of Messina (Italy). An informed consent form was signed by each patient. The MMSE (Mini-Mental State Examination) score was used as inclusion criteria for the enrollment of the patients. After the confirmation of the diagnosis, all patients were evaluated for gender, age, schooling, estimated age of onset of the disorder, marital status and on the MMSE scores. The use of any medications and drugs was estimated: all patients had been receiving them for at least three months before the assessment. All cognitive and clinical valuations were carried out by the same examiners. The Alzheimer Disease's was diagnosed according to the National Institute on Aging-Alzheimer's Association criteria. Each subject was evaluated at a baseline time (time T0) and 3 months later (time T1). At the first evaluation, time T0, the neuropsychological assessment of patients exhibited MMSE scores of 23.4_6.69 for the MCI group. At T1, 3 MCI patients exhibited a conversion from MCI to AD.

(2) *EEG registration*: The EEGs were recorded during the resting state of the patient, in accordance with the 10–20 International System (Fig. 25.3, 19 electrodes: Fp1, Fp2, F3, F4, C3, C4, P3, P4, O1, O2, F7, F8, T 3, T 4, T 5, T 6, Fz, Cz, Pz). The linked earlobe (A1–A2) was used as reference. The sampling rate was 1024 and a 50 Hz notch filter was applied. Before the recordings, patients and caregivers were interviewed about the duration and quality of the sleep of the last night and about the last meal timing and content. The EEGs were recorded in the morning. During the acquisition stage, the patients kept their eyes closed but remained awake. In order to prevent the sleepiness of the

(a) (b)

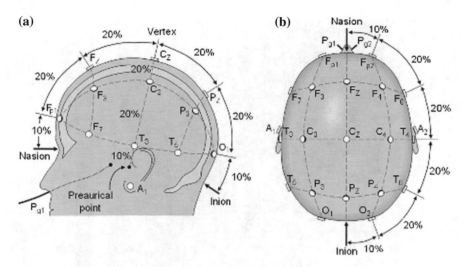

Fig. 25.3 The 10–20 International System seen from left (**a**) and above (**b**) the scalp. *A* ear lobe, *C* central, *Pg* nasopharyngeal, *P* parietal, *F* frontal, *Fp* frontopolar, *O* occipital

patient, the technician kept the subject alert, by calling her/him name. In fact, observing the EEGs, the recordings did not show any sleep patterns. The EEG duration was about 5 min.

(3) *EEG preprocessing*: each n-channel EEG was processed through a sliding temporal window (5 s width). After the down sampling operation, the sampling rate was 256 Hz, so one window includes $N = 5 \times 256 = 1280$ samples. This window was stored on a computer as a $n \times N$ (channels $\times$ samples) matrix. The EEG was splitted into m non-overlapping windows, where m is the number of windows and is function of the length of the EEG recording. Once the wavelet coherence was calculated, it was partitioned into the six EEG sub-bands: delta (0–4 Hz), theta (4–8 Hz), alpha 1 (8–10 Hz), alpha 2 (8–12 Hz), beta 1 (12–18 Hz) and beta2 (18–30 Hz). Every sub-band refers to a different functional and physiological area of the brain.

25.3 Results

• *Analysis of the coherence matrices*

Figures 25.5 and 25.6 show how the average coherence evolves from time T0 to T1 in the six sub-bands: each sub-band is characterized by a 2D representation (Fig. 25.4) where the pixel i, j correspond to the average coherence value of the pairs of electrodes i, j and is represented with a coloration from dark blue (0) to dark red (1).

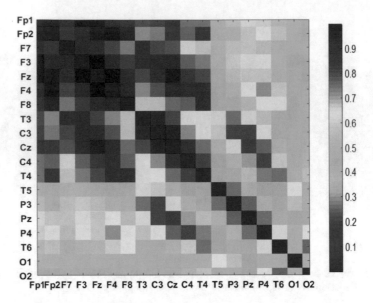

Fig. 25.4 2D representation of the average coherence for a sub-band. Each pixel relates to an average coherence value estimated for the couple of electrodes i, j. The coherence value is ranged between 0 (*dark blue*) and 1 (*dark red*)

- *As regards MCI converted to AD*: in Patient AD-03, at time T0, parietal-occipital zone was characterized by high coherence values in all sub-bands and in particular in the alpha 2 sub-band; at time T1, the coherence moved in the left frontal temporal area identify by Fp1, F7, F3, T3 electrodes, in all sub-bands. In Patient AD-71 (Fig. 25.5) and Patient AD-51 at time T0, coherence was uniformly distributed in all areas and all sub-bands; at time T1 we detected a significantly reduction of coherence and concentrated in the frontal zone especially in alpha 2 and beta sub-bands.
- *As regards MCI not converted to AD*: Patient MCI-23 and Patient MCI-30 show a similar behavior to the Patient AD-03 described above: coherence values moved from parietal-occipital zone to the frontal temporal area in all sub-bands. Patient MCI-41, Patient MCI-57 do not have significant changings from time T0 to T1. Coherence remained between the frontal temporal electrodes. Figure 25.6 shows the behavior of Patient MCI-41. As regards Patient MCI-72 we observed an increase of coherence from time T0 to time T1 in all areas and sub-bands.

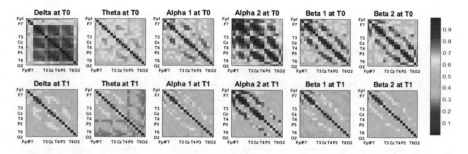

Fig. 25.5 2D representation of the average coherence of the six sub-bands at time T0 and time T1 for the patient AD-71

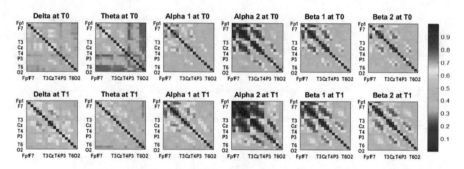

Fig. 25.6 2D representation of the average coherence of the six sub-bands at time T0 and time T1 for the patient MCI-41

- *Boxplots Analysis*

  Boxplots (Fig. 25.7) show the estimated parameters coherence values of MCI (blue box) and AD (red box) patients at time T0 and T1.

- *As regards MCI converted to AD*: the boxplots showed a reduction of the median of the coherence in all sub-bands, except delta band in Patient AD-51 and beta2 band in Patient AD-03. Moreover, a strong reduction, not only of medians but also of the range of coherence can be observed in Patient AD-03 (delta band), Patient AD-51 (theta, alpha(s), beta(s) bands), Patient AD-71 (every bands).

- *As regards MCI not converted to AD*: the median of coherence was mostly constant even if Patient MCI-72 showed an increase of the median in all bands, Patients MCI-57 a decrease in alpha(s) and beta(s) bands, Patient MCI-30 a strong reduction of median and range of coherence in the delta band.

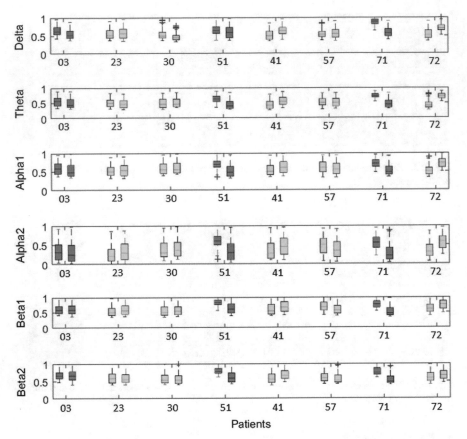

Fig. 25.7 Boxplot of coherence at time T0 and T1. For each patient, the *boxplot on the left* represents the values assumed at time T0 whereas the *boxplot on the right* represents the values assumed at time T1. The *blue boxplots* are associated to patients who were diagnosed MCI at time T0, whereas the *red ones* are associated to AD patients. Each box consists of the median (*central mark*), the 25th and 75th percentiles (*the edges of the box*); the whiskers extend to most extreme data points not considered outliers

25.4 Conclusion

In this work, we proposed a novel approach to carry out a longitudinal study of EEG recordings of subjects affected by MCI. We investigated the changings of the cortical connectivity due to the progression of MCI through wavelet coherence. We monitored 8 patients affected by MCI at time T0 and we studied their follow up at time T1 (3 months later): three of them showed a conversion to AD, five remained MCI. The wavelet coherence method (described in Sect. 25.2.1) was used. The 2D representation of the average coherence and the box plot representation provided useful tools to observe which regions of the brain are damaging and more generally, which the level of cerebral coherence is at time T0 and T1. To sum up we observed a significant mutation of coherence in the AD group rather than MCI. Future works consider a bigger database, including Healthy Controls, and using a hierarchical clustering approach.

Acknowledgements This work was funded by the Italian Ministry of Health, project code: GR-2011-02351397.

References

1. Gauthier, S., et al.: Mild cognitive impairment. The Lancet **367**(9518), 1262–1270 (2006)
2. Mammone, N., Bonanno, L., De Salvo, S., Bramanti, A., Bramanti, P., Ieracitano, C., Campolo, M., Morabito, F.C., Adeli, H.: Hierarchical clustering of the electroencephalogram spectral coherence to study the changes in brain connectivity in Alzheimer's disease. In: Proceedings of International Joint Conference on Neural Networks (2016)
3. Morabito, F.C., Campolo, M., Labate, D., Morabito, G., Bonanno, L., Bramanti, A., de Salvo, S., Marra, A., Bramanti, P.: A longitudinal EEG study of Alzheimer's disease progression based on a complex network approach. Int. J. Neural. Syst. **25**(2),1550005(1–18) (2015)
4. Jeong, J.: EEG dynamics in patients with Alzheimer's disease. Clin. Neurophysiol. **115**(7), 1490–1505 (2004)
5. Sankari, Z., Adeli, H., Adeli, A.: Intrahemispheric, interhemispheric, and distal EEG coherence in Alzheimer's disease. Clin. Neurophysiol. **122**(5), 897–906 (2011)
6. Sankari, Z., Adeli, H., Adeli, A.: Wavelet coherence model for diagnosis of Alzheimer disease. Clin. EEG Neurosci. **43**(4), 268–278 (2012)
7. Grinsted, A., Moore, J.C., Jevrejeva, S.: Application of the cross wavelet transform and wavelet coherence to geophysical time series. Nonlinear Processes Geophys. **11**(5/6), 561–566 (2004)
8. Torrence, C., Compo, G.: A practical guide to wavelet analysis. Bull. Am. Meteorol. Soc. **79**, 61–78
9. Lachaux, J.-P., et al.: Estimating the time-course of coherence between single-trial brain signals: an introduction to wavelet coherence. Neurophysiol. Clin. Neurophysiol. **32**(3), 157–174 (2002)
10. Morabito, F.C., et al.: Monitoring and diagnosis of Alzheimer's disease using noninvasive compressive sensing EEG. In: SPIE Defense, Security, and Sensing. International Society for Optics and Photonics (2013)
11. Sankari, Z., Adeli, H.: Probabilistic neural networks for diagnosis of Alzheimer's disease using conventional and wavelet coherence. J. Neurosci. Methods **197**(1), 165–170 (2011)
12. Klein, A., et al.: Conventional and wavelet coherence applied to sensory-evoked electrical brain activity. IEEE Trans. Biomed. Eng. **53**(2), 266–272 (2006)
13. Kay, Steven M.: Modern Spectral Estimation. Prentice-Hall, Englewood Cliffs, NJ (1988)
14. Daubechies, I.: Ten lectures on wavelets. In: CBMS-NSF Regional Conference Series in Applied Mathematics, p. 61. SIAM, Philadelphia, PA (1992)

Chapter 26
On the Classification of EEG Signal by Using an SVM Based Algorithm

Valeria Saccá, Maurizio Campolo, Domenico Mirarchi,
Antonio Gambardella, Pierangelo Veltri
and Francesco Carlo Morabito

Abstract In clinical practice, study of brain functions is fundamental to notice several diseases potentially dangerous for the health of the subject. Electroencephalography (EEG) can be used to detect cerebral disorders but EEG study is often difficult to implement, taking into account the multivariate and non-stationary nature of the signals and the invariable presence of noise. In the field of Signal Processing exist many algorithms and methods to analyze and classify signals reducing and extracting useful information. Support Vector Machine (SVM) based algorithms can be used as classification tool and allow to obtain an efficient discrimination between different pathology and to support physicians while studying patients. In this paper, we report an experience on designing and using an SVM based algorithm to study and classify EEG signals. We focus on Creutzfeldt-Jakob disease (CJD) EEG signals. To reduce the dimensionality of the dataset, principal component analysis (PCA) is used. These vectors are used as inputs for the SVM classifier with two classification classes: pathologic or healthy. The classification accuracy reaches 96.67% and a validation test has been performed, using unclassified EEG data.

Keywords Classification · SVM · Early detection

26.1 Introduction

The Electroencephalography (EEG) is a non-invasive analysis used to study brain activity by recording cerebral waves placing electrodes on the scalp. The EEG signal is a random process (as shown in Fig. 26.1) derived by neurons activity and is

V. Saccá · D. Mirarchi · A. Gambardella · P. Veltri (✉)
Department of Surgical and Medical Science, University Magna Graecia Catanzaro,
Catanzaro, Italy
e-mail: veltri@unicz.it

M. Campolo · F.C. Morabito
Department of Civil Engineering, Energy, Environment and Materials,
University Mediterranea Reggio Calabria, Reggio Calabria, Italy

© Springer International Publishing AG 2018   271
A. Esposito et al. (eds.), *Multidisciplinary Approaches to Neural Computing*,
Smart Innovation, Systems and Technologies 69,
DOI 10.1007/978-3-319-56904-8_26

Fig. 26.1 Example of an
EEG signal

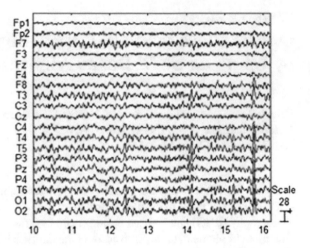

composed by several waves. Each waves reflects the state of cerebral health and in case of diseases the EEG analysis shows several abnormalities. The identification of these abnormalities enables the physician to estimate the disease and its stage, in order to simplify the clinical diagnosis. But EEG study has many difficulties, e.g. signal dimensions and noise presence, thus that it is often difficult to implement an automatic abnormality detection [1]. There is the necessity to have efficient and accurate methods to support signals elaboration. In the literature there are many classification algorithms used to analyze EEG signals (e.g., [3–11] and [12–14]). We experienced in analysing and classifying EEG signals in particular to investigate Alzheimer's disease and epilepsy [15, 16]. Also onset signals have been detected in [17] for ECG signals. In this paper, we report on using SVM based classifier to analyze EEG signals. The aim is to define an algorithm able to: (i) define a classification tool trained by means of pathological and healthy patients and (ii) perform test by using blindly healthy and non healthy signals. We used EEG signals related to CreutzfeldtJakob disease (CJD) healthy patients. Data signals derived by real patients conditions and have been extracted from available clinical dataset. We focus on CJD disease as a rapidly progressive one, characterized by the accumulation of an abnormal protein in the brain [18]. EEG signals exhibit several characteristic in considered pathological condition, depending on the stage of the disease. The here presented classification algorithm is thus based on the identification of the periodic sharp wave complexes (PSWC) that represent the hallmark EEG finding in patients with CJD. We designed and implemented a classifier model of EEG signals, considering the Creutzfeldt-Jakob disease (CJD) based on SVM as classification algorithm. The here proposed method has been testing on clinical data provided by University Magna Graecia clinicians group and the obtained results will be used to optimize the process and for the model tuning.

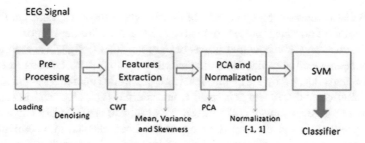

Fig. 26.2 SVM classifier block framework

26.1.1   Related Works

SVM is a machine learning based method and it has been largely used recent as kernel for classification tools. E.g., in [19] has been performed a comparison among SVM and ANN (artificial neural network), to classify eyes blinking showing better accuracy of the SVM based tool. Recently SVM algorithm has mainly used in EEG classification to study several brain diseases, as for instance in [3–9]. In [3] an SVM classifier for EEG signals is presented and used to detect the onset of epileptic seizures. Features are generated for both seizure and not seizure activity and a RBF kernel has been chosen, with optimal results. Moreover the EEG classification can be used to investigate the brain tumors, as in [8], and to detect drowsiness onset while driving [10]. In the cited papers, a spectral analysis method has been applied for extracting generic features by the signals. Nevertheless, the spectral analysis, i.e. the FFT method, is based on simple functions (i.e. sinusoids) and is not suitable for complex signals as EEG ones. We derive features by using temporal frequency analysis, using a continue wavelet transform, that is closest with non-stationary signals, as the EEG [20], by using similar approach than the ones reported in [21, 22] (Fig. 26.2).

26.2   Preprocessing and Data Features Extraction

The aim of this paper is to define a classifier to support decision in diagnosis analyzing EEG signals. Data set needs to be preprocessed and normalized to train SVM based classifier methods. Sixty EEG signals have been gathered from clinical database referring to several patients. In particular, thirty signals relate to healthy subjects and thirty ones refer to CJD patients. These signals have been processed by using the following workflow: (i) preprocessing phase, aiming to reduce artifacts and noise; (ii) features extraction phase, by using a wavelet transformation; (iii) principal component analysis (PCA) and normalization phase, aiming to remove the redundant data; (iv) SVM phase, to generate the classifier model. The input data set generated for the SVM is represented by an N × M matrix where the rows represent the EEG signals and the columns represent the extracted features. We here report on how to extract features and thus how matrix has been created. EEG signals are gathered by using nineteen electrodes placed on the patient scalp. Figure 26.1 shows an example of EEG signals derived by the electrodes. Each electrode generates a signal containing information and artifacts generated from external (i.e., electrical interference) or internal (e.g., eyes movement) sources. To eliminate artifacts EEG signals must be preprocessed. In our data set each signal has been cleaned in a semiautomatic way, by selecting and cutting the artifact of interest to improve the signal quality with a good accuracy, without using a filter. Each signal has been reduced from ten minutes recording to couple of minutes. We have been supported by physicians of University Magna Graecia medical school. The preprocessed signals are then manipulated through a feature extraction process. Extracting the features aims to reduce the data complexity and to simplify the sequel information process. An analysis in time domain has been carried out, using the Continuous Wavelet Transform (CWT). The CWT is an effective tool in signal processing, in which the signal to be analyzed is matched and convolved with the wavelet basis function at continuous time and frequency increments. As result the original signal is expressed as a weighted integral of the continuous basis wavelet function [20]. Choosing the wave type is an important algorithm phase. Indeed, EEG signal is a random process derived by neurons asynchronous activity, which leads a particular form. In this paper, the wavelet "mexh" has been considered, because is very close to the type of the requested format wave. Mean, variance and skewness features have been evaluated for each signal. Each signal captured by each electrode has been divided in twenty-four epochs of five seconds and the CWT has been applied to evaluate the aforementioned features for each signal. Each signal has been divided in three bands, thus calculating three values for mean, variance and skewness. Finally, the mean of these three values has been evaluated for each signal, obtaining twelve features for each EEG signal. Since each EEG is generated by using 19 electrodes, the final number of features extracted for each subject is equal to 228 features. For the experienced dataset, we used 60 subjects, thus that the matrix generated by the preprocessing and features extraction phases is a 60 rows and 228 columns N × M matrix. We consider an M+1 column, used to distinguish healthy from non healthy patients, assigning 1 to pathological patients and 0 to healthy ones, obtaining in such a case a (60 × 229) matrix. Thus, latter column in the right part of Table 26.1 is used to distinguish healthy from non healthy patients. To apply an SVM module and thus to obtain a classifier, data needs to be reduced in size and also needs to be treated in order to obtain more homogeneous values. Indeed, the more large is the dataset, the more high is the probability of obtaining errors or misclassification. Similarly, the more values are non-homogeneous, it is more likely that SVM classifier is not able to discriminate between the considered classes. To reduce the size of the dataset, a method of Blind Source Separation (which is the separation of a set of source signals from a set of mixed signals) has been used, to improve SVM performance (as in [24, 25]). We used Principal Component Analysis (PCA) to reduce the number variables (representing features). In this case, PCA has reduced the features by 228 to 59. To make values more homogeneous a normal-

Table 26.1 The training vector and classes: Patient Id is the subject identifier while health status may be 0 or 1 (healthy or non healty). The vector has been divided in groups of ten

Patient Id	Healthy Status	Patient Id	Healthy Status	Patient Id	Healthy Status	Patient Id	Healthy Status	Patient Id	Healthy Status	Patient Id	Healthy Status
1	1	11	1	21	1	31	0	41	0	51	0
2	1	12	1	22	1	32	0	42	0	52	0
3	1	13	1	23	1	33	0	43	0	53	0
4	1	14	1	24	1	34	0	44	0	54	0
5	1	15	1	25	1	35	0	45	0	55	0
6	1	16	1	26	1	36	0	46	0	56	0
7	1	17	1	27	1	37	0	47	0	57	0
8	1	18	1	28	1	38	0	48	0	58	0
9	1	19	1	29	1	39	0	49	0	59	0
10	1	20	1	30	1	40	0	50	0	60	0

ization algorithm has been applied and values have been mapped into a $(-1, +1)$ range. Finally, the algorithms have been implemented and run by using the Matlab programming environment and LibSVM [26].

26.3   SVM Based Classifier

Data set are used to train and develop the SVM based classifier. The training data is used to generate a model used to predict the target values of the test data [27]. We used leave-one-out as training method [28]. It consists in calculating the model with the exclusion of one object at a time and predicting its value. Starting from a data set, it works as follows: (i) remove one element from the data set; (ii) define a prediction model using the data (less the one removed); (iii) predict and assign the removed element to a class by using the defined model; (iv) repeat the procedure for all elements. The workflow of the Leave-one-out algorithm is reported in Fig. 26.3. To complete the algorithm, the implementation requires the definition of a *kernel function*. We used an RBF kernel function because it is most appropriate in the biomedical signal processing and requires the setting of two parameters: *boxconstraint* or C and γ. γ defines how far the influence of a single training example reaches, while the boxconstrain C controls the classification and the misclassification, due to data overfitting. In this case, C and γ have the same value equal to 1. We implemented an RBF based kernel with the parameters defined above. The latter column of the data matrix has been used as training vector. Results are used to define a confusion matrix, which is used to compare the predicted elements with the original classes. Also, by using the confusion matrix, we calculated four parameters: true positives (TP), true negatives (TN), false positives (FP) and false negatives (FN). These values allow to calculate the accuracy, the specificity and the sensibility.

Fig. 26.3 Leave one out
algorithm workflow

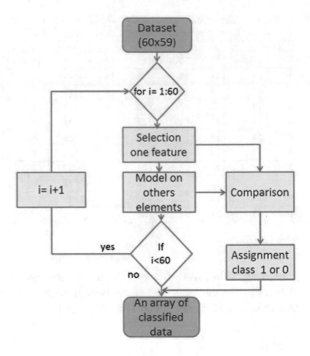

26.4   Experimental Results

We used the algorithm reported above to define an SVM based classifier. The used
data set consists of 60 available EEG signals. The performance of the SVM classifier
can be used to test and produce diagnosis of CJD, considering EEG track with PSWC
as hallmark of disease. Using the 60 signals in the dataset, the accuracy of our classifier is 96.67%, the specificity is 93.33% and the sensibility is 100%. These results
can be refined making additional changes, such as increase the signals number or
introduce new features. In this paper to improve the classifier performance, we used
9 additional EEG signals of subjects suffering from CJD. These signals have been
preprocessed and added to the dataset. The SVM classifier has been trained by using
different values of C and γ. By setting C and γ both to 0.4, and using the whole set
of data, the value of accuracy is 97.10%, the sensitivity 96.67% while the specificity
97.44%. The addition of these new signals has improved the accuracy and sensitivity
of the classifier, while the specificity is slightly lower.

26.5   Conclusion

CJD diagnosis is currently performed by using both EEG signals as well as clinical and healthy status. Nevertheless, it is still difficult to make reliable diagnosis.
We defined a methodology based on a classification tool that uses Support Vector

Machine (SVM) technique. First of all EEG signals have been elaborated by using wavelet based mapping function and by evaluating statistical features. PCA and normalization are used to reduce data heterogeneity. For the training phase, a leave-one-out based algorithm has been implemented and though the confusion matrix, we reached a 96.67% accuracy. The performance of our SVM classifier confirms the classification ability and a candidate as decision support system for EEG analysis in case of CJD suspect cases.

Acknowledgements Authors thank Rocco Cutellé for his support and experiments in denoising and preprocessing signals, and Umberto Aguglia for furnishing supports for EEG signals.

References

1. Bhuvaneswari, P., Kumar, J.S.: Support vector machine technique for EEG signals. Int. J. Comput. Appl. **63**(13) (2013)
2. Li, Y., Wen, P., et al.: Classification of EEG signals using sampling techniques and least square support vector machines. In: Rough Sets and Knowledge Technology. Springer (2009)
3. Shoeb, A.H., Guttag, J.V.: Application of machine learning to epileptic seizure detection. In: Proceedings of the 27th International Conference on Machine Learning (2010)
4. Kumari, R., Jose, J.P.: Seizure detection in EEG using time frequency analysis and SVM. In: 2011 International Conference on Emerging Trends in Electrical and Computer Technology (ICETECT) (2011)
5. Panda, R., Khobragade, P.S., Jambhule, P.D., Jengthe, S.N., Pal, P.R., Gandhi, T.K.: Classification of EEG signal using wavelet transform and support vector machine for epileptic seizure diction. In: International Conference on Systems in Medicine and Biology (ICSMB) (2010)
6. Li, S., Zhou, W., Yuan, Q., Geng, S., Cai, D.: Feature extraction and recognition of ICTAL EEG using EMD and SVM. Comput. Biol. Med. (Elsevier) **43**(7) (2013)
7. Temko, A., Thomas, E., Marnane, W., Lightbody, G., Boylan: EEG-based neonatal seizure detection with support vector machines. Clin. Neurophysiol. (Elsevier) **122**(3) (2011)
8. Murugesan, M., Sukanesh, R.: Towards detection of brain tumor in electroencephalogram signals using support vector machines. Int. J. Comput. Theory Eng. (IACSIT Press) **1**(5) (2009)
9. Sabeti, M., Boostani, R., Katebi, S.D., Price, G.W.: Selection of relevant features for EEG signal classification of schizophrenic patients. Biomed. Signal Process. Control (Elsevier) **2**(2) (2007)
10. Yeo, M.V.M., Li, X., Shen, K., Wilder-Smith, E.P.V.: Can SVM be used for automatic EEG detection of drowsiness during car driving. Saf. Sci. (Elsevier) **47**(1) (2009)
11. Shoker, L., Saeid, S., Alex, S.: Distinguishing between left and right finger movement from EEG using SVM. In: 27th Annual International Conference of the Engineering in Medicine and Biology Society, IEEE-EMBS (2005)
12. Hazarika, N., Chen, J.Z., Ah Chung, T., Sergejew, A.: Classification of EEG signals using the wavelet transform. Digit. Signal Process. Proc. (IEEE) **1** (1997)
13. Subasi, A., Erelebi, E.: Classification of EEG signals using neural network and logistic regression. Comput. Methods Programs Biomed. **78** (2005)
14. Subasi, A., Alkana, A., Koklukayab, E., Kemal Kiymika, M.: Wavelet neural network classification of EEG signals by using AR model with MLE preprocessing. Neural Netw. (Elsevier) **18** (2005)
15. Labate, D., Palamara, I., Mammone, N., Morabito, G., La Foresta, F., Morabito, F.C.: SVM classification of epileptic EEG recordings through multiscale permutation entropy. Neural Netw. (IEEE) (2013)

16. Morabito, F.C., Campolo, M., Labate, D., Morabito, G., Bonanno, L., Bramanti, A., De Salvo, S., Marra, A., Bramanti, P.: A longitudinal EEG study of Alzheimer's disease progression based on a complex network approach. Int. J. Neural Syst. **25**(02) (2015)
17. Vizza, P., Curcio, A., Tradigo, G., Indolfi, C., Veltri, P.: A framework for the a trial fibrillation prediction in electrophysiological studies. Comput. Methods Programs Biomed. (Elsevier) (2015)
18. Wieser, H.G., Schindler, K., Zumsteg, D.: EEG in Creutzfeldt–Jakob disease. Clin. Neurophysiol. (Elsevier) **117**(5) (2006)
19. Rajesh, S., Brijil, C., Arun, K., Santosh, Jayashree: Comparison of SVM and ANN for classification of eye events in EEG. J. Biomed. Sci. Eng. (Scientific Research Publishing) **4**(01) (2011)
20. Adeli, H., Zhou, Z., Dadmehr, N.: Analysis of EEG records in an epileptic patient using wavelet transform. J. Neurosci. Methods (Elsevier) **123** (2003)
21. Inuso, G., La Foresta, F., Mammone, N., Morabito, F.C.: Brain activity investigation by EEG processing: wavelet analysis, Kurtosis and Renyi's entropy for artifact detection. Inf. Acquis. (IEEE) (2007)
22. Inuso, G., La Foresta, F., Mammone, N., Morabito, F.C.: Wavelet-ICA methodology for efficient artifact removal from electroencephalographic recordings. Neural Netw. (IEEE) (2007)
23. Martnez, A.M., Kak, A.C.: PCA versus LDA. IEEE Trans. Pattern Anal. Mach. Intell. **23**(2) (2001)
24. Gursoy, Subast: A comparison of PCA, ICA and LDA in EEG signal classification using SVM. In: 2008 IEEE 16th Signal Processing, Communication and Applications Conference (2008)
25. Subasi, A., Gursoy, M.I.: EEG signal classification using PCA, ICA, LDA and support vector machines. Expert Syst. Appl. (Elsevier) **37**(12) (2010)
26. Chang, C.-C., Lin, C.-J.: LIBSVM: a library for support vector machines. ACM Trans. Intell. Syst. Technol. (TIST) **2**(3) (2011)
27. Hsu, C.-W., Chang, C.-C., Lin, C.-J., et al.: A practical guide to support vector classification (2003)
28. Weston, J.: Leave-One-Out Support Vector Machines (IJCAI) (1999)

Chapter 27
Preprocessing the EEG of Alzheimer's Patients to Automatically Remove Artifacts

Nadia Mammone

Abstract Alzheimer's disease (AD) is a neurological degenerative disorder that causes the impairment of memory, behaviour and cognitive abilities. AD is considered a cortical disease because it causes the loss of functional connections between the cortical regions. Electroencephalography (EEG) consists in recording, non-invasively, the electrical potentials produced by neuronal activity. EEG is used in the evaluation of AD patients because they show peculiar EEG features. The EEG traces of AD patients usually exhibit a shift of the power spectrum to lower frequencies as well as reduced coherence between the cortical areas. This is the reason why AD is defined as "disconnection disorder". However, the correct interpretation of the EEG can be very challenging because of the presence of "artifacts", undesired signals that overlap to the EEG signals generated by the brain. Removing artifacts is therefore crucial in EEG processing. Recently, the author contributed to develop an automatic EEG artifact rejection methodology called Enhanced Automatic Wavelet Independent Component Analysis (EAWICA) which achieved very good performance on both simulated and real EEG from healthy subjects (controls). The aim of the present paper is to test EAWICA on real EEG from AD patients. According to the expert physician's feedback, EAWICA efficiently removed the artifacts while saving the diagnostic information embedded in the EEG and not affecting the segments that were originally artifact free.

Keywords Electroencephalography · Automatic artifact rejection · Alzheimer's disease · EAWICA · AWICA · Independent component analysis · Wavelet · Entropy · Kurtosis

N. Mammone (✉)
IRCCS Centro Neurolesi Bonino-Pulejo, Via Palermo c/da Casazza, SS. 113, Messina, Italy
e-mail: nadiamammone@tiscali.it; nadia.mammone@irccsme.it

© Springer International Publishing AG 2018   279
A. Esposito et al. (eds.), *Multidisciplinary Approaches to Neural Computing*,
Smart Innovation, Systems and Technologies 69,
DOI 10.1007/978-3-319-56904-8_27

27.1 Introduction

Alzheimer's Disease (AD) is a neurological degenerative disorder that causes the impairment of memory, behaviour and cognitive abilities. The average survival time of AD patients is only 5–8 years after they are diagnosed with AD [8] because this pathology is still intractable. At the moment, only medications that help to control the symptoms of this neurological disorder are available. In the future, if AD early detection will be possible, medical treatment is expected to be more effective, if engaged before the brain has been significantly damaged. Whether the scientific community will succeed in developing treatments that are able to stop or at least slow down the disease's evolution, or just medication that help to manage only the symptoms of AD will be available in the near future, the goal must be diagnosing AD as early as possible. If fact, whatever the goal, slowing down or managing the symptoms, early diagnosis would increase the probability of success of the treatment. To this purpose, a non invasive, well tolerated and low-cost methodology should be developed so that the high-risk population, Mild Cognitive Impaired (MCI) subjects, people aged over 65, etc., can undergo a long-term and periodic monitoring to assess their own brain's health state. Given the requirements that such a monitoring system should fulfill (non-invasive, well tolerated and low-cost), it will be likely based on Electroencephalography (EEG), which should be processed to extract relevant diagnostic parameters. The EEG consists in recording, at the scalp, the electrical potentials generated by the brain. EEG abnormalities arise in the brain of AD patients because of the anatomical and functional degeneration of the cerebral cortex. One of the effects is a disconnection between the different areas of the cortex, such disconnection affects the behaviour of the cortical electrical activity and therefore affects the EEG. The EEG signals of AD patients are characterized by a shift of the power spectrum to lower frequencies as well as reduced coherence between the cortical areas [8], probably due to functional disconnection caused by the death of neurons. As AD affects EEG, an EEG-based diagnostic tool should be able to process the EEG automatically and extract relevant information about the effects of the disease on the cerebral electrical activity, for example, through the mapping of descriptive features [15, 17] or the analysis of EEG complexity [6]. Unfortunately, EEG traces are often corrupted by artifacts, which are signals generated by artifactual sources (Electromyography (EMG), ocular artifacts, Electrocardiographic signals (ECG), 50 Hz/60 Hz noise, etc.). Artifactual signals mix to the signals generated by neuronal activity and therefore corrupt the EEG. In this way, the cortical electrical activity might be completely or partially clouded and the subsequent EEG processing might provide incorrect results. Preprocessing the EEG in order to remove artifacts is almost always necessary, especially when the EEG is meant to undergo automatic processing. There are two possible approaches to artifact rejection: (1) artifactual segments (epochs) suppression; (2) artifacts extraction and manipulation in order to reconstruct the corrupted EEG epoch in the best possible way, which would unavoidably cause some information loss, but would allow to save the artifactual epoch. Several algorithms have been proposed in the literature so far. Huang et al. [7] introduced

a Swarm Intelligence Optimization (SIO)-based algorithm to deal with ocular artifacts and subsequently extract P300 information. Bedoya et al. [4] introduced a fuzzy
clustering approach that estimates the similarity between time series. Mateo et al.
[18] proposed to use artificial neural networks to remove eye blinks and movements.
Independent Component Analysis (ICA) has widely employed in artifactual signals
extraction. The author contributed in the development of automatic algorithms based
on ICA and higher order statistics to automatically detect the presence of artifacts
[9, 12, 13, 16]. De Vos et al. [5] focused on the evaluation of ICA-based algorithms
that automatically remove artifacts in neonatal EEG, in order to subsequently process
the cleaned EEG with the aim of detecting seizures. Liao et al. [10] implemented in
hardware and tested an online recursive ICA (ORICA) for eye blink artifact rejection.
After that, they introduced a real-time system that suppresses eye blink in real time
[7]. Bartels et al. [3] proposed an algorithm which jointly exploits ICA and Support
Vector Machines (SVM) to remove artifacts from the EEG recordings. An algorithm
based on the joint use of ICA and continuous wavelet transformation (CWT) was proposed by Bagheri et al. [2] to eliminate ECG artifacts from EEG recordings. ICA and
wavelet denoising (WD), were introduced by Akhtar et al. [1] to preprocess the EEG
and reduce artifacts' effects. Zeng et al. [19] proposed a technique that applies ICA to
extract the artifactual independent components and then denoises them by Empirical
Mode Decomposition (EMD). Recently, the author contributed in the development
of a technique, called Enhanced Automatic Wavelet ICA (EAWICA), which automatically rejects artifacts from the EEG signals with no epoch suppression [11, 14].
As previously discussed, artifact rejection is crucial when the EEG is recorded to
diagnose neurological disorders like AD. The present paper analyzes the effects of
automatic artifact rejection on the EEG of AD patients through EAWICA. EAWICA
has never been tested on the EEG of patients affected by neurological disorders. The
performance of a method for the automatic EEG artifact rejection may significantly
depend on the specific application and the topic of removing artifacts from the EEG
of Alzheimer is mostly unexplored. The paper is organized as follows: Sect. 27.2.1
describes how the EEG was recorded, Sect. 27.2.2 summarizes EAWICA technique,
Sect. 27.3 reports the results and Sect. 27.4 discusses the conclusions.

27.2 Methodology

27.2.1 EEG Recording

The EEG was acquired according to the 10–20 International System (19 channels,
electrode montage: Fp1, Fp2, F3, F4, C3, C4, P3, P4, O1, O2, F7, F8, T3, T4, T5,
T6, Fz, Cz, and Pz), with linked earlobe (A1–A2) reference and 1024 Hz sampling
rate, applying a notch filter at 50H. The EEG was filtered in the range 0.5–64 Hz and
then downsampled to 256 Hz. The patient kept his/her eyes closed throughout the
recording. The recording lasted 168 s. EEG acquisition is a totally non-invasive and
comfortable procedure. It was performed according to a specific protocol approved

by the Ethics Committee of the IRCCS Neurolesi Center (Messina, Italy). The patient
signed an approved informed consent form.

27.2.2 Enhanced Automatic Wavelet-ICA for EEG Artifact Rejection

EAWICA is an algorithm for the automatic extraction, identification and suppression of the wavelet-independent components (WICs) associated to artifactual activity. Higher order statistics, kurtosis and Renyi's Entropy (RE), perform the automatic
artifactual components detection. First of all, the EEG is split into the four major EEG
sub-bands, delta (0–4 Hz), theta (4–8 Hz), alpha (8–12 Hz) and beta+gamma (12–64
Hz) and the wavelet components (WCs) are extracted. The WCs that are highly likely
to be artifactual are automatically marked by kurtosis and RE. They are processed
by ICA and the wavelet independent components (WICs) are extracted. WICs are
processed by kurtosis and RE, the epochs that are marked as artifactual are then sup-
pressed. Inverse ICA and inverse Wavelet Transform are then applied to reconstruct
the EEG. The method is described in detail in [14].

27.2.3 Power Spectral Density

The EEG signals were band-pass filtered between 0.5 and 64 Hz and then downsampled to 256 Hz. The EEG was then partitioned into Nw non-overlapping windows
(width = 5s). Given an EEG window under analysis and given a x-th EEG channel
(electrode), the corresponding time series will be herein denoted as x. The Power
Spectral Density (PSD) is defined as the Fourier Transform of the autocorrelation
function of x:

$$PSD_{xx} = F\{ACF_{xx}(t)\} \tag{27.1}$$

In order to estimate the effects of artifact rejection in the frequency domain, the PSD
was estimated, in the whole frequency range 0.5–64 Hz. The PSD was estimated
within different subareas of the brain: the whole scalp (averaging the EEG signals of
all the electrodes); the frontal, temporal, central, parietal and occipital areas (averaging the EEG signals of the electrodes belonging to the given sub-area under analysis).

27.3 Results

Figures 27.1 and 27.2 show the six artifactual epochs of the EEG recording, before
and after artifact rejection by EAWICA. All artifacts are related to bad electrode
contact or electrode movement, except Artifact 6, which was generated by ocular

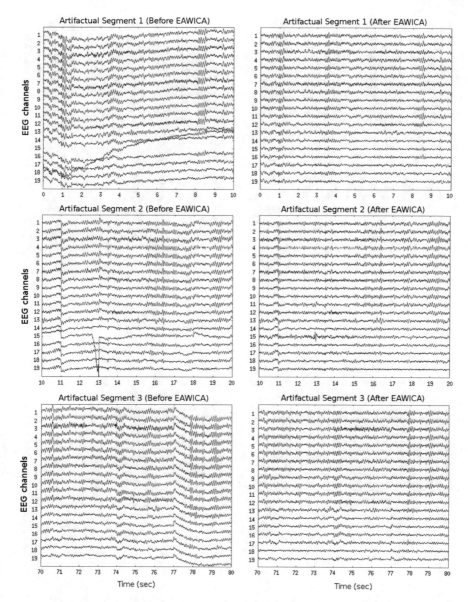

Fig. 27.1 Artifactual segments 1, 2 and 3 of the EEG recording before and after artifact rejection by EAWICA

movements. EAWICA successfully detected and reduced all the artifacts, as we can see in Figs. 27.1 and 27.2. According to the visual evaluation of the expert physician, the diagnostic information was not corrupted by EAWICA whereas artifacts were significantly reduced. Furthermore, the segments that were originally artifact free,

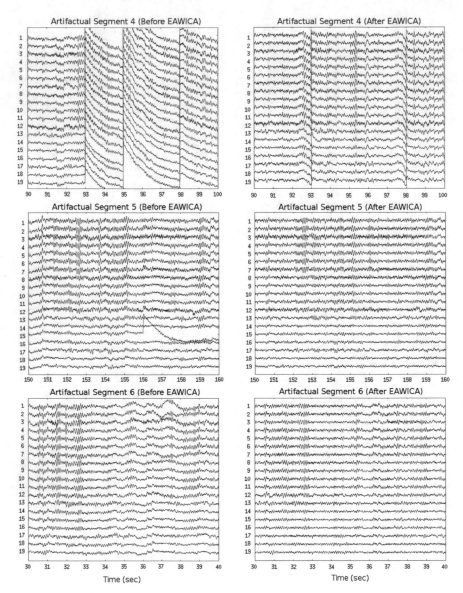

Fig. 27.2 Artifactual segments 4, 5 and 6 of the EEG recording before and after artifact rejection by EAWICA

were not distorted by EAWICA. Figure 27.3 shows the Power Spectral Density of the EEG (computed as described in Sect. 27.2.3), before and after artifact rejection, in every specific sub-area of the scalp (Frontal, Temporal, Central, Parietal, Occipital) as well as in the whole scalp. We can see that the two PSD profiles (before and after artifact rejection) show a similar trend even though PSD looks slightly lower in

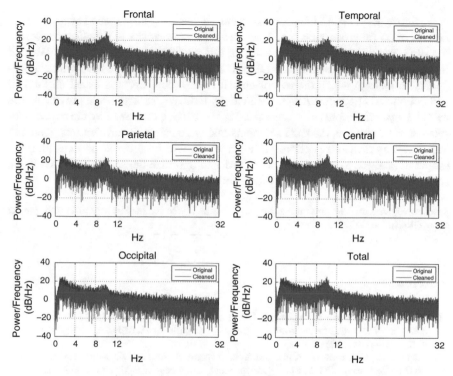

Fig. 27.3 Power Spectral Density of the EEG in the different sub-areas of the scalp (Frontal, Temporal, Central, Parietal, Occipital) and in the whole scalp (Total). Every subplot shows the PSD before (*blue*) and after (*magenta*) artifact removal through EAWICA

delta and theta band after artifact rejection. This is due to the effect of suppressing the bad contact artifact, which can occur even though the EEG was recorded in a resting state, and which typically affects the lower frequency ranges. In conclusion, EAWICA successfully reduced the artifacts while saving the diagnostic information embedded in the EEG and not affecting the artifact free EEG data segments. Future research will be focused on extending the proposed analysis to a large dataset in order to optimize EAWICA parameters' settings for Alzheimer's EEG.

27.4 Conclusions

Alzheimer's disease (AD) is a neurological degenerative disorder characterized by a loss of functional connections between the cortical regions. The Electroencephalography (EEG) of such patients exhibit peculiar features, however, the standard visual inspection of the EEG cannot reveal deep diagnostic information about the disease and advanced EEG processing algorithms are needed if we aim to early diagnose AD in the future. Advanced EEG processing must necessarily be preceded by proper arti-fact rejection preprocessing step because artifacts, unwelcome signals that corrupt the EEG, might lead to wrong results. In this paper Enhanced-Automatic Wavelet Independent Component Analysis (EAWICA), a recent automatic EEG artifact rejec-tion methodology which the author contributed to develop, was tested on the real EEG of one AD patient. EAWICA efficiently reduced artifacts while preserving the useful diagnostic information embedded in the EEG, as assessed by the expert neu-rologist. Furthermore, the EEG segments that were originally artifact free, were not influenced by artifact rejection. Future research will be focused on extending the proposed analysis to a large dataset in order to optimize the settings of EAWICA's parameters for Alzheimer's EEG processing.

Acknowledgements The present research was funded by the Italian Ministry of Health, project code: GR-2011-02351397.

References

1. Akhtar, M.T., Mitsuhashi, W., James, C.J.: Employing spatially constrained ICA and wavelet denoising, for automatic removal of artifacts from multichannel EEG data. Signal Process. **92**(2), 401–416 (2012)
2. Bagheri, H., Chitravas, N., Kaiboriboon, K., Lhatoo, S., Loparo, K.: Automated removal of EKG artifact from EEG data using independent component analysis and continuous wavelet transformation. IEEE Trans. Biomed. Eng. **61**(6), 1634–1641 (2014)
3. Bartels, G., Shi, L.C., Lu, B.L.: Automatic artifact removal from EEG—a mixed approach based on double blind source separation and support vector machine. Conf. Proc. IEEE Eng. Med. Biol. Soc. 5383–5386 (2010). doi:10.1109/IEMBS.2010.5626481
4. Bedoya, C., Estrada, D., Trujillo, S., Trujillo, N., Pineda, D., Lopez, J.D.: Automatic compo-nent rejection based on fuzzy clustering for noise reduction in electroencephalographic signals. In: 2013 XVIII Symposium of Image, Signal Processing, and Artificial Vision (STSIVA), pp. 1–5. IEEE (2013)
5. De Vos, M., Deburchgraeve, W., Cherian, P.J., Matic, V., Swarte, R.M., Govaert, P., Visser, G.H., Van Huffel, S.: Automated artifact removal as preprocessing refines neonatal seizure detection. Clin. Neurophysiol. **122**(12), 2345–2354 (2011)
6. Ferlazzo, E., Mammone, N., Cianci, V., Gasparini, S., Gambardella, A., Labate, A., Latella, M., Sofia, V., Elia, M., Morabito, F., Aguglia, U.: Permutation entropy of scalp EEG: a tool to investigate epilepsies: suggestions from absence epilepsies. Clin. Neurophysiol. **125**(1), 13–20 (2014)
7. Huang, K.J., Liao, J.C., Shih, W.Y., Feng, C.W., Chang, J.C., Chou, C.C., Fang, W.C.: A real-time processing flow for ICA based EEG acquisition system with eye blink artifact elimination. In: 2013 IEEE Workshop on Signal Processing Systems (SiPS), pp. 237–240. IEEE (2013)
8. Jeong, J.: EEG dynamics in patients with Alzheimer's disease. Clin. Neurophysiol. **115**(7), 1490–1505 (2004)
9. La Foresta, F., Inuso, G., Mammone, N., Morabito, F.C.: PCA-ICA for automatic identification of critical events in continuous coma-EEG monitoring. Biomed. Signal Process Control **4**, 229–235 (2009)
10. Liao, J.C., Shih, W.Y., Huang, K.J., Fang, W.C.: An online recursive ICA based real-time mul-tichannel EEG system on chip design with automatic eye blink artifact rejection. In: 2013 International Symposium on VLSI Design, Automation, and Test (VLSI-DAT), pp. 1–4. IEEE (2013)

11. Mammone, N., La Foresta, F., Morabito, F.C.: Automatic artifact rejection from multichannel scalp EEG by wavelet ICA. IEEE Sensors J. **12**(3), 533–542 (2012)
12. Mammone, N., Morabito, F.C.: Independent component analysis and high-order statistics for automatic artifact rejection. In: 2005 International Joint Conference on Neural Networks, pp. 2447–2452. Montral, Canada (2005)
13. Mammone, N., Morabito, F.C.: Enhanced automatic artifact detection based on independent component analysis and Renyi's entropy. Neural Netw. **21**(7), 1029–1040 (2008)
14. Mammone, N., Morabito, F.C.: Enhanced automatic wavelet independent component analysis for electroencephalographic artifact removal. Entropy **16**(12), 6553–6572 (2014)
15. Mammone, N., Morabito, F.C., Principe, J.C.: Visualization of the short term maximum lyapunov exponent topography in the epileptic brain. In: Proceedings of 28th Annual International Conference on IEEE Engineering in Medicine and Biology Society (EMBC), pp. 4257–4260. New York City, USA (2006)
16. Mammone, N., Morabito, F.: Analysis of absence seizure EEG via permutation entropy spatio-temporal clustering. In: Proceedings of International Joint Conference on Neural Networks (IJCNN), pp. 1417–1422 (2011)
17. Mammone, N., Principe, J., Morabito, F., Shiau, D., Sackellares, J.C.: Visualization and modelling of STLmax topographic brain activity maps. J. Neurosci. Methods **189**(2), 281–294 (2010)
18. Mateo, J., Torres, A.M.: Eye interference reduction in electroencephalogram recordings using a radial basic function. IET Signal Process. **7**(7), 565–576 (2013)
19. Zeng, H., Song, A., Yan, R., Qin, H.: EOG artifact correction from EEG recording using stationary subspace analysis and empirical mode decomposition. Sensors **13**(11), 14839–14859

Chapter 28
Smell and Meaning: An OERP Study

Sara Invitto, Giulia Piraino, Arianna Mignozzi, Simona Capone, Giovanni Montagna, Pietro Aleardo Siciliano, Andrea Mazzatenta, Gianbattista Rocco, Irio De Feudis, Gianpaolo F. Trotta, Antonio Brunetti and Vitoantonio Bevilacqua

Abstract The purpose of this work is to investigate the olfactory response to a neuter and a smell stimulation through Olfactory Event Related Potentials (OERP). We arranged an experiment of olfactory stimulation by analyzing Event Related Potential during perception of 2 odor stimuli: pleasant (Rose, 2-phenyl ethanol $C_2H_4O_2$) and neuter (Neuter, Vaseline Oil CH_2). We recruited 15 adult safe non-smokers volunteers. In order to record OERP, we used VOS EEG, a new device dedicated to odorous stimulation in EEG. After the OERP task, the subject filled a visual analogic scale, regarding the administered smell, on three dimensions: pleasantness (P), arousing (A) and familiarity (F). We performed an artificial neural network analysis that highlighted three groups of significant features, one for each amplitude component. Three neural network classifiers were evaluated in terms of accuracy on both full and restricted datasets, showing the best performance with the latter. The improvement of the accuracy rate in all VAS classifications was: 13.93% (A), 64.81% (F), 9.8% (P) for P300 amplitude (Fz); 16.28% (A), 49.46% (F), 24% (P) for N400 amplitude (Cz, Fz, O2, P8); 110.42% (A), 21.19% (F), 24.1% (P) for N600 amplitude (Cz, Fz). Main results suggested that in smell presentation we can observe the involvement of slow Event-Related-Potentials, like N400 and N600, ERP involved in stimulus encoding.

S. Invitto (✉) · A. Mignozzi
Human Anatomy and Neuroscience Lab, Department of Biological and Environmental
Sciences and Technologies, University of Salento, Lecce, Italy
e-mail: sara.invitto@unisalento.it

G. Piraino
Istituto Santa Chiara, Lecce, Italy

S. Capone · G. Montagna · P.A. Siciliano
CNR Microsystemic and Microelectronics, Lecce, Italy

A. Mazzatenta
Physiology and Physiopathology Secty, Neuroscience, Imaging and Clinical Science
Department, University of Chieti, Chieti, Italy

G. Rocco · I. De Feudis · G.F. Trotta · A. Brunetti · V. Bevilacqua (✉)
Department of Electrical and Information Engineering, Polytechnic of Bari, Bari, Italy
e-mail: vitoantonio.bevilacqua@poliba.it

© Springer International Publishing AG 2018
A. Esposito et al. (eds.), *Multidisciplinary Approaches to Neural Computing*,
Smart Innovation, Systems and Technologies 69,
DOI 10.1007/978-3-319-56904-8_28

Keywords Olfactory Event-Related Potentials · Artificial Neural Network · Classification

28.1 Introduction

The sense of smell is used by many animal species to keep track of a prey, predators or find out for mating where in humans evoke memories. Although we can discriminate hundreds of different smells, we have an adequate vocabulary to describe them.

Analyzing the sensory processing is necessary to keep in mind that the perceptive stimulus is related with the space in which it is perceived; although this report is clear as regards the hearing and the sight, smell needs further clarification. It seems necessary considering not only the structural aspects of the stimuli and the induced neural activity but also the functional aspects of different olfactory perceptions. Scientific evidence has found that our olfactory system, in addition to detecting the stimuli, seems to be able to discriminate and identify the same.

For what concerns the processes of stimuli detection, it can be defined, by the application of rigorous statistical criteria, accurate detection thresholds of the olfactory stimulus [1] that represent the standardization of the human ability to detect the stimulus. It is also important to note that the ability of olfactory detection, in addition to have good intrinsic characteristics, may be strengthened by means of appropriate training; the improvement of such ability correlates, especially in women, with a reduction of detection thresholds [2]. Even the discrimination capability detects excellent skills, because the individuals seem able not only to discriminate differences in concentration, but also changes in the structure of detected stimuli [3].

Finally, the olfactory system is specialized in the identification of olfactory cues; in humans, this process becomes even more complex as the olfactory identification seems to be correlated with the linguistic identification. Interestingly, familiarity with a smell could produce a best olfactory identification, but not a linguistic identification [4]. For this reason, generally olfactory stimuli, once identified by the sensory point of view, are recognized and described linguistically in terms of pleasantness and agreeableness. The correlation between the perceptual and linguistic identification is not detected only from a behavioral point of view but also from a neural point of view; a partial overlap was detected between the neural networks deputies to the olfactory perception and those used olfactory imagery [5].

In the light of these concepts, it is clear how the olfactory perception is an active process able to detect, discriminate and identify specific olfactory molecules. From the neural perspective, it is possible to notice the spatial location of an olfactory stimulus activated around the time (Figs. 28.1 and 28.2), so even the localization of a visual and acoustic stimulus, identifying this area as a center of integration and transformation of sensory information into space coordinates. Further studies also

Fig. 28.1 Mapping of Rose odor segmentation in different temporal ranges

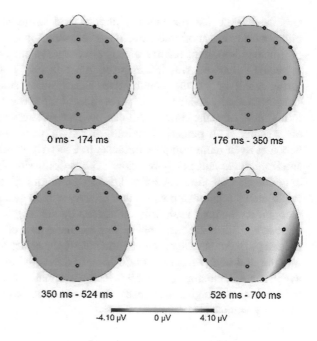

0 ms - 174 ms 176 ms - 350 ms

350 ms - 524 ms 526 ms - 700 ms

-4.10 μV 0 μV 4.10 μV

Fig. 28.2 Mapping of Neutral odor segmentation in different temporal ranges

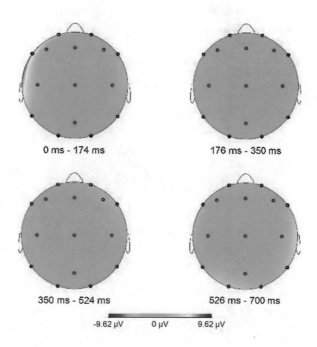

0 ms - 174 ms 176 ms - 350 ms

350 ms - 524 ms 526 ms - 700 ms

-9.62 μV 0 μV 9.62 μV

have identified that the central, parietal and occipital areas are designated for olfactory identification and selectively recruited for visual imagery processes [6].

Another important feature of our olfactory system is related to the ability of a molecule detection, which can reach the olfactory epithelium through the nose (ortho-nasal) or via the mouth (retro-nasal); some studies have found a double sensory modality related to the perception of olfactory stimuli [7], so it is possible to highlight how the same smell can produce different neural responses depending on whether it is perceived in ortho-nasal or post-nasal way [8]. This process becomes even more complex because, in everyday situations, we find ourselves to smell mixtures and not pure odors. This second situation is typical, however, of experimental situations. A lot of knowledge on olfaction has to be still get mainly regarding its relationship with other sensory stimuli as audio, visual, taste [9]. Odor recognition memory is slightly influenced by the length of retention intervals. This is observed for short intervals (a few minutes) as well as longer retention period (years) [10]. Another approach to understand the olfaction links to other senses is to monitor and analyze human brain activity during odors perception. By measuring event-related potentials (ERPs) by electroencephalography, it is possible to measure the electrophysiological response of brain to a specific event, i.e. the presentation of olfactory stimuli [11] (see Fig. 28.3).

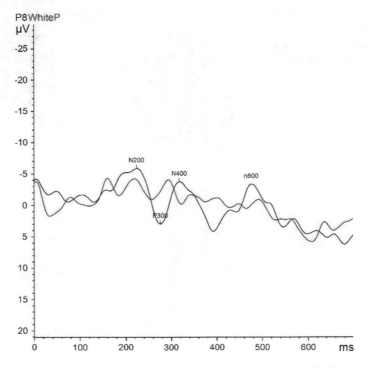

Fig. 28.3 Olfactory event related potentials: PEA (*Black Line*) and neuter (*Red Line*)

In this paper, we used a new device to record olfactory event-related potentials, the US2017127971 (A1) [12]. The invention consists in the creation of a new measuring system of event-related brain potentials using electroencephalogram during the olfactory stimulation in a controlled, automated and synchronized acquisition of the EEG signal. The measuring system, of new conception, has been implemented through the design and implementation of an olfactometer interfaced to an EEG instrumentation. In particular, a two channels olfactometer, interfacing EEG, allows the occurrence of ERP via the controlled administration of two different chemo-olfactory stimuli to the individual during EEG analysis. The subjects had to identify some odorous substances for the definition of experiments, interesting for the study of brain activity under stimulation by pleasant and neutral odors. We use this approach to understand an important effect of smell on perception and to investigate new paradigm of smell administration, as well as to implement new technology and new devices for olfactory cognition, built through microelectronic systems. Aim of this study is to understand how the olfactory system can process an olfactory stimulus in a simple recognition task. Through these results we aim to investigate and to implement innovative applications of cognitive neuroscience in order to improve olfactory basic knowledge.

28.2   Method

Subjects: we recruited 15 adult safe volunteers (mean age 24.4 years old; s.d 4.4); the subjects were non-smoking university students. The sample of recruited volunteers had normal hearing, normal or corrected-to-normal vision, and a right manual dominance. The subjects recruited had no previous experience of EEG and cognitive tasks. None of them had previously taken part in such experiments. All participants gave written informed consent according to the Helsinki declaration. The study was approved inside the 'Olfactory Protocol' approved by ASL Lecce Ethical Committee.

Smell Tools: we arranged an experiment of olfactory stimulation by analyzing Event Related Potential during perception of two stimuli: pleasant smell of rose (2-phenyl ethyl alcohol—PEA C8H10O) and neuter the control vaseline oil (CnH (2n + 2)). The experimental concentration of PEA is 20 μL in 10 mL of Vaseline oil. A dilution ratio was suitably considering in order to realize easily perceptible pleasant, odors perception. The odorous solutions were put into 20 mL glass vials sealed with septum till the exposure time to the volunteers.

EEG: during the smell presentation task, we have recorded data coming from an EEG 16 Channels of Brain AMP-Brain Vision Recorder, and Galvanic Skin Reflex. We considered, in EEG tracks, the Event Related Potentials for averaged Neuter and Rose Smell triggers. The EEG signal was recorded by 16 channels recording electrodes, obtained by Brain Vision Recorder apparatus (Brain Vision). A further electrode was positioned above the left eyebrow for electro-oculogram recording.

The ERPs analysis was obtained using the Brain Vision Analyzer and the time off-line analysis was from 100 ms pre-stimulus to 500 ms post-stimulus with 1000 ms baseline-correction. Thereafter, trials with blinks and eye movements were rejected since horizontal electro-oculogram with an Independent Component Analysis (ICA).

An artifact rejection criterion of 60 V was used at all other scalp sites to reject trials with excessive EMG or other transient noise. The sampling rate was 256 Hz. After a transformation and re-segmentation of data with the Brain Vision Analyzer, the artifact-free EEG tracks corresponding to the affordance object, marked by the motor response, were averaged in each case to extract the main waveforms of the P300, N400 and N600. We performed a semi-automatic peak detection with the mean of the maximum area for the different components of the ERP waves.

Task: subjects were seated in a comfortable chair, 100 cm from the black cave, and performed only one task during the experiment: the subject had to breath in black cave, and try to perceive the changes in odor, without naming or reporting them in any way.

VOS EEG [12]: synchronization, along with the program of olfactory stimulation, allows us to study in details the ERP according to the triggers that will be used and be related to the stimulus and interfaced digitized output from the flow meter. The olfactory triggers are modulated as a function of the variables intensity and duration of stimulation. Our system of odorous stimuli presentation allows to measure the OERPs under olfactory stimuli in a controlled, automated mode synchronized to the acquisition of the EEG signal. The system is based on the dynamic headspace sampling method. Aliquots of samples of odorous substances (at a given concentration in solution—vaseline solvent) are placed inside 20 ml sealed by septa PTFE gas-tight glass vial; the volatile compounds of the substance, passed in vapor phase, are extracted by means of a flow of air taken from the environment. An acoustically isolated micro-pump conveys the air flow, adjusted and kept constant (100 ml/min) by a flow controller, through an appropriate combination of solenoid valves into two vials, where two different odors sources S1 and S2 are contained; solenoid mini valves with a low level of noise emissions have been used in the olfactometer to limit the possible influences on the individual caused by pneumatic or auditory cues. Rise time of the olfactorystimulus is about 2 s. In our experiment S1 is Rose (Rose 2 μl and 10 ml of Vaseline Oil) and Neutral (2 μl and 10 ml of Vaseline Oil). The presentation is conveyed through a plexiglas tube in a black cave that avoids the presentation of visual stimuli. The sequence of stimulus presentation S1 (for 3 s) and S2 (for 3 s) is random with an inter-stimulus period of 60 s during which the individual is exposed only to air. The air flow, after collecting the volatile components of the substances S1 and S2, transfers them to an exposure open box (a cave), where the subject will have to put his head. The cave has been appropriately obscured to avoid visual constraints. The duration of the task session was 25 min.

VAS: the subjects, after the EEG task, had to fill in a 5-points visual analogic scale on three dimensions: Pleasant, Familiar and Arousing dimension.

28.3 Data Analysis and Results

28.3.1 EEG Datasets

Dataset concerned amplitude and latency of P300, N400 and N600 components of the ERP waves. The collected data were the observations, in 15 attributes corresponding to 15 electrodes (F1, F2, ..., F15), on 15 individuals split into 2 classes related to the specific smell of experiment (smell1 and smell2). The correspondence between features and EEG channels is shown in Table 28.1.

28.3.2 Data Analysis

Former analysis performed the significance of features to discriminate between individuals of the 2 classes and then, by using the two-tailed Student's t-distribution, 15 features were selected as an input of an Artificial Neural Network (ANN) pattern recognition strategy [13]. For each data collection, both the amplitude and the latency of P300, N400 and N600 were computed, for each index i, the average of the two classes for the i-th feature (Eq. 28.1).

$$\mu_{i,c} = \frac{1}{14} \sum_{j=1}^{14} x_{i,j} \quad \text{with} \quad i = 1, 2, \ldots, 15 \quad \text{and} \quad c = 1, 2 \tag{28.1}$$

Then, it was used the Student's t-distribution to determine whether there was a statistically significant difference between the means of 2 classes [13].

The inferential statistical analysis revealed that only P300 and N400 in latency showed some significant features. The significant features are shown with the corresponding p-value in Table 28.2.

Table 28.1 Mapping of features and EEG channels

Feat	F1	F2	F3	F4	F5	F6	F7	F8	F9	F10	F11	F12	F13	F14	F15
EEG	Fp1	Fp2	F3	F4	C3	C4	P7	P8	O1	O2	F7	F8	Cz	Pz	Fz

Table 28.2 Significant features with the corresponding p-value, for each ERP wave component

ERP	Significant features	p-value < 0.05	ERP	Significant features	p-value < 0.05
P300 latency	F7	0.0321	N400 latency		
	F12	0.0042		F8	0.0163
	F14	0.0229			

28.3.3 Binary Neural Classifier

The measure of the goodness and the representativeness of the subsets restricted to the significant features extracted in the previous step (Table 28.2) was directly related to their discrimination ability between the two classes (smell1 Rose and smell2 Neuter). For the purpose, 10 neural networks classifiers were designed [14–16] trained and tested on each input pattern. In order to evaluate the classifier performances, the following indexes were used: Accuracy (Eq. 28.4), Area Under the ROC Curve (AUC), Sensitivity (Eq. 28.2) and Specificity (Eq. 28.3). The mean values calculated on the ten neural networks were considered.

$$sensitivity = \frac{TP}{TP + FN} \tag{28.2}$$

$$specificity = \frac{TN}{TN + FP} \tag{28.3}$$

$$accuracy = \frac{TP + TN}{TP + FP + FN + TN} \tag{28.4}$$

In Table 28.3, there is the confusion matrix showing the number of false positives (FP), false negatives (FN), true positives (TP), and true negatives (TN). The input data for the neural network were the six ERP wave components (amplitude and latency of P300, N400 and N600, independently). In training, it was used the following settings: random initialization of synaptic weights with values between ±1, learning rate of 0.3 and momentum of 0.2, method of cross-validation with a training time of 500 epochs.

28.3.4 Results

The comparison of the results in classification for each ERP wave component (Table 28.2) on the full dataset and on the restricted number of significant features one, shows that only P300 and N400 latency produced the best classification performances for the corresponding restricted datasets cases, as shown in Table 28.4.

Table 28.3 A confusion matrix allows visualization of the classification algorithm performances

		Expert advices		Total
		Smell 1	Smell 2	
Prediction	Smell 1	TP	FP	Class 1 (TP + FP)
	Smell 2	FN	TN	Class 2 (FN + TN)
	Total	Smell 1 (TP + FN)	Smell 2 (FP + TN)	

Table 28.4 Performances of P300 and N400 latency, using the full data set input and the dataset restricted to the statistically significant features

ERP	Dataset	Number of neurons in hidden layer	Avg accuracy	Avg AUC	Avg sensitivity	Avg specificity
P300 latency	**[F1, F2, ..., F15]**	8	50.00	0.58	0.46	0.50
	[F8]	1	**68.67**	**0.73**	**0.65**	**0.72**
N400 latency	**[F1, F2, ..., F15]**	8	57.00	0.59	0.34	0.79
	[F7, F12, F14]	2	**78.00**	**0.86**	**0.71**	**0.86**
	[F7, F12]	2	**89.33**	**0.90**	**0.82**	**0.97**

On average, the smell 2 group showed, for each significant feature, higher values of N400 Latency and lower values of P300 Latency than the smell1 group.

The location of electrodes corresponding to the statistically significant features of the ERP wave components represent the Regions of Interest (ROIs) of the specific phenomenon. In the case of P300 latency, the ROI was composed by F8; in the case of N400 latency the ROI covers the area composed by F7, F12, F14. In order to improve the classification performances reducing the number of features to be analyzed, we looked for a smaller set of features. In the second case, it was obtained a fewer set concluding that the significant features for discriminating individuals in the two classes could be reduced to F7, F12 (Table 28.4). A different approach it was tried to find out if amplitude could contain knowledge about the phenomenon. On the same datasets (P300, N400, P600 amplitude and latency) it was performed the Kruskal-Wallis test obtaining the following results: for the latency in each ERP component, the values of all channels had medians which do not differ significantly from each other; instead, there were significant variations between the medians of amplitudes values. The Dunn post hoc highlighted channel pairs with significant differences; they are shown for each ERP component in Table 28.5. The features extracted from the pairs obtained after the previous step were the ones with the greatest number of occurrences weighted with the p-value summary. The result was the following: Fz for the P300 amplitude; Cz, Fz, O2, P8 for the N400 amplitude and Cz, Fz for the N600 amplitude.

In order to discriminate between the grades in each VAS (pleasant, familiar and arousing), we designed feed-forward error back propagation neural networks. In this case, the neural classifiers were trained, validated and tested with one-against-all strategy to perform a multi-classification (5 classes). Their performances on the full datasets and the datasets limited to the significant features, were compared (Table 28.6). The values of accuracy in classification of the three VAS were always higher in the case of dataset restricted to the selected significant features. For P300 and N400 amplitude there was a significant improvement in the *familiar VAS* classification, as well as for N600 amplitude in *arousing VAS* classification.

Table 28.5 Dunn post hoc results: channel pairs with significant differences with p-value summary

P300 amplitude			N400 amplitude			N600 amplitude		
P8	Fz	*	Fp1	Cz	*	Fp1	Cz	*
O1	Fz	*	Fp1	Fz	***	Fp1	Fz	***
O2	Fz	*	Fp2	Cz	**	Fp2	Cz	**
F7	Fz	**	Fp2	Fz	****	Fp2	Fz	***
F8	Fz	**	F3	P8	***	F3	P8	*
Pz	Fz	*	F3	O2	**	P7	Cz	*
			F4	P8	**	P7	Fz	***
			F4	O2	*	P8	Cz	***
			P7	Cz	**	P8	Fz	****
			P7	Fz	***	O1	Cz	**
			P8	Cz	****	O1	Fz	****
			P8	Fz	****	O2	Cz	**
			O1	Cz	**	O2	Fz	****
			O1	Fz	****	F7	Cz	**
			O2	Cz	****	F7	Fz	****
			O2	Fz	****	F8	Cz	*
			F7	Cz	**	F8	Fz	***
			F7	Fz	***	Pz	Fz	**
			F8	Cz	**			
			F8	Fz	****			
			Pz	Fz	**			

p-value summary: * $p < 0.05$; ** $p < 0.01$; *** $p < 0.001$; **** $p < 0.0001$

Table 28.6 Accuracy rate gain of P300, N400 and N600 amplitude in VAS discrimination, using the full dataset input and the dataset confined to the statistically significant features

ERP	Compared sets of features	Arousing VAS accuracy gain (%)	Familiar VAS accuracy gain (%)	Pleasant VAS accuracy gain (%)
P300 Amplitude	[F1, F2, …, F15] versus [Fz]	**3.83** (31.33–27.5)	**17.5** (44.5–27)	**4.33** (48.5–44.17)
N400 Amplitude	[F1, F2, …, F15] versus [Fz, Cz, O2, P8]	**5.67** (40.5–34.83)	**16.16** (48.83–32.67)	**6.16** (31.83–25.67)
N600 Amplitude	[F1, F2, …, F15] versus [Fz, Cz]	**26.5** (50.50–24)	**7.84** (44.84–37)	**4.5** (23.17–18.67)

28.4   Discussion and Conclusion

Three groups of significant features were identified, one for each amplitude component: Fz for the P300 amplitude; Cz, Fz, O2, P8 for the N400 amplitude, Cz, Fz for the N600 amplitude. Furthermore, using three neural classifiers, we checked the performance in accuracy with full set and restricted one, always getting the best performance with the latter.

An important result found in this work is that in both P300 and N400 amplitude there is a significant improvement in the familiar classification, while the N600 amplitude enhances the arousing classification. Our results suggest that, in a presentation where the threshold of detection is low, we can observe the involvement of slow event related potentials, like N400 and N600, ERP involved in language and in a semantic meaning [17]. It happens probably because an encoding processing for naming identification of familiar and arousing stimuli occurs. The olfactory system is specialized in the identification of olfactory cues; this process becomes even more 'cognitive' when the olfactory identification seems to be correlated with the linguistic identification that is when the stimulus seems less identifiable. Even for the olfactory condition, as is the case for the other sensory conditions relating to the mental imagery, the ideational aspects play an important role in the modulation of amplitude and latency aspects of the event related potentials [18–20].

Interestingly, familiarity and arousing with a smell could have a best olfactory identification in slow latency.

Acknowledgements 'Università del Salento—publishing co-funded with '5 for Thousand Research Fund'.

References

1. Young, J., Shykind, B., Lane, R., Priddy, L., Ross, J., Walker, M., Williams, E., Trask, B.: Odorant receptor expressed sequence tags demonstrate olfactory expression of over 400 genes, extensive alternate splicing and unequal expression level. Genome Biol. **4**, 1–15 (2003)
2. Dalton, P., Doolittle, N., Breslin, P.: Gender specific induction of enhanced sensitivity to odors. Nat. Neurosci. 199–200 (2002)
3. Laska, M., Ayabe-Kanamura, S., Hubener, F., Saito, S.: Olfactory discrimination ability for aliphatic odorants as a function of oxygen moiety. Chem. Senses 189–197 (2000)
4. Lawless, H., Egen, T.: Association to odors: interference, mnemonic and verbal labelling. J. Exp. Psychol.: Hum. Learn. Memory **3**, 52–59 (1977)
5. Djordjevic, J., Zatorre, R., Petrides, M., Boyle, J., Gotam, M.: Functional neuroimaging of odor imagery. NeuroImage 791–801 (2004)
6. Porter, J., Anand, T., Johnson, B., Khan, R., Sobel, N.: Brain mechanisms for extraction spatial information from smell. Neuron 581–592 (2005)
7. Rozin, P.: "Taste smell confusions" and the duality of the olfactory sense. Percept. Psychophys. 397–401 (1982)

8. Small, D., Gerber, J., Mak, Y., Hummel, T.: Differential neural responses evoked by orthonasal versus retronasal odorant perception in humans. Neuron 593–605 (2005)
9. Ademoye, O., Ghinea, G., Oluwakemi, A.: Olfaction-enhanced multimedia: perspectives and challenges. Multimedia Tools Appl. **55**(3), 601–626 (2011)
10. Zucco, G.M.: Anomalies in cognition: olfactory memory. Eur. Psychol. 77–86 (2003). doi:10.1027/1016-9040.8.2.77
11. Kim, Y., Watanuki, S.: Characteristics of electroencephalographic responses induced by a pleasant and an unpleasant odor. J. Physiol. Anthropol. Appl. Hum. Sci. **22**(6), 285–291 (2003)
12. Invitto, S., Capone, S., Montagna, G., Siciliano, P.A.: US2017127971 (A1)—Method and system for measuring physiological parameters of a subject undergoing an olfactory stimulation (2017)
13. Menolascina, F., Tommasi, S., Paradiso, A., Cortellino, M., Bevilacqua, V., Mastronardi, G.: Novel data mining techniques in aCGH based breast cancer subtypes profiling: the biological perspective. In: IEEE Symposium on Computational Intelligence and Bioinformatics and Computational Biology, 2007. CIBCB'07, pp. 9–16. IEEE (2007)
14. Bevilacqua, V., Mastronardi, G., Menolascina, F., Pannarale, P., Pedone, A.: A novel multi-objective genetic algorithm approach to artificial neural network topology optimisation: the breast cancer classification problem. In: International Joint Conference on Neural Networks (IJCNN), Vancouver, BC, Canada, pp. 1958–1965. IEEE (2006)
15. Bevilacqua, V., Brunetti, A., Triggiani, M., Magaletti, D., Telegrafo, M., Moschetta, M.: An optimized feed-forward artificial neural network topology to support radiologists in breast lesions classification. In: Proceedings of the 2016 on Genetic and Evolutionary Computation Conference Companion, pp. 1385–1392. ACM (2016)
16. Bevilacqua, V., Cassano, F., Mininno, E., Iacca, G.: Optimizing feed-forward neural network topology by multi-objective evolutionary algorithms: a comparative study on biomedical datasets. In: Advances in Artificial Life, Evolutionary Computation and Systems Chemistry, pp. 53–64. Springer International Publishing (2015)
17. Olofsson, J.K., Hurley, R.S., Bowman, N.E., Bao, X., Mesulam, M.M., Gottfried, J.: A designated odor-language integration system in the human brain. J. Neurosci. **34**(45), 14864–14873 (2014)
18. Invitto, S., Faggiano, C., Sammarco, S., De Luca, V., De Paolis, L.: Haptic, virtual interaction and motor imagery: entertainment tools and psychophysiological testing. Sensors **16**(3), 394 (2016)
19. Gilbert, A.N., Crouch, M., Kemp, S.E.: Olfactory and visual mental imagery. J. Mental Imagery **22**(3&4), 137–146 (1998)
20. Gibbs, J., Raymond, W., Berg, E.: Mental imagery and embodied activity. J. Mental Imagery (2002)

Chapter 29
Automatic Detection of Depressive States from Speech

Aditi Mendiratta, Filomena Scibelli, Antonietta M. Esposito,
Vincenzo Capuano, Laurence Likforman-Sulem,
Mauro N. Maldonato, Alessandro Vinciarelli and Anna Esposito

Abstract This paper investigates the acoustical and perceptual speech features that differentiate a depressed individual from a healthy one. The speech data gathered was a collection from both healthy and depressed subjects in the Italian language, each comprising of a read and spontaneous narrative. The pre-processing of this dataset was done using Mel Frequency Cepstral Coefficient (MFCC). The speech samples were further processed using Principal Component Analysis (PCA) for correlation and dimensionality reduction. It was found that both groups differed with respect to the extracted speech features. To distinguish the depressed group from the healthy one on the basis the proposed speech processing algorithm the Self Organizing Map (SOM) algorithm was used. The clustering accuracy given by SOM's was 80.67%.

F. Scibelli · V. Capuano · A. Esposito
Dipartimento di Psicologia, Università della Campania "Luigi Vanvitelli", Caserta, Italy

A. Mendiratta
School of Computing Science and Engineering, VIT University, Vellore, India
e-mail: aditi.mendiratta2012@vit.ac.in

A. Esposito (✉)
International Institute for Advanced Scientific Studies (IIASS), Vietri sul Mare, Italy
e-mail: iiass.annaesp@tin.it

A.M. Esposito
Istituto Nazionale di Geofisica e Vulcanologia, Sez. di Napoli Osservatorio Vesuviano, Naples, Italy
e-mail: antonietta.esposito@ingv.it

L. Likforman-Sulem
Telecom ParisTech, Paris, France
e-mail: laurence.likforman@telecom-paristech.fr

M.N. Maldonato
Dipartimento di Scienze Umane, Università della Basilicata, Potenza, Italy
e-mail: mauro.maldonato@unibas.it

A. Vinciarelli
School of Computing Science, University of Glasgow, Glasgow, UK
e-mail: alessandro.vinciarelli@glasgow.ac.uk

© Springer International Publishing AG 2018
A. Esposito et al. (eds.), *Multidisciplinary Approaches to Neural Computing*,
Smart Innovation, Systems and Technologies 69,
DOI 10.1007/978-3-319-56904-8_29

Keywords Speech analysis · Depression feature extraction · MFCC · PCA · Self Organizing Maps (SOM)

29.1 Introduction

It was found that most of the identified acoustic differences between depressed and healthy speech are attributed to changes in F0 frequency values and F0 related measures, formants' frequencies, power spectral densities, Mel Frequency Cepstral (MFCC) coefficients, speech rate, and glottal parameters such as jitter and shimmer [1, 4, 20–24]. Recently, it has been observed that the depressed speech exhibits a "slow" auditory dimension [19] and it is perceived as sluggish with respect to the respective healthy samples. Esposito et al. [4] reported that this effect can be primarily attributed to lengthened empty pauses (no significant differences between filled pause durations and consonant/vowel lengthening), and shortened phonation time (less and shorter clauses), whose distribution in a dialogue significantly differs between depressed and healthy subjects. Such measures can be used to develop algorithms that can be implemented in automatic diagnostic tools for the diagnosis of different degrees of depressive states and in general for embedding in ICT interfaces socially believable and contextual information [5]. For this reason, our goal is to extend the earlier research work in Esposito et al. [4] and propose a way through which the detection of such features and measures can be automated so that the depressed speech could be distinguished from the healthy one. The proposed method is based on MFCC speech processing algorithm [16, 22, 26] and the SOM [9, 27, 28] clustering method adopted for the accuracy analysis. This approach is motivated by the fact that MFCC encoding captures subtle variations in the speech [14], structuring the data according to their properties. The algorithm implemented by SOM is able to identify such structures clustering similarities together. The authors have tested the algorithm on seismic signals obtaining valuable performance [8, 9]. The paper is structured as follows. The descriptions of the used database are presented in Sect. 29.2; detailed descriptions of the given dataset processing using MFCC and PCA are presented in Sects. 29.3 and 29.4 respectively. Section 29.5 reports our clustering results followed by conclusions in Sect. 29.6.

29.2 Database

Read and spontaneous speech narratives, collected from healthy and depressed Italian participants were used for the proposed research. For the read narratives, participants were asked to read the tale "The North Wind and the Sun" which is a standard phonetically balanced short folk tale composed of approximately six sentences. For the spontaneous narratives, participants were asked to narrate the daily activities performed by them during the past week. The depressed patients were recruited, with the help of psychiatrists in the Department of Mental Health (Dipartimento di salute mentale) at Caserta (Italy) general Hospital, the Institute for Mental Health (Istituto di Igiene Mentale) at Santa Maria Capua Vetere (Italy) general Hospital, the Centre for Psychological Listening (Centro di Ascolto Psicologico) in Aversa (Italy) and a private psychiatric office. Consent forms were collected from all participants, who were then administered the Beck Depression Inventory Second Edition, BDI II [4] assessed for the Italian language by Ghisi et al. [11]. BDI scores were calculated for both depressed and healthy subjects. A total of 24 sets of recordings were collected, 12 from healthy (or typical speech) and 12 from depressed patients. Each set further contains two types of recordings, i.e. read and spontaneous narratives. On an average, the duration of each set ranges from 150 (lower duration bound) to 300 s (approximatively). Therefore, 150 s from every set is selected and divided into 15 speech waveforms of 10 s. In this selection, the first 130 s belong to the spontaneous speech and the last 20 s to the read-narrative speech.

The recordings were made using a clip-on microphone (Audio-Technica ATR3350), with external USB sound card. Speech was sampled at 16 kHz and quantized at 16 bits. For each subject, the recording procedure did not last more than 15 min. The demographic description of each subject involved in the experiment is reported in Esposito et al. [4].

29.2.1 Analysis of the Database

The BDI-II scores in the range of 0–12 are from control subjects. Table 29.1 illustrates the BDI-II score distributions of depressed subjects which are significantly higher than the matched control group. The scores are from a mild/moderate to a severe depression degree. A T-Student test (one and two tailed testing hypothesis) for independent samples suggested that the factors contributing to the discrimination between healthy and depressed speech [4] are:

- The total duration of speech pauses (empty, filled and vowel lengthening taken all together) is significantly longer for depressed subjects compared to healthy ones
- The total duration of empty pauses is significantly longer for depressed subjects compared to healthy ones
- The clause duration is significantly shorter for depressed subjects compared to healthy ones
- The clause frequency is significantly lower for depressed subjects compared to healthy ones

Table 29.1 BDI-II score distributions of depressed patients with respect to age groups

Depression degree	Age (years)	BDI-II score
Mild + Moderate	21–30	21–25
	41–50	16–20
	51–60	16–20
	61–70	26–30
Severe	31–40	36–40
	41–50	31–35
	51–60	31–35

29.3 Mel Frequency Cepstral Coefficients (MFCC)

From the above information, it was decided to process the speech data through an MFCC pre-processing algorithm, since it has been shown that this kind of pre-processing is the most accurate in extracting perceptual features from speech [22, 26, 14]. The length of recordings obtained from the participants ranged from approximatively 200–360 s. In order to account for the same amount of data for each participant, only 150 s of speech were selected from the beginning of each recording. Each of these 150 s of speech waves has been divided into 15 segments of 10 s each giving a total of T = 360 speech signals ((12 depressed + 12 healthy samples) * 15 segments). Mel-frequency cepstral coefficients (MFCC) are the results of a cosine transform of the real logarithm of the short-term energy spectrum presented on a Mel-frequency scale. The MFCC algorithm is based on human hearing perception which is able to process the details having low frequency ranges below 1000 Hz. Higher frequency ranges are instead grouped more coarsely. In other words, MFCC is based on known variation of the human ear's critical bandwidth and it is generally used to obtain the best parametric perceptual representation of acoustic signals [26, 30]. The MFCC algorithm exploits two kinds of filters that are spread linearly and logarithmically according to the frequencies in the signal below or above 1000 Hz [22] respectively. A subjective pitch of pure tones is present on the Mel Frequency Scale to capture significant characteristic of the speech perceptual features. Thus for each tone with an actual frequency t measured in Hz, a subjective pitch is measured on the so called 'Mel Scale'. The Mel frequency scale produces a linear frequency spacing below 1 kHz and a logarithmic spacing above. The extraction of the MFCC coefficients is made according to the steps illustrated in Fig. 29.1. For sake of clarity, these steps are shortly described in the following.

29.3.1 Pre-processing and FFT

The speech signal is passed through a first order filter that increases high frequencies energy to emphasize these frequencies according to equation $S'(\alpha) = S$

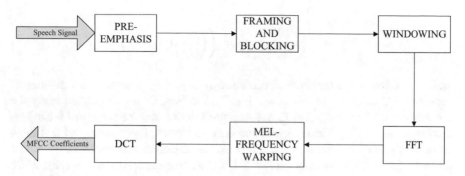

Fig. 29.1 MFCC processing

$(\alpha) - A \times S(\alpha - 1)$, where $S'(\alpha)$ is the output of the filtered speech signal $S'(\alpha)$, $A = 0.97$ is the pre-emphasis coefficient, and α is the sample index. The pre-emphasised speech signal is segmented into small frames of TW length, with a shift of TS (all in ms). In our case, adjacent frames are separated by M = 640 overlapped samples (M < N) [16, 22]. A Hamming window [10] of 100 ms was applied according to Eqs. (29.1) and (29.2):

$$S'_w(\alpha) = S'(\alpha) \times W(\alpha) \qquad (29.1)$$

$$W(\alpha) = 0.54 - 0.46 \; \cos\left(\frac{2\pi\alpha}{N-1}\right), \quad 0 \le \alpha \le N - 1 \qquad (29.2)$$

where N is the number of samples in each window, $S'_w(\alpha)$ the output, and $S'(\alpha)$ the input of the windowing process. The windowing was applied to any α-th speech frame of 10 s, $1 \le \alpha \le T = 360$ speech frames. The windowed signal is then Fast Fourier Transformed (FFT, [22]).

29.3.2 Mel Frequency Warping and DCT

A set of triangular filters are used to compute a weighted sum of the FFT spectral components so that the output approximates to a set of Mel-frequencies (Eq. (29.3))

$$F(Mel) = 2595 * \log(1 + f/700) \qquad (29.3)$$

The amplitude of a given filter over the Mel scale is represented as m_j, where $1 \le j \le NC$. NC is the number of filterbanks (30 channels in our case). The cepstral parameters ($c\tau$) are calculated from the filterbank amplitudes m_j using the Eq. (29.4):

$$c_\tau = \sqrt{\left(\frac{2}{N_C}\right)} \sum_{j=1}^{N_C} m_j \cos\left(\left(\frac{\pi\tau}{N_C}\right)(j-0.5)\right) \qquad (29.4)$$

where τ is index of the cepstral coefficients, $1 \leq \tau \leq x$, and x the number of cepstral coefficients (5 in this case). Finally, the MFCCs are calculated using the discrete cosine transform (DCT) and cepstral liftering routine through Linear Prediction Analysis [30]. Through trial and error processes, it was observed that there were no significant differences in the SOM classification accuracy trained over a dataset of 12 MFCCs versus one of 5 MFCCs, (for each 10 s of speech), while considering less than 5 MFCCs the SOM classification accuracy decreased.

Figures 29.2 and 29.3 report the MFCC processing of a 10 s speech wave for a depressed (Fig. 29.2) and healthy (Fig. 29.3) subject respectively. The figures are intended to show that such processing is able to capture the frequency and duration of clauses and empty pauses. Indeed, it is possible to see that empty pauses are clearly more frequent in the depressed speech producing a different MFCC encoding. In each figure, the topmost subfigure is the plot of the original 10 s speech wave. The middle one displays the energy of the same speech after a 30 channel Mel-frequency processing. On the x-axis is the time, and on y-axis the number of Mel-frequency. The different colours indicate the amount of energy, for a given sample at a given Mel-frequency filterbank. The bottommost subplot represents the MFCC encoding. When comparing the middle and bottom subplots of Figs. 29.2 and 29.3 it can clearly be seen that the energy of the depressed speech is lower as compared to a typical speech in the given time-frame distribution.

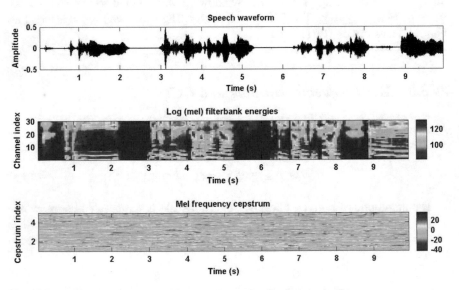

Fig. 29.2 MFCC processing on a 10 s speech sample of a depressed subject

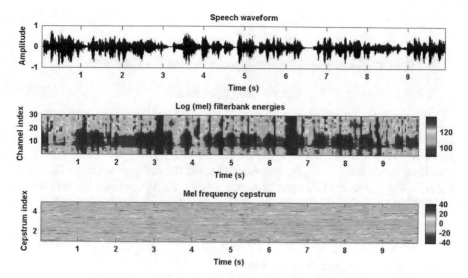

Fig. 29.3 MFCC processing on a 10 s speech sample of a typical person

29.3.3 Principal Component Analysis (PCA)

Principal Component Analysis is a common dimension reduction method applied for feature extraction in speech recognition [13]. PCA maps m-dimensional input data to n-dimensional one, where n $\leq$ m. The method assumes that features that best describe the data are in the directions along which the variations of the data are the largest [12, 29]. Given F feature vectors each of H cepstral coefficients represented as x_{ij}, x_{ij}, $1 \leq i \leq H, 1 \leq j \leq F$, the PCA processing is given by the Eqs. (29.5) and (29.6):

$$\nu_{ij} = x_{ij} - \overline{x_i}, 1 \leq i \leq H, 1 \leq j \leq F \tag{29.5}$$

$$\overline{x_i} = \frac{1}{F} \Sigma_{j=1}^{F} x_{ij} \tag{29.6}$$

where ν_{ij} is the new jth—centered vector data for PCA and $\overline{x_i}$ is the mean of the original dataset containing ith MFCCs. Usually, PCA analysis contains only a single covariance matrix. However, we had to compute P covariance matrices, for each 10 s of speech, as given by Eqs. (29.7) and (29.8).

$$P = \frac{Total\,sample\,inputs\,of\,PCA}{Cepstral\,coefficient\,per\,speech\,sample} = \frac{H}{x} \tag{29.7}$$

$$Cov_i = \frac{1}{F} \sum_{j=1}^{F} \nu_{ij}\nu_{ij}^T, 1 \leq i \leq P \tag{29.8}$$

The principal components are obtained by solving the equation:

$$\lambda_i(y_i) = Cov_i(y_i), 1 \leq i \leq P \tag{29.9}$$

where $\lambda \geq 0$ and $y \in v_{iF}$. The dimensionality reduction step is performed by keeping only the eigenvectors corresponding to the K largest eigenvalues (K $\leq$ P). The resultant values are stored in the matrix $Y_K = [y_1\, y_2 \ldots y_K]$ where $y_1, \ldots, y_K$, are eigenvectors and $\lambda_1, \ldots, \lambda_k$, are eigenvalues of the covariance matrix Cov_r (r $\in$ [1, K]). The reduced PCA transformation matrix Y_K is obtained by solving the Eq. (29.10)

$$z_r = Y_K^T \nu_{rj}, 1 \leq r \leq K, 1 \leq j \leq F \tag{29.10}$$

where z_r denotes the transformed vector.

The first subplot of Fig. 29.4 shows the distribution of the MFCC coefficients before applying the PCA algorithm. The data points have the largest variation along the x-axis. The second subplot shows the reduced dataset with data points correlated to the corresponding mean values of the original MFCC values. The dataset reduced from P = 360 vectors (each of 248 features) to K = 75 vectors. Both the depressed and healthy data, represented as principal MFCC coefficients, have been plotted together. As mentioned above, features discriminating between depressed and typical speech are the *total duration of speech pauses (empty, filled and vowel lengthening taken all together), the total clause durations, and the clause frequency*. These features are not the ones used for our clustering with SOM, since our speech samples were processed through the MFCC algorithm. However, it is

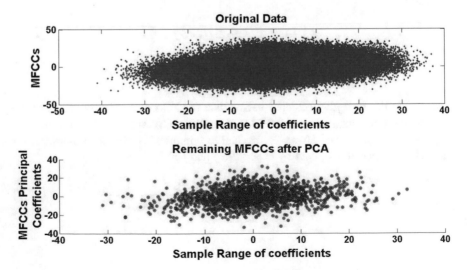

Fig. 29.4 Plots of the obtained (*top*) and PCA reduced MFCC coefficients (*bottom*)

possible that MFCC coefficients encode these parameters. Figures 29.2 and 29.3 support this hypothesis since the MFCC coefficients extracted from depressed speech (Fig. 29.2) display lower energy as compared to those extracted from healthy speech waves (Fig. 29.3). This indirectly suggests more silences and longer empty pauses.

29.4   Self Organizing Map (SOM) Clustering

The SOMs carry out a nonlinear projection of the input data space to a set of units (neural network nodes) on a two-dimensional grid. The grid contains μ neurons given $\mu = R \times C$, where R and C are the number of rows and columns of the SOM grid respectively. Each neuron has an associated weight vector which is randomly initialized. During the training, the weight vector associated to each neuron is likely to become the center of a cluster of input vectors. In addition, the weight vectors of adjacent neurons (neighboring neurons) move close to each other to fit a high-dimensional inputs space into the two dimensions of the network topology [8, 9, 28].

29.5   Results

The main goal of this research was to automatically discriminate between depressed and healthy speech. To this aim, the final MFCC dataset after the PCA reduction was fed into a U_{RC} SOM using the MATLAB Neural Network Toolbox [2]. The R and C values were taken each equal to 6 making a grid of 36 neurons. After training the SOM for 600 epochs, clusters of input vectors with similar MFCC-PCA reduced coefficients are formed on the grid, as illustrated in Fig. 29.5. Figure 29.5 represents the resultant coefficient hits per neuron—i.e. the number of coefficients that cluster in a given neuron of the SOM. The class of a node corresponds to the majority class of the samples clustered in it. Generally, a cluster centre is a neuron that holds a high density of coefficient hits and is closest to all the remaining neurons in that particular cluster as compared to any other neurons in the same cluster. The centre of a cluster of neurons that collects the majority of hits from each class (in our case two classes: depressed and typical speech) is chosen as the neuron containing the maximum number of hits for a given class whose neighbouring neurons also attract the majority of hits from the same class. To this aim, the neuron 13th is a practical option for depressed speech and the 24th for typical speech.

Figure 29.5 represents a statistical analysis of the SOM clustering of the entire dataset containing 50% of depressed speech and 50% of control/typical speech feature coefficient hits. The x-axis represents the μ neurons (in this case $\mu = 36$) on the SOM grid and the y-axis the number of hits for the healthy (red line) and depressed (blue line) speech. The neurons in Fig. 29.5 are not pure classes of only

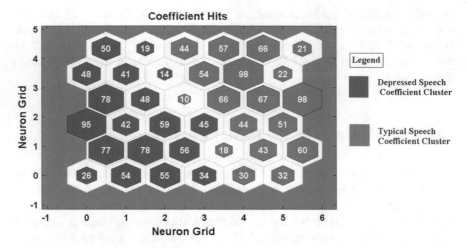

Fig. 29.5 The resulting SOM. The healthy (*orange*) and depressed (*blue*) MFCC features groups into two different clusters along the matrix diagonal. Neuron numbers on the grid must be read from *left* to *right*, *bottom* to *top*

one type of hit (as in the ideal case of 100% accuracy). They contain hits from both typical and depressed speech, as it can be seen in Fig. 29.6. In the real life scenarios, it is quite possible that a small number of typical speech coefficients hit a cluster which have a majority of depressed coefficients and vice versa. Therefore, for the neuron 13 in Fig. 29.5—all the coefficients hits are 95. However, when the SOM output is analysed through the Matlab routine "nctool" that allows to quantify the hits in each neuron, it appears that out of 95 hits, 91 belong to the depressed speech (true hits) and 4 to the typical speech (false hits). This is illustrated in Fig. 29.6 where the 13th neuron shows 91 rather than 95 hits. The same reasoning applies to the neuron 24th and all the remaining neurons. To obtain a realistic clustering accuracy, the testing procedures were repeated three times, using the Rand Measure technique [18] and exploiting three different sets of input vectors, randomly chosen with different proportions of depressed and healthy feature coefficients. The three sets of input vectors were accordingly defined as:

1. Two sets of input vectors (extracted from the 60% of the original dataset), containing 75% of depressed speech features and 25% of healthy ones;
2. Two sets of input vectors (extracted from the 40% of the original database) containing the 12.5% of depressed speech and the 87.5% of healthy one;
3. Two sets of input vectors (the entire dataset) containing an equal amount of depressed and healthy speech (50% each).

The mean performance accuracy from the resultant SOM clustering on each of the three sets, according to the numbers of SOM hits is reported on Table 29.2.

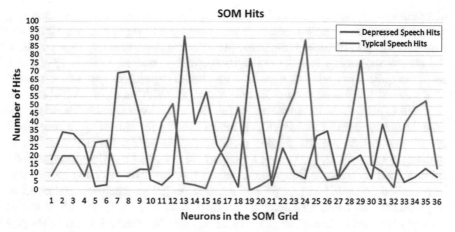

Fig. 29.6 Analysis of the result after clustering of the entire dataset. The x axis represents neurons and the y axis the total number of coefficient hits

Table 29.2 SOM classification results on three different sets of input data

Dataset for classification using SOMs			Accuracy (%)
% of the original data set	% of depressed speech used in each run	% of healthy speech used in each run	
60	75	25	83.34
40	12.5	87.5	78.67
100	50	50	80
Mean accuracy			80.67

29.6 Discussion and Conclusion

There are many parameters in the speech of depressed people that show significant differences compared to a healthy reference group [1, 4, 20, 22–24]. In this study, these parameters were automatically extracted from a dataset of healthy and depressed speech waves by using the MFCC speech processing algorithm [10, 16]. Further correlation of the processed speech was performed by using the PCA algorithm to reduce the data dimensionality. The findings in literature suggest that features discriminating depressed healthy speech are produced by the framing of speech pauses that are elongated, and duration and frequency of speech clauses which are shortened and less frequent for depressed subjects [4, 15]. It is possible that these features are captured by processing speech waves through the MFCC algorithm and through the PCA concept to select from MFCC coefficients those that show the greatest variability with respect to the variance of the data.

In this context, it was found that the combination of MFCC and PCA is a powerful technique for the automatic feature extraction of depressed speech features since, by using a SOM clustering algorithm on such processed data the discrimination accuracy of 80.67% (see Table 29.2) was obtained. The clustering was performed on a small database of 24 recordings (12 depressed and 12 healthy subjects). Despite of these limitations, the discrimination accuracy was far above the chance suggesting that the extracted automatic features d (Sects. 29.2.1 and 29.5) are quite descriptive of depressed and healthy speech despite of the amount of data used for the automatic feature extraction. With more of such data, it is expected to achieve an improvement of the discrimination accuracy. Therefore, the combination of MFCC and PCA is a robust process for extracting features from speech and SOMs provide a good platform for clustering. Similar results were obtained by Kiss et al. [15] using a Support Vector Machine Classification algorithm trained on a larger Hungarian database and tested on the same Italian data. The method presented in this paper for the same Italian database resulted in an improvement of the discrimination accuracy with respect to the classification accuracy reported in Kiss et al. [15]. This study can be extended to a multi-lingual speech database for detecting depression in a language independent way with a larger dataset.

Currently there is a huge demand for complex autonomous systems able to assist people on several needs, ranging from long term support of disordered health states (including caring of elders with motor-skills restrictions) to mood and communicative disorders. Provisions of support have been made either through the monitoring and detection of changes in the physical, and/or cognitive, and/or social daily functional activities, as well as in offering therapeutic interventions [3, 17, 25]. According to the World Health Organization (WHO) at the least 25% of people visiting family doctors live with depression. As reported on the WHO website, (http://www.euro.who.int/en/health-topics/noncommunicable-diseases/mental-health/news/news/2012/10/depression-in-europe). This number is projected to increase and place considerable burdens on national health care institutions in terms of medical, and social care costs associated to the assistance of such people. Voice Assistive Computer Interfaces able to detect depressive states from speech can be a solution to this problem because they can provide an automated on-demand health assistance reducing the abovementioned costs. However, speech is intrinsically complex, and emotional speech is even more [7] requiring the need of an holistic approach that account for several factors including personality traits [27], social and contextual information and cultural diversities [5]. *"The goal is to provide experimental and theoretical models of behaviors for developing a computational paradigm that should produce [ICT interfaces] equipped with a human level [of] automaton intelligence"* ([6], p. 48).

Acknowledgements The patients, healthy subjects (with typical speech) and doctors (psychiatrists) are acknowledged for their involvement and contribution towards this research. The International Institute for Advanced Scientific Studies (IIASS) and Professor Ferdinando Mancini (President, IIASS), is acknowledged for having supported the first author during her internship.

References

1. Alpert, M., Pouget, E.R., Silva, R.R.: Reflections of depression in acoustic measures of the patient's speech. J. Affect. Disord. **66**, 59–69 (2001)
2. Beale, M.H., Hagan, M.T., Demuth, H.B.: Neural network toolbox. User's Guide, The Mathworks Inc., 7–39 (2010)
3. Cordasco, G., Esposito, M., Masucci, F., Riviello, M.T., Esposito, A., Chollet, G., Schlögl, S., Milhorat, P., Pelosi, G.: Assessing voice user interfaces: the assist system prototype. In: Proceedings of 5th IEEE international Conference on Cognitive Info Communications, Vietri sul Mare, 5–7 Nov, pp. 91–96 (2014)
4. Esposito, A., Esposito, A.M., Likforman-Sulem, L., Maldonato, N.M., Vinciarelli, A.: On the significance of speech pauses in depressive disorders: results on read and spontaneous narratives. In: Esposito, A., et al. (eds.) Springer SIST series on Recent Advances in Nonlinear Speech Processing, vol. 48, pp. 73–82 (2016)
5. Esposito, A., Jain, L.C.: Modeling social signals and contexts in robotic socially believable behaving systems. In Esposito, A., Jain, L.C. (eds.) Toward Robotic Socially Believable Behaving Systems Volume II—"Modeling Social Signals" Springer International Publishing Switzerland, ISRL series 106, pp. 5–13 (2016)
6. Esposito, A., Esposito, A.M., Vogel, C.: Needs and challenges in human computer interaction for processing social emotional information. Pattern Recogn. Lett. **66**, 41–51 (2015)
7. Esposito, A., Esposito, A.M.: On the recognition of emotional vocal expressions: motivations for an holistic approach. Cogn. Process. J. **13**(2), 541–550 (2012)
8. Esposito, A.M., D'Auria, L., Angelillo, A, Giudicepietro, F., Martini, M.: Predictive analysis of the seismicity level at Campi Flegrei volcano using a data-driven approach. In: Bassis, et al. (eds.) Recent Advances of Neural Network Models and Applications, Springer Series in Smart Innovation, Systems and Technologies, vol. 19, pp. 133–145 (2014)
9. Esposito, A.M., D'Auria, L., Angelillo, A, Giudicepietro, F., Martini, M.: Waveform variation of the explosion-quakes as a function of the eruptive activity at Stromboli volcano. In: Bassis, et al. (eds.) Neural Nets and Surroundings, Springer Series in Smart Innovation, Systems and Technologies, vol. 19, pp. 111–119 (2013)
10. Gupta, S., Jaafar, J., Ahmad, W.F., Bansal, A.: Feature extraction using MFCC. Signal Image Process. (SIPIJ) **4**(4), 101–108 (2013)
11. Ghisi, M., Flebus, G.B., Montano, A., Sanavio, E., Sica, C.: Beck Depression Inventory-II. Manuale Italiano. Firenze, Organizzazioni Speciali (2006)
12. Jackson, J.E.: A User's Guide to Principal Components, p. 592. Wiley (1991)
13. Jolliffe, I.T.: Principal Component Analysis, 2nd edn. pp. 299–316. Springer (2002)
14. Kakumanu, P., Esposito, A., Gutierrez-Osuna, R., Garcia, O.N.: A comparison of acoustic coding models for speech-driven facial animation. Speech Commun. **48**(6), 598–615 (2006)
15. Kiss, G.C., Tulics, M.G., Sztahó, D., Esposito, A., Vicsi, K.: Language independent detection possibilities of depression by speech. In: Esposito, A., et al. (eds.) Springer SIST series on Recent Advances in Nonlinear Speech Processing, vol. 48, pp. 103–114 (2016)
16. Kopparapu, K.S., Laxminarayana, M.: Choice of Mel filter bank in computing MFCC of a resampled speech. In: IEEE International Conference on Information Sciences Signal Processing and their Applications (ISSPA 2010), Malaysia 10–13 May, pp. 121–124 (2010)
17. Maldonato, N.M., Dell'Orco, S.: Making decision under uncertainty, emotions, risk and biases. In: Bassis, S., Esposito, A., Morabito, F.C. (eds.) Advances in Neural Networks: Computational and Theoretical Issues, SIST Series 37, pp. 293–302. Springer International Publishing Switzerland (2015)
18. Manning, C.D., Raghavan, P., Schütze, H.: Introduction to Information Retrieval: Evaluation of Clustering, pp. 349–356. Cambridge University Press (2008)
19. Marazziti, D., Consoli, G., Picchetti, M., Carlini, M., Faravelli, L.: Cognitive impairment in major depression. Eur. J. Pharmacol. **626**, 83–86 (2010)

20. Moore, E., Clements, M., Peifer, J., Weisser L.: Investigating the role of glottal parameters in classifying clinical depression. In: Proceedings of the 25th Annual International Conference of the IEEE Engineering in Medicine and Biology Society, vol 3, pp. 2849–2852 (2003)
21. Moore, E., Clements, M.A., Peifer, J.W., Weisser, L.: Critical analysis of the impact of glottal features in the classification of clinical depression in speech. IEEE Trans. Biomed. Eng. **55**, 96–107 (2008)
22. Muda, L., Begam, M., Elamvazuthi, I.: Voice recognition algorithms using Mel Frequency Cepstral Coefficient (MFCC) and Dynamic Time Warping (DTW) techniques. J. Comput. **2** (3), 138–143 (2010)
23. Mundt, J.C., Snyder, P.J., Cannizzaro, M.S., Chappie, K., Geralts, D.S.: Voice acoustic measures of depression severity and treatment response collected via interactive voice response (IVR) technology. J. Neurolinguist. **20**, 50–64 (2007)
24. Mundt, J.C., Vogel, A.P., Feltner, D.E., Lenderking, W.R.: Vocal acoustic biomarkers of depression severity and treatment response. Biol. Psychiatry **72**, 580–587 (2012)
25. Rosser, B.A., Vowles, K.E., Keogh, E., Eccleston, C., Mountain, G.A.: Technologically-assisted behaviour change: a systematic review of studies of novel technologies for the management of chronic illness. Telemed. Telecare **15**(7), 327–338 (2009)
26. Tiwari, V.: MFCC and its applications in speaker recognition. Int. J. Emerg. Technol. 19–22 (2010)
27. Troncone, A., Palumbo, D., Esposito, A.: Mood effects on the decoding of emotional voices. In: Bassis, S., et al. (eds.) Recent Advances of Neural Network Models and Applications, SIST 26, pp. 325–332. International Publishing Switzerland (2014)
28. Vesanto, J., Alhoniemi, E.: Clustering of the self-organizing map. IEEE Trans. Neural Netw. **11**(3), 586–600 (2000)
29. Viszlay, P., Pleva, M., Juhár, J.: Dimension reduction with principal component analysis applied to speech supervectors. J. Electr. Electron. Eng. **4**(1), 245–250 (2011)
30. Young, S., Evermann, G., Gales, M., Hain, T., Kershaw, D., Liu, X., Moore, G., Odell, J., Ollason, D., Povey, D., Valtchev, V., Woodland, P.: The HTK Book (for HTK Version 3.4.1). Engineering Department, Cambridge University, pp. 56–80 (2006)

Chapter 30
Effects of Gender and Luminance Backgrounds on the Recognition of Neutral Facial Expressions

Vincenzo Capuano, Gennaro Cordasco, Filomena Scibelli,
Mauro Maldonato, Marcos Faundez-Zanuy and Anna Esposito

Abstract In this study we challenged the universal view of facial emotion perception evaluating the effects of gender and different luminance backgrounds on the recognition accuracy of neutral facial expressions. To this aim, we applied the Ekman standard paradigm for assessing the human ability to decode neutral facial expressions reproduced on black, white and grey backgrounds and portrayed by male and female actors. The exploited stimuli consisted of 10 different neutral faces (5 females) selected from the Dutch Radboud database (Langner et al. Cogn Emot, 2010 [21]) where luminance backgrounds were changed in black, grey and white. The resulted 30 stimuli were assessed by 31 subjects (16 females) who were asked to tag each of them with one of the six primary emotion labels. The data analysis demonstrates a significant gender effect where neutral male faces are less accurately decoded than females ones. On the other hand, no effects of luminance backgrounds have been identified.

Keywords Emotions · Emotional recognition · Neutral facial expressions ·
Socio-cultural influence · Gender effects

G. Cordasco · A. Esposito (✉)
International Institute for Advanced Scientific Studies (IIASS), Vietri sul Mare, Italy
e-mail: iiass.annaesp@tin.it

V. Capuano · G. Cordasco · F. Scibelli · A. Esposito
Dipartimento di Psicologia, Università della Campania "Luigi Vanvitelli", Caserta, Italy
e-mail: gennaro.cordasco@unicampania.it

M. Maldonato
Dipartimento di Scienze Umane, Università della Basilicata, Potenza, Italy
e-mail: mauro.maldonato@unibas.it

M. Faundez-Zanuy
Pompeu Fabra University, Barcelona, Spain
e-mail: faundez@tecnocampus.cat

© Springer International Publishing AG 2018
A. Esposito et al. (eds.), *Multidisciplinary Approaches to Neural Computing*,
Smart Innovation, Systems and Technologies 69,
DOI 10.1007/978-3-319-56904-8_30

315

30.1 Introduction

For more than a century the research on emotions recognition has been guided by the idea of an innate motor program generating our emotional facial expressions through the movements of a particular set of muscular facial actions (AU—Action Units). This research sustains that observers from different cultures can read these facial actions easily. The experimental evidences of facial emotional expressions as *spontaneous by products of innate motor mechanisms* [9, 18, 25], have been debated recently by several studies showing that the AU movements are affected by different contextual information[1] such as descriptions of social situations [5], voices, body postures, visual scenes [1, 3, 4, 10, 29] and other emotional faces [23, 24]. These findings point out that *"even there exists a universal substrate tying the expression and decoding of emotional faces, this process can be less influential than it was previously hypothesized in regulating social emotional exchanges"* [4]. Therefore, in order to properly recognize facial emotional expressions, and in general interlocutors' emotional states, several cognitive processes must be at play, involving the ability to decode intra personality, emotion regulation strategies, personal experiences, see [14, 33] and inter contextual information [14].

Though several studies have found a relationship between colors and emotions [19, 28, 31], to our knowledge, only few have investigated the relationship between colors and emotional facial expressions [7] and no studies have verified the influence of different "image backgrounds" on the proper recognition of facial emotional expressions.

The present study evaluate the emotional recognition performance of 31 (15 males) observers using 10 neutral (i.e., no emotional) facial expressions (5 women), selected from the Radboud Faces Database [21] and administered with 3 different luminance backgrounds (white, black, grey). The aim was to verify whether simple physical contextual information (like e.g. the background of a picture) and the gender of the actor might affect the human ability to decode neutral facial expressions.

30.2 Materials and Procedure

30.2.1 Participants

We recruited 31 white Caucasian western observers (15 males) aged between 22 and 30 years (M = 26.35, SD = ±2.90) with a minimal experience of other cultures (as assessed by the screening questionnaire reported in the appendix of this paper as *Supplemental Material*) and normal or corrected-to-normal vision.

---

[1]Context information is intended here as including the physical, social, individual, and organizational aspects of context surrounding the emotional facial expression that has to be recognized.

The study was exempt from ethical approval procedures since it involved health subjects which volunteered their participation. All observers filled and signed a consent form declaring their voluntary participation and authorized the researchers to use the collected data for scientific purposes.

30.2.2   Stimuli

10 different neutral facial expressions (5 males) were selected from the Radboud Faces Database [21] and photo edited in order to create 3 different luminance background conditions (see Fig. 30.1 for an example): white (rgb^2 = 255, 255, 255), grey (rgb = 128, 128, 128) and black (rgb = 0, 0, 0). A total of 30 stimuli were obtained and randomly presented to the observers through the Superlab^3 software. Each stimulus was presented for 3 s, after a 1 s fixation point. The observers were asked to categorize the stimuli among the six Ekman universal emotions (happiness, fear, sadness, anger, surprise, disgust) plus a "no emotion" and "other" as possible label responses.

30.2.3   Procedure

An 8 Alternative Emotion Forced-Choice (8AEFC—happiness, fear, sadness, anger, surprise, disgust, other, no emotion) was created: the observers were instructed according to the Ekman paradigm [30].

Observers established familiarity with the task by reading detailed instructions provided through the Superlab software. Each observer viewed and categorized the 30 stimulus' trials presented across 1 experimental session with a constant viewing distance of around 68 cm, with images subtending 14.8° (vertical) and 9.4° (horizontal) of visual angles, reflecting the average size of a human face during natural social interactions [17]. The stimuli appeared in the central visual field of a computer screen. Each observer was asked to label the stimuli (forced choice). All observers remained naïve to the presentation order of the stimuli (which were randomly displayed) and the aims of the experiment.

---

^{2}RGB is an additive color model based on three primary colors: red, green, and blue. Each color is represented by three integers ranging in the numerical set of [0, 255], and representing the intensity of each primary color [32].

^{3}www.superlab.com/. Copyright 2012 Cedrus Corporation.

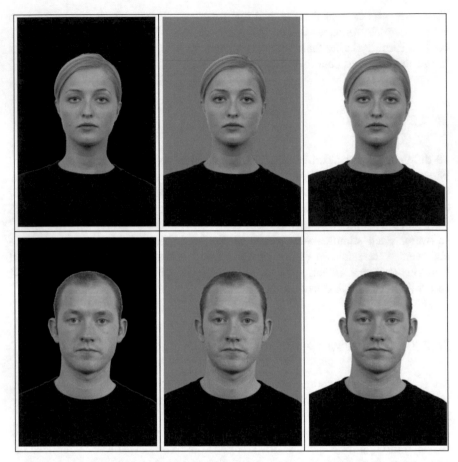

Fig. 30.1 An example of 2 stimulus' identities (a female and a male), presented in the 3 different luminance backgrounds: *black* (rgb 0, 0, 0), *grey* (rgb 128,128, 128) and *white* (rgb 255, 255, 255)

30.3 Results

A 2 × 2 × 3 repeated measures ANOVA was performed in order to verify the influence of gender and luminance backgrounds on the observers' recognition's performance. In particular, the considered analysis took into account as a between subject variable the participants' gender, and as within subject variables the gender of stimuli and the luminance backgrounds (black, grey and white). The confidence level was set to alpha = 0.05. Table 30.1 presents the emotional labels (in percentages) attributed to each stimulus, according to the stimulus gender and its luminance backgrounds (F-Black, M-Black, F-Grey, M-Grey, F-White, M-White).

Table 30.1 Emotional labels (in percentages) attributed to each stimulus, according to the stimulus gender and its luminance backgrounds (F-Black, M-Black, F-Grey, M-Grey, F-White, M-White). Reported are both the percentages of correct answers (associated with the neutral labels) and the percentages of confusions with other emotions

%		F-Black	M-Black	F-Grey	M-Grey	F-White	M-White
Female−subjects	Neutral	62.50	47.50	53.75	47.50	57.50	42.50
	Sadness	15.00	12.50	12.50	11.25	18.75	12.50
	Fear	1.25	3.75	1.25	5.00	0.00	2.50
	Happiness	3.75	7.50	3.75	7.50	3.75	5.00
	Disgust	1.25	3.75	1.25	1.25	1.25	0.00
	Anger	10.00	12.50	10.00	15.00	8.75	18.75
	Surprise	2.50	2.50	2.50	2.50	1.25	6.25
	Others	3.75	10.00	15.00	10.00	8.75	12.50
Male−subjects	Neutral	45.33	46.67	48.00	40.00	49.33	34.67
	Sadness	12.00	9.33	9.33	8.00	12.00	9.33
	Fear	8.00	4.00	5.33	5.33	1.33	9.33
	Happiness	8.00	16.00	6.67	13.33	5.33	13.33
	Disgust	9.33	6.67	10.67	4.00	12.00	5.33
	Anger	9.33	10.67	9.33	14.67	6.67	16.00
	Surprise	1.33	2.67	5.33	8.00	8.00	10.67
	Others	6.67	4.00	5.33	6.67	5.33	1.33
Total subjects	Neutral	54.19	47.10	50.97	43.87	53.55	38.71
	Sadness	13.55	10.97	10.97	9.68	15.48	10.97
	Fear	4.52	3.87	3.23	5.16	0.65	5.81
	Happiness	5.81	11.61	5.16	10.32	4.52	9.03
	Disgust	5.16	5.16	5.81	2.58	6.45	2.58
	Anger	9.68	11.61	9.68	14.84	7.74	17.42
	Surprise	1.94	2.58	3.87	5.16	4.52	8.39
	Others	5.16	7.10	10.32	8.39	7.10	7.10

Luminance Background Effects

No luminance background effects ($F_{(2, 58)} = 1.66$, $p = 0.19$) were found on the correct recognition of neutral faces. When participants labeled the neutral faces with another emotional category, a significant difference was found for surprise ($F(2;58) = 3942$; $p = 0.025$). A post-hoc comparison using the Bonferroni test indicated that surprise was significantly misrecognized (p-value $= 0.034$) in the white ($M = 1.327$, $SD = 0.059$) rather than the black ($M = 1.112$, $SD = 0.039$) background condition. The graph in Fig. 30.2 illustrates emotional labels (in percentages) attributed to each stimulus, according to its luminance background (Black, Grey, White).

Gender Effects

The gender of stimuli significantly affects the correct recognition of neutral faces ($F_{(1, 29)} = 8.332$; $p = 0.007$). In particular, female neutral stimuli ($M = 3.637$,

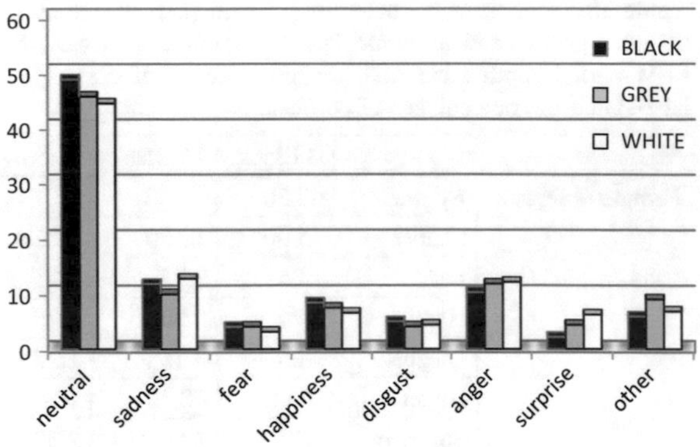

Fig. 30.2 Emotional labels (in percentages) attributed to each stimulus, according to its luminance background (*Black*, *Grey*, *White*). Reported are both the percentages of correct answers (associated with the neutral labels) and the percentages of confusions with other emotions

Fig. 30.3 Emotional labels (in percentages) attributed to each stimulus, according to its gender (Female, Male). Reported are both the percentages of correct answers (associated with the neutral labels) and the percentages of confusions with other emotions

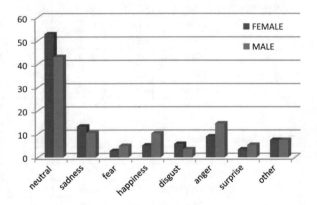

SD = 0.245) have been recognized significantly better than male ones (M = 3.157, SD = 0.224). When observers mis-recognized neutral faces, the gender of the stimuli significantly affected the mis-recognition. In particular, neutral faces portrayed by a male actor were significantly more than female faces mis-decoded as happiness (F (1, 29) = 5.791, p = 0.023) and anger (F (1, 29) = 5.824, p = 0.022).

The gender of the observers did not affected the recognition accuracy (F (1, 29) = 0.815, p = 0.377). The graph in Fig. 30.3 illustrates emotional labels (in percentages) attributed to each stimulus, according to its gender.

30.4 Conclusion

The intent of this work was to contribute to the research investigating on the human ability to recognize and correctly decode emotional facial expressions.

In particular, it was investigated whether different luminance backgrounds and the gender of the stimuli affect the ability of an observer to correctly decode neutral facial expressions.

In spite of the results reported in Lambrecht et al. [20], who recently found no significant gender-specific effects on the recognition of emotions through visual and audiovisual stimuli, our data reveals a significant effect of the gender. In particular neutral facial expressions portrayed by female actresses are decoded significantly better than those portrayed by male actors. In addition, observers confuse more a neutral facial expression with a happy, and/or angry face when the expression is portrayed by a male rather than a female.

Our data adds a socio-cultural effect (the gender of the faces) to recent findings showing that the recognition of emotional facial expressions is affected by social situation descriptions [5], voices [1], body postures [3], visual scenes [8, 29], and other emotional faces [23, 24].

The automatic recognition of faces (and in particular emotional faces) under gross environmental variations and in real time is still a largely unsolved problem [6]. However, humans do not seem to be affected by these limitations since our data did not show any effect of the three different luminance backgrounds.

In summary, our data adds to the abovementioned findings (and only for neutral faces), a socio-cultural effect due to the gender of the stimuli, and excludes a physical effect due to white, black, and grey luminance backgrounds.

In the fascinating research debate investigating whether emotional facial expressions are *learned communication patterns* [15, 16, 22, 27] or *spontaneous expressions of innate mechanisms* [9, 18, 25, 26] our results seem to suggest that facial emotional expressions "are social messages dependent upon context and personal motives" [11, 12]. This complex "dynamic situated" attempt [2, 34] to account for variabilities affecting the human ability to decode emotional facial expressions need further investigations and more data.

These further investigations should account of information of the physical, organizational, cultural, and social environment in which the facial expression is manifested, the characteristics of the individual experiencing and/or recognizing the emotion itself (understood as his personality, his habitual way to express and regulate emotions) and her/his degree of sensitivity (her/his ability to read emotional signals).

Supplemental Material

Observer Questionnaire

Each potential observer completed the following (culture appropriate) questionnaire. We selected only participants that answered to the following questions as reported below[4]:

---

[4]The signed squares represents the accepted answers: vision (normal vision, glasses, contact lenses), with no learning disabilities, that they are no synaesthete, answering "no" to all questions for participation in the experiment.

Please answer to the following questions

1. Vision

Do you have normal (uncorrected) vision, do you wear glasses or contact lenses, or are you visually impaired?–
    ☑ Normal vision ☑ Glasses ☑ Contact Lenses ☐ Visually Impaired

2. Laterality

Since the left part of the brain control the right side of the body, and the right part of the brain controls the left side, it is important to know such things as handedness of subjects before running an experiment. Are you left, right-handed or ambidextrous (use both hands equally)?
    ☐ Left-handed ☑ Right-handed ☐ Ambidextrous

3. Learning disabilities

The term learning disability is used to refer to individuals who show a discrepancy between their apparent capacity to learn and their level of achievement. Specific learning disabilities include difficulties with:

• reading (dyslexia)
• writing (dysgraphia)
• speech and language (dysphasia or aphasia)
• maths (dyscalcula)
• motor-planning (dyspraxia)

A learning disability can also be a difficulty with information processing, such as visual and auditory perception (e.g. difficulty recognizing shapes). If you are aware of any learning disability that you may have please specify this here. Do you suffer from learning disability?
    ☐ Yes ☑ No

4. Synaesthesia

Synaesthetes experience a fascinating phenomenon called synaesthesia, a "joining together" of two senses that are normally experienced separately. There are many, many types of synaesthesia. For example, synaesthetes may perceive colours when they read numbers, letters, or words; the colours may be superimposed on the graphemes themselves or merely perceived in their "mind's eye". Synaesthetes may "hear colours", "see sounds", or feel touch sensations on their own bodies when they observe others being touched. Likewise, the days of the week or even numbers may appear in a specific location, form, or shape.
    Take a look at the following questions and statements. If your answer to any of them is either "yes" or "sometimes", then you may be a synaesthete.

• I experience colours when I look at written numbers.
• I experience colours when I look at written letters.
• I experience colours when I look at written words.

- Each number/letter/word has a specific colour.
- I associate numbers to colours.
- I associate letters to colours.
- I associate words to colours.
- I experience touch on my own body when I look at someone else being touched (i.e., I feel touch sensations on my own body when I observe them on another person's body).
- I experience touch on my own body when I look at something else being touched (i.e., I feel touch sensations on my own body when I observe them on objects).
- I experience touch in response to body postures.
- Do these experiences have specific locations (i.e., on your body, on words or objects, in front of your eyes) or not (i.e., you just "know" or they feel as though they are in your "mind's eye")?
- Do you think about ANY of the following being arranged in a specific pattern in space (i.e., in a line, a circle, etc.)? ALPHABET, CALENDAR YEAR, DAYS OF THE WEEK, WEEKS, TIME, NUMBERS (NUMBER LINE)
- Do you think about numbers/letters/words as having personalities or genders?
- Do you experience colours in response to: SOUNDS, MUSIC, VOICES, TOUCH. If your answer to any of those questions is YES or SOMETIMES, you may be a synaesthete.

    Are you a synaesthete? ☐ Yes ☑ No

5. Hearing

    Do you have normal hearing, corrected hearing (e.g. cochlear implant) or is your hearing impaired (e.g. tinnitus)?
    ☑ Normal ☐ Corrected ☐ Impaired

References

1. Aviezer, H., et al.: Angry, disgusted, or afraid? Studies on the malleability of emotion perception. Psychol. Sci. **19**(7), 724–732 (2008)
2. Barrett, L.F.: Variety is the spice of life: a psychological construction approach to understanding variability in emotion. Cogn. Emot. **23**(7), 1284–1306 (2009)
3. Barrett, L.F., Mesquita, B., Gendron, M.: Context in emotion perception. Curr. Dir. Psychol. Sci. **20**(5), 286–290 (2011)
4. Capuano, V., Riviello, M.T., Cordasco, G., Mekyska, J., Faundez-Zanuy, M., Esposito, A.: Assessing natural emotional facial expressions: an evaluation of the I.Vi.T.E. database. In: von Dieter Mehnert, H., Kordon, U., Wolff, M. (eds.) Rüdiger Hoffmann zum, vol. 65, pp. 249–255. Geburtstag Systemtheorie Signalverarbeitung Sprachtechnologie, TUD press. Kartoniert (2013)
5. Carroll, J.M., Russell, J.A.: Do facial expressions signal specific emotions? Judging emotion from the face in context. J. Pers. Soc. Psychol. **70**(2), 205 (1996)
6. Cheon, Y., Kim, D.: Natural facial expression recognition using differential-AAM and manifold learning. Pattern Recogn. **42**(7), 1340–1350 (2009)

7. da Pos, O., Green-Armytage, P.: Facial expressions, colours and basic emotions. JAIC-J. Int. Colour Assoc. **1** (2012)
8. de Gelder, B., Meeren, H.K., Righart, R., Van den Stock, J., van de Riet, W.A., Tamietto, M.: Beyond the face: exploring rapid influences of context on face processing. Prog. Brain Res. **155**, 37–48 (2006)
9. Ekman, P., Cordaro, D.: What is meant by calling emotions basic. Emot. Rev. **3**(4), 364–370 (2011). doi:10.1177/1754073911410740
10. Esposito A, Capuano V, Mekyska J, Faundez-Zanuy M: A naturalistic database of thermal emotional facial expressions and effects of induced emotions on memory. In: Esposito, A., et al. (eds.) Behavioural cognitive systems, LNCS, vol. 7403, pp. 158–173. Springer, Heidelberg (2012). ISBN: 978-3-642-34583-8 e-ISBN 978-3-642-34584-5
11. Esposito, A.: The situated multimodal facets of human communication. In: Coverbal Synchrony in Human-Machine Interaction, pp. 173–202 (2013a)
12. Esposito, A.: Emotional expressions: communicative displays or psychological universals?. In: COST Strategic Workshop on Social Robot Booklet-Perception. www.cost.eu/download/38170, p. 32 (2013b)
13. Esposito, A., Esposito, A.M., Vogel, C.: Needs and challenges in human computer interaction for processing social emotional information. Pattern Recogn. Lett. **66**, 41–51 (2015a)
14. Esposito, A., Palumbo, D., Troncone, A.: Effects of narrative identities and attachment style on the individual's ability to categorize emotional voices. In: Bassis S. et al. (eds.) Advances in Neural Networks: Computational and Theoretical Issues, Springer International Publishing Switzerland, SIST Series, vol. 37, pp. 265–272 (2015b)
15. Esposito, A., Jain, L. C.: Modeling Emotions in Robotic Socially Believable Behaving Systems Toward Robotic Socially Believable Behaving Systems, vol. I, pp. 9–14. Springer (2016)
16. Fridlund, A. J.: The new ethology of human facial expressions. In: The Psychology of Facial Expression, vol. 112, pp. 103–129 (1997)
17. Hall, E.: The hidden dimension. Garden City, NY: Doubleday (1966)
18. Izard, C.E.: Innate and universal facial expressions: evidence from developmental and cross-cultural research. Psychol. Bull. **115**(2), 288–299 (1994)
19. Joosten, E., Van Lankveld, G., Spronck, P.: Influencing player emotions using colors. J. Intell. Comput. **3**(2), 76–86 (2012)
20. Lambrecht, L., Kreifelts, B., Wildgruber, D.: Gender differences in emotion recognition: impact of sensory modality and emotional category. Cogn. Emot. **28**(3), 452–469 (2014)
21. Langner, O., Dotsch, R., Bijlstra, G., Wigboldus, D.H.J., Hawk, S.T., van Knippenberg, A.: Presentation and validation of the Radboud Faces Database. Cogn. Emot. (2010)
22. Maldonato, N.M., Dell'Orco, S.: Making decision under uncertainty, emotions, risk and biases. In: Bassis, S., Esposito, A., Morabito, F.C. (eds.) Advances in Neural Networks: Computational and Theoretical Issues, SIST series, vol. 37, pp. 293–302. Springer International Publishing, Switzerland (2015)
23. Masuda, T., Ellsworth, P.C., Mesquita, B., Leu, J., Tanida, S., Van de Veerdonk, E.: Placing the face in context: cultural differences in the perception of facial emotion. J. Pers. Soc. Psychol. **94**(3), 365–381 (2008). doi:10.1037/0022-3514.94.3.365
24. Masuda, T., Gonzalez, R., Kwan, L., Nisbett, R.E.: Culture and aesthetic preference: comparing the attention to context of East Asians and Americans. Pers. Soc. Psychol. Bull. **34**(9), 1260–1275 (2008). doi:10.1177/0146167208320555
25. Matsumoto, D., Keltner, D., Shiota, M. N., Frank, M. G., O'Sullivan, M.: What's in a face? Facial expressions as signals of discrete emotions. In: Lewis, M., Haviland, J.M., Feldman Barrett, L. (eds.) Handbook of Emotions, pp. 211–234. Guildford Press, New York (2008)
26. Matsumoto, D., Seung Hee, Y., Fontaine, J.: Mapping expressive differences around the world: the relationship between emotional display rules and individualism versus collectivism. J. Cross Cult. Psychol. **39**(1), 55–74 (2008). doi:10.1177/0022022107311854
27. Mehu, M., Scherer, K.R.: A psycho-ethological approach to social signal processing. Cogn. Process. **13**(2), 397–414 (2012)

28. Oberascher, L., Gallmetzer, M.: Colour and emotion. In: Proceedings of AIC 2003 Bangkok: Color Communication Management, pp. 370–374 (2003)
29. Righart, R., de Gelder, B.: Rapid influence of emotional scenes on encoding of facial expressions: an ERP study. Soc. Cogn. Affect. Neurosci. 3(3), 270–278 (2008)
30. Rosenberg, E.L., Ekman, P.: Conceptual and methodological issues in the judgment of facial expressions of emotion. Motiv. Emot. 19(2), 111–138 (1995)
31. Simmons, D.R.: Colour and emotion. In: New Directions in Colour Studies, pp. 395–414 (2011)
32. Tkalcic, M., Tasic, J.F.: Colour spaces: perceptual, historical and applicational background. In: Paper Presented at the Eurocon (2003)
33. Troncone, A., Palumbo, D., Esposito, A.: Mood effects on the decoding of emotional voices. In: Bassis, S., et al. (eds.) Recent Advances of Neural Network Models and Applications, SIST, vol. 26, pp. 325–332. International Publishing, Switzerland (2014)
34. Wilson-Mendenhall, C.D., Barrett, L.F., Barsalou, L.W.: Situating emotional experience. Front. Hum. Neurosci. 7 (2013)

Chapter 31
Semantic Maps of Twitter Conversations

Angelo Ciaramella, Antonio Maratea and Emanuela Spagnoli

Abstract Twitter is an irreplaceable source of data for opinion mining, emergency communications, or fact sharing, whose readability is severely limited by the sheer volume of *tweets* published every day. A method to represent and synthesize the information content of conversations on Twitter in form of semantic maps, from which the main topics and the main orientations of tweeters may easily be read, is proposed hereafter. After a preliminary grouping of *tweets* in conversations, relevant keywords and Named Entities are extracted, disambiguated and clustered. Annotations are made using extensive knowledge bases and state-of-the-art techniques from Natural Language Processing and Machine Learning. The results are in form of coloured graphs, to be easily interpretable. Several experiments confirm the high understandability and the good adherence to tackled topics of the mapped conversations.

Keywords Semantic processing · Natural language processing · Clustering · Twitter mining

31.1 Introduction

All technologies supporting social interaction are recently experiencing a rapid growth and microblogging in particular is being credited as a privileged channel for emergency communications, due to its growing ubiquity, fastness, and cross-platform accessibility.

A. Ciaramella (✉) · A. Maratea · E. Spagnoli
Department of Science and Technology, University of Naples "Parthenope",
Isola C4, Centro Direzionale, 80143 Napoli (na), Italy
e-mail: angelo.ciaramella@uniparthenope.it
URL: http://www.scienzeetecnologie.uniparthenope.it

A. Maratea
e-mail: antonio.maratea@uniparthenope.it

E. Spagnoli
e-mail: emanuela.spagnoli@uniparthenope.it

© Springer International Publishing AG 2018   327
A. Esposito et al. (eds.), *Multidisciplinary Approaches to Neural Computing*,
Smart Innovation, Systems and Technologies 69,
DOI 10.1007/978-3-319-56904-8_31

Twitter[1] is a microblogging service launched in 2006 where every user can publish messages up to 140 characters long, called *tweets*, which are visible on a public board or through third-party applications [10]. As of December 2015, Twitter declares 320 millions of monthly active users and it is deemed to have more than one billion of registered users and 100 millions of daily active users in more than 35 languages.

Registered users can read and *post* tweets, while unregistered users can only read them. Both can access Twitter through the website interface, SMS or mobile device applications. Users can group posts together by topic or type by use of *hashtags*, that is words or phrases prefixed with a "#"). To share a message from another user with his own followers, a user can *retweet* the tweet.

Twitter offers a fast and easy way to retrieve, produce and spread facts and personal opinions in form of short messages and has been used for trying to regulate disaster response in earthquakes events or security threats [3, 5], as well as a source of data for analyzing political orientation [4].

In this paper a method for generating semantic maps of twitter conversations is proposed; its purpose is to obtain an intuitive graphic representation of the content of a conversation, from which the main topics and the main orientations of the *tweeters* may easily be read. A tool named *TweetCheers* has been developed, leveraging powerful technologies as BabelNet, Babelfy, Wordnet, Wikipedia and Wikirelate! Only the English language has been tested at the moment.

The paper is organized as follows: in Sect. 31.2 each logical block used by *TweetCheers* is presented; in Sect. 31.3 the global architecture of *TweetCheers* is described in detail; in Sect. 31.4 several experiments are shown and results discussed.

31.2 Building Blocks

The building blocks used by *TweetCheers* and the underlying ideas are outlined hereafter.

31.2.1 WordNet

WordNet is a large lexical dictionary [7] inspired by current psycholinguistic theories on human lexical memory. It contains about 155,000 nouns, verbs, adjectives, and adverbs in English organized into synonym sets (*synsets*), each representing one lexical concept. Formally it is a semantic network (acyclic graph). Words have a definition and are wedged in an hypernym/hypnoym (is-a) taxonomy, that allows to compute their semantic relatedness for example through their Lowest Common Hypernym (LCH) or Least Common Subsumer (LCS).

---

[1]http://www.twitter.com.

31.2.2   Wikipedia and WikiRelate!

Wikipedia is a free-access, open Internet encyclopedia, supported and hosted by the non-profit Wikimedia Foundation. While the WordNet taxonomy is well-structured and trustful, it does not provide information about the named entities (e.g. Henry Potter, Rolling Stones, P450). Wikipedia, on the opposite side, provides a large corpus of information about entities and concepts, and gives a knowledge base for computing word relatedness (including named entities) in a more comprehensive fashion than WordNet [11].

In [11] the authors use Wikipedia for computing semantic relatedness in a tool called *WikiRelate!* In particular, the online encyclopedia and its folksonomy is used for computing the relatedness of words. The authors show that Wikipedia outperforms WordNet when applied to the largest available benchmark datasets.

31.2.3   BabelNet and Babelfy

Babelnet[2] is a very large, wide-coverage multilingual semantic network [9]. The resource is automatically built by a methodology that integrates lexicographic and encyclopedic knowledge from WordNet and Wikipedia. It represents an *encyclopedic dictionary*, that provides concepts and named entities lexicalized in many languages and connected with large amounts of semantic relations.

Babelfy[3] is a unified graph-based approach to Entity Linking and Word Sense Disambiguation based on a loose identification of candidate meanings coupled with a densest subgraph heuristic which selects high-coherence semantic interpretations [8].

31.2.4   TextRank

TextRank is a graph-based ranking model for text processing that was the foundation of the Google PageRank algorithm and that was extended to weighted graphs in [6]. TextRank can be applied for unsupervised keyword or sentence extraction. Lexical units are identified as vertex and edge is created if the corresponding lexical units co-occur within a window of N words. The vertices added to the graph can be restricted with syntactic filters, which select only lexical units of a certain type (i.e. only nouns or verbs).

---

[2]http://www.babelnet.org.
[3]http://www.babelfy.org.

31.2.5 Fuzzy Relational Data Clustering

Cluster analysis or clustering is the task of grouping a set of objects so that objects in the same group (called a cluster) are more similar (in some sense) among them than to those in other groups (clusters). Fuzzy c-means (FCM) is a method of clustering which allows one object to belong to two or more clusters with a different membership value [1]. In fields like management, industrial engineering, and social sciences, relational data, that is data that describe objects specifying their pairwise similarities instead of feature measurements, are frequently encountered. In this case a robust non-Euclidean Fuzzy Relational data Clustering (RFC) algorithm can be used, where the relational data are derived as measures of dissimilarity between the objects [2].

31.3 TweetCheers

The 140 characters limitation of each *tweet* presents a challenge because, in the effort to compress concepts as much as possible, users abuse of slang words, abbreviations, links, acronyms and fancy hashtags, none of which can have a unique, clear or timely definition. Moreover *tweets* cannot be sampled exhaustively and each sampled stream is full of heterogeneous topics, spam, incidental or groundless conversations—ultimately noise. For these reasons, a tool able to automatically map in a coloured graph the main topics in a stream of tweets, pruning most noise, is precious to have an overview of the information content of sampled conversations.

Aim of *TweetCheers* is to generate semantic maps from a number of *tweets*, preliminary grouped into conversations. Figure 31.1 sketches its general architecture.

First of all many streams of data are gathered querying the Twitter API with a set of predefined keywords, called *seeds*. Tweets are then linked to form a number of conversations, using the *in_reply_to* field of the auxiliary information given by the API. Once conversations are built, these are processed as a whole unit of text by the logical blocks described in Sect. 31.2. Ultimately, a semantic map is obtained by collecting information from *keywords*, *hashtags* and two *agglomerating* approaches, one on the *Named Entities* and another on the *Wikicategories*. The seeds words are the root nodes of the map, and the other nodes are added as the various blocks process the data. One separate map is obtained for each seed. Nodes corresponding to the hashtags are obtained parsing the conversations, and they are added to the map creating a link to the seed word to which the conversation belongs.

Nodes corresponding to non-seed keywords are obtained through TextRank, that is scoring each word in each conversation and picking only the most relevant. They are added to the map creating a link to the seed word to which the conversation belongs.

In parallel to these blocks, a Babelfy disambiguation and Part Of Speech Tagging (POS) process is started, aiming to contextualize the conversations relative to

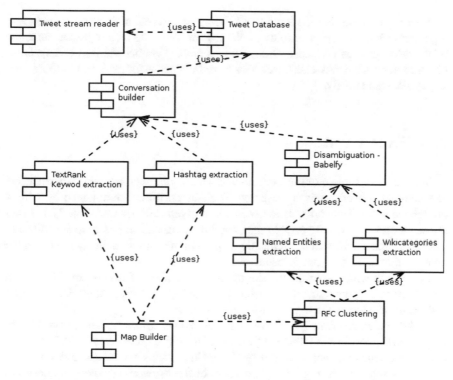

Fig. 31.1 Component diagram of Tweetcheers architecture

each seed. For each conversation, a list of N Named Entities and relative $M \geq N$ Wikipedia Categories is extracted by Babelfy, and for each pair of entities (e_i, e_j) the similarity $s_{ij}^{(E)}$ is computed using Wikirelate! while for each pair of categories (c_i, c_j) the similarity $s_{ij}^{(C)}$ is computed using Lowest Common Ancestor (LCA). The matrix $S^{(E)} = [s_{ij}]_{N \times N}^{(E)}$ is the similarity matrix of entities and the matrix $S^{(C)} = [s_{ij}]_{M \times M}^{(C)}$ is the similarity matrix of categories. Using as distance between categories the steps to the LCA has two drawbacks: when the set of categories is crowded it is inefficient and retrieved ancestors are often too generic. To avoid the drawbacks a filtering in the list of categories is performed, so to exclude the ones that have too many subcategories. An RFC clustering is then performed on both the Entities and the Wikicategories (with 2 or 3 as the number of clusters). The centroids (entitites and categories with the lowest average distance among all elements in the cluster) are considered the most relevant topics in each conversation and are linked to the seed node in each map; the other elements in each cluster are linked to their centroid, so to obtain a snowflake schema.

The semantic map is obtained connecting all nodes generated, with a different color for each node type and a weight representing the distance from each cluster centroid.

Seed words represent the topic that is being monitored, and are reported in *red*, together with the cluster centroids of Named Entities. The arcs represent semantic relatedness between concepts. *Orange* nodes are the Named Entities (NEs); *black* nodes represent the Wikicategories. *Green* and *violet* are for extracted hashtags and keywords, respectively.

31.4 Experiments

In this section a sample of the experimental results obtained applying TweetCheers to some real conversations is presented. The conversations are obtained by using a predefined set of keywords, called *seeds*, in a given time frame. Table 31.1 shows the seeds and time frames used in the experiments. Experiments were conceived to emphasize the stability of the semantic map, to highlight the main arguments in the conversations and its dynamic behaviour.

In the following experiments, the minimum number of tweets considered in a conversation is 7, the number of clusters for both NEs and Wikicategories are 2 or 3 and the dimension of the n-grams for Textrank is 2.

In the first experiment 70000 tweets and 50 conversations gathered from June to August 2015, considering the seed "violence" have been used. In Fig. 31.3 a particular of the semantic map obtained by TweetCheers clustering NEs and Wikicategories is visualized. In Fig. 31.2 the clusters of the keywords (violet) and hashtags (green) are highlighted, respectively. The center of the keywords cluster is "family" and the main concepts are: "women", "children", "gunmen", "gender", "agitation". The words of the hashtags are: "justice", "islam", "ukraine", "nigeria".

From these clusters, can be observed that they contain very general conversation concepts and they are not easily interpretable. Instead, in Fig. 31.3 the main extracted clusters are related to "violence against women", "eastern Christianity" and "Mad Max", an American movie remade on May 2015. In Table 31.2 the concepts and the semantic similarities of the "eastern Christianity" cluster are summarized.

In a new experiment the number of conversations for the seed "violence" is expanded to 160. Although, the main concepts of the keywords cluster are unchanged, some words are added: "utopia", "player", "peace", "songs". Also for the hashtags cluster some added words are: "russia", "fiction", "bcc live".

Moreover, Fig. 31.4 shows the results obtained clustering NEs and Wikicategories. Here, a new subgraph can be observed in the semantic map: it is dedicated to

Table 31.1 Twitter conversations

Keywords	# Tweets	# Conversations	Period
Violence	70000	50 − 160	06.2015 − 08.2015
Isis	130000	400 − 500	05.2015 − 09.2015

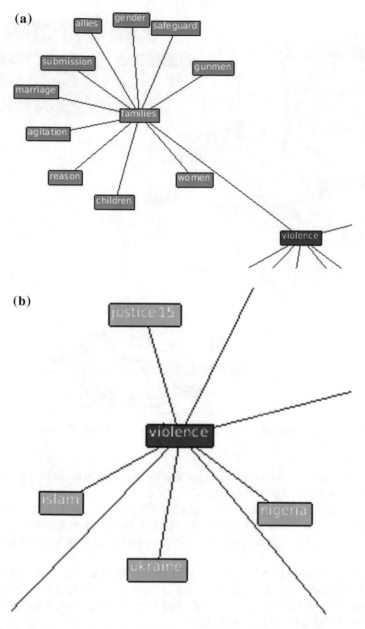

Fig. 31.2 Semantic map obtained considering the keyword "violence" (50 conversations): **a** keywords cluster; **b** hashtags cluster

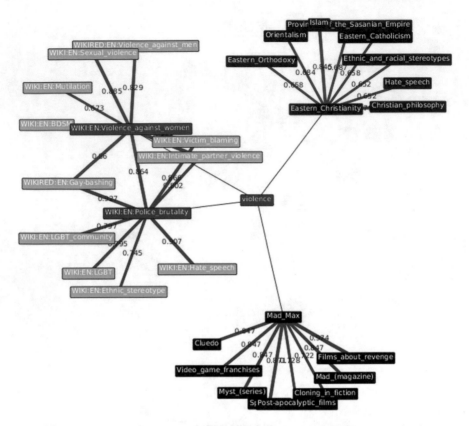

Fig. 31.3 Semantic map obtained considering the keyword "violence" (50 conversations)

"Al Qaeda" (see Table 31.3 for details), a heavily discussed worldwide theme, which seems to replace the cluster "Mad Max".

Also in this case, the clusters of keywords and hashtags contain very general conversation concepts, but they are useful to complete the interpretability of the semantic map. In the next experiments, for shortness and clarity, these clusters are not shown.

In the second experiment, the seed "ISIS" is considered. 130000 tweets with 400 and 500 conversations, gathered from May to September 2015, were collected.

Figures 31.5 and 31.6 contain the obtained semantic maps before and after a specific event: "ISIS threats against the Pope".

In the first case the main extracted concepts are related to "Arab World", "Barack Obama", "Gulf War", "Organizations affiliated with Al-Qaeda" and "Areas controlled by the Islamic State of Iraq and the Levant". In the second case the main extracted concepts are related to "Al Jazeera", "Arab World", "Organizations affiliated with Al-Qaeda", "Treaties of Syria" and "Sovereigns of Vatican City" (see

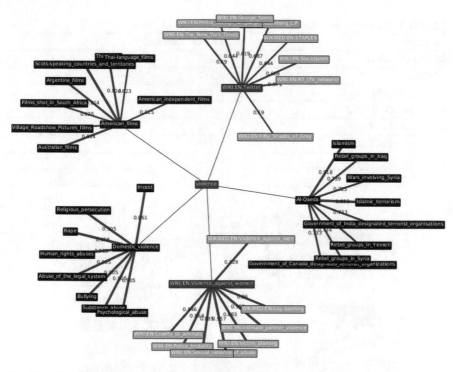

Fig. 31.4 Semantic map obtained considering the keyword "violence" (160 conversations)

Table 31.2 Twitter conversations: seed "violence"; cluster "eastern Christianity"

Concepts	Semantic similarity
Islam	0.845
Christian philosophy	0.750
Province of the sasanian empire	0.687
Orientalism	0.684
Eastern orthodoxy	0.658
Eastern catholicism	0.658
Ethnic and racial stereotypes	0.652
Hate speech	0.652

Table 31.4 for details). It can be observed that in the latter a novel cluster emerges about the Vatican State due to the approaching the Jubilee: it replaces the unimportant clusters of the previous semantic map.

Table 31.3 Twitter conversations: seed "violence"; cluster "Al Qaeda"

Concepts	Semantic similarity
Islamic terrorism	0.933
Islamism	0.918
Rebel groups in Iraq	0.739
Rebel groups in Syria	0.734
Government of India …	0.733
Government of Canada …	0.733
Rebel groups in Yemen	0.732
Wars involving Syria	0.725

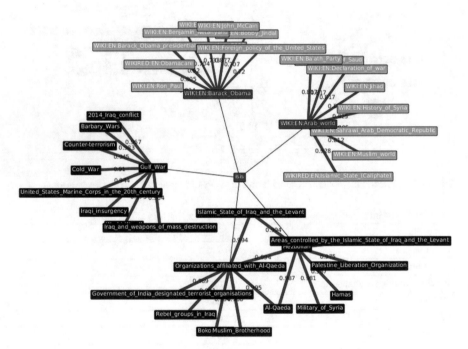

Fig. 31.5 Semantic map obtained considering the keyword "ISIS" (before ISIS threats against the Pope)

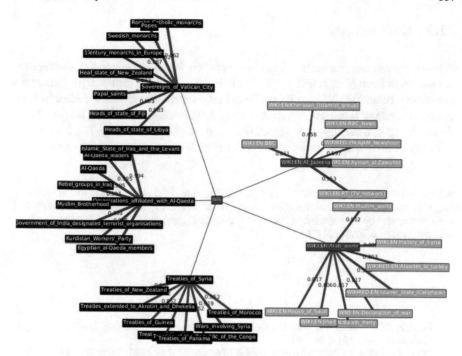

Fig. 31.6 Semantic map obtained considering the keyword "ISIS" (after ISIS threats against the Pope)

Table 31.4 Twitter conversations: seed "ISIS"; cluster "Sovereigns of Vatican City"

Concepts	Semantic similarity
Popes	0.982
Roman Catholic monarchs	0.962
Papal saints	0.938
Swedish monarchs	0.887
Twetury monarchs in Europe	0.885
Heaf state of New Zealand	0.883
Heads state of Fiji	0.883
Heads state of Libya	0.883

31.5 Conclusions

A tool for generating semantic maps representing information content of Twitter conversations in form of coloured graphs has been proposed and tested. Using extensive knowledge bases (i.e. Wikipedia and Wordnet), recent techniques for Natural Language Processing (i.e. Babelnet, Babelfy) and agglomeration techniques (i.e., Relational Fuzzy Clustering), the information content of conversations has been effectively synthesized, in an easily readable form. Future work is in modeling explicitly the temporal evolution of maps.

Acknowledgements The authors thank Agostino Perasole, Michele Crivellari and Antonio Lo Regio for the help during their bachelor's degree at the University of Naples "Parthenope". This work was funded by the University of Naples "Parthenope" (project "Sostegno alla ricerca individuale per il triennio 2015–2017").

References

1. Bezdek, J.C., Ehrlich, R., Full, W.: FCM: the fuzzy c-means clustering algorithm. Comput. Geosci. **10**(2–3), 191–203 (1984)
2. Dave, R.N., Sen, S.: Robust fuzzy clustering of relational data. IEEE Trans. Fuzzy Syst. **10**(6), 713–727 (2002)
3. Hughes, A.L., Palen, L.: Twitter adoption and use in mass convergence and emergency events. In: Proceedings of the 6th International ISCRAM Conference pp. 1–10 (2009)
4. Jungherr, A.: Analyzing Political Communication with Digital Trace Data: The Role of Twitter in Social Science Research. Springer (2015)
5. McLean, H.: Social Media Benefits and Risks in Earthquake Events. Encycl. Earthq. Eng. pp. 1–8 (2015)
6. Mihalcea, R., Tarau, P.: Textrank: bringing order into texts. Association for Computational Linguistics (2004)
7. Miller, G.A., Beckwith, R., Fellbaum, C., Gross, D., Miller, K.J.: Introduction to wordnet: an on-line lexical database. Int. J. Lexicogr. **3**(4), 235–244 (1990)
8. Moro, A., Raganato, A., Navigli, R.: Entity linking meets word sense disambiguation: a unified approach. Trans. Assoc. Comput. Linguist. **2**, 231–244 (2014)
9. Navigli, R., Ponzetto, S.P.: Babelnet: Building a very large multilingual semantic network. In: Proceedings of the 48th Annual Meeting of the Association For Computational Linguistics pp. 216–225 (2010)
10. Stone, B.: Things a Little Bird Told Me: Confessions of the Creative Mind. Grand Central Publishing, USA (2014)
11. Strube, M., Ponzetto, S.P.: Wikirelate! computing semantic relatedness using wikipedia. In: Proceedings of AAAI'06 2, pp. 1419–1424 (2006)

Chapter 32
Preliminary Study on Implications of Cursive Handwriting Learning in Schools

Andreu Comajuncosas, Marcos Faundez-Zanuy, Jordi Solé-Casals and Marta Portero-Tresserra

Abstract The aim of this study is to describe a new database acquired in two different elementary schools of Barcelona province. The study assessed the effect of the type of handwriting learning in general writing performance. In the first school, classical cursive handwriting is learnt while the second one substitutes this skill for keyboarding and print-script handwriting. Analyses in two different groups of age (8–9 and 11–12 years old) for both schools have been performed. A set of 14 different handwriting tasks has been acquired for each student using an Intuos Wacom series 4 tablet plus ink pen and specific software to conduct the analysis. The results revealed that cursive handwriting might improve the handwriting performance by increasing the speed of writing and drawing.

Keywords Cursive handwriting · Print-script handwriting · Biometrics · Learning · Elementary school

A. Comajuncosas (✉) · M. Faundez-Zanuy
Fundació Tecnocampus, Pompeu Fabra University, Av. Ernest Lluch 32,
08302 Mataró, Spain
e-mail: comajunc@tecnocampus.cat

M. Faundez-Zanuy
e-mail: faundez@tecnocampus.cat

J. Solé-Casals
University of Vic – Central University of Catalonia, Vic, Spain
e-mail: jordi.sole@uvic.cat

M. Portero-Tresserra
Pompeu Fabra University, Barcelona, Spain
e-mail: marta.portero@upf.edu

© Springer International Publishing AG 2018
A. Esposito et al. (eds.), *Multidisciplinary Approaches to Neural Computing*,
Smart Innovation, Systems and Technologies 69,
DOI 10.1007/978-3-319-56904-8_32

32.1 Introduction

Many elementary schools teach children how to write beginning with uppercase print-script letters (block letters), then introducing lowercase print-script, and finally cursive (or joined) handwriting. In some cases cursive handwriting is taught before print-script [1], and in some other cases cursive handwriting is discarded, teaching only print-script [2–4].

In recent years there has been a controversy about the convenience of teaching cursive handwriting, or discarding it in order to devote this learning time to more useful skills, as keyboarding. Some qualitative arguments have been presented in favor of both approaches, but we consider that the decision should be made based on experimental evidences. Main arguments for both approaches are the following.

Arguments in favor of learning cursive handwriting [5–7]:

- Cursive handwriting helps to develop fine motor skills.
- Cursive handwriting implicates higher coordination.
- Words are more clearly separated.
- Cursive handwriting requires a higher effort and activation of brain areas related to writing.

Arguments against learning cursive handwriting [3, 4]:

- Print-script is easier to learn.
- Print-script facilitates the reading of books and printed documents.
- Cursive is more difficult to read.
- Time spent on learning cursive can be devoted to learn other skills.
- There are other ways to develop fine motor skills than writing cursive.

32.2 Method

In order to achieve experimental evidence in favor, or against, learning cursive handwriting, we have set up an experiment consisting in acquiring handwriting samples in two different schools and at two different ages (8–9 and 11–12 years old). The first school (we will call it 'Cursive') follows the classic approach of teaching cursive letters while the second one (we will call it 'Print') teaches only print-script handwriting. Samples have been acquired in two classes for every school and every age, making a total number of almost 200 students.

Figure 32.1 shows the template used to acquire samples, which includes 14 different tasks. Both assignments on drawing and writing have been included, in order to compare the students' ability to draw and write. Text is in uppercase print-script, which has been learnt by all students.

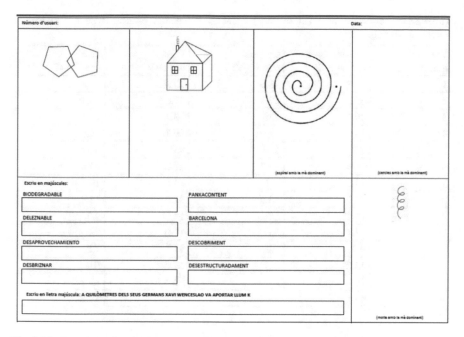

Fig. 32.1 Template of tasks (*pentagons*, *house*, *spiral*, *circles*, *spring*, *words*, and *sentence*)

A graphics tablet and specific software programmed by us (freely available on request) has been used for sample acquisition. In this way, not only the writing outcome has been stored, but also the pen inclination, all the on-air pen movements, and pen pressure over paper, everything as a function of time. All these data allow calculating various parameters which can be useful to assess the writing quality. The most immediate parameter is time required to carry out every task. In-air and on-surface time can be distinguished.

The hypothesis is that if cursive handwriting learning produces a better writing skill, then these students will require less time to perform the tasks than those who just learned print-script writing. On the other hand, if both types of students require the same time, conclusion is that cursive handwriting doesn't cause a better hand control to draw and write.

32.3   Results

Time periods to perform each task have been compared; mainly, between 'Cursive' and 'Print' students, for both age groups, and also between younger and older students, for every school. This last comparison is expected to produce an obvious result, and it is used as a control.

Table 32.1 On-surface time (in seconds) spent as a function of task, type of handwriting schooling, and age

Task number	Cursive, age 8–9	Cursive, age 11–12	Print, age 8–9	Print, age 11–12
1	22.2	18.0	23.5	19.8
2	35.5	32.1	37.6	34.1
3	29.2	22.9	28.1	21.8
4	15.7	12.0	19.7	14.0
5	7.7	5.8	9.9	6.1
6	20.4	13.2	20.2	13.9
7	14.1	9.3	14.8	10.1
8	22.0	13.6	21.6	15.2
9	13.1	8.9	13.9	9.7
10	12.5	8.6	13.0	9.2
11	9.9	7.2	11.1	7.2
12	13.0	8.8	13.6	9.0
13	19.4	13.8	20.4	13.7
14	48.8	33.2	51.1	35.2

Fig. 32.2 Time improvement for older students. 'Cursive' school

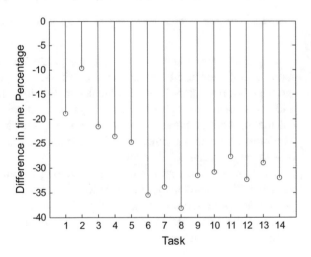

The most significant parameter has been on-surface time (t_down), disregarding on-air time for each task. As expected, older students required less time to complete all tasks, in both schools; between 10 and 40% less, depending on the task. Time periods, in seconds, are summarized in Table 32.1, and time improvement for older students is shown in Figs. 32.2 and 32.3.

On-surface time comparison for both schools shows that, for most of the tasks, students who have learnt cursive handwriting have required less time than students who have learnt only print-script. Results are shown in Figs. 32.4 and 32.5.

Fig. 32.3 Time improvement for older students. 'Print' school

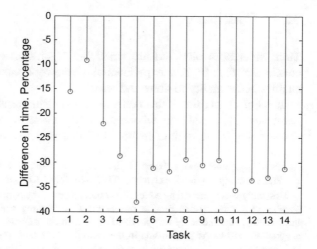

Fig. 32.4 Time improvement for 'cursive' students. Age 8–9

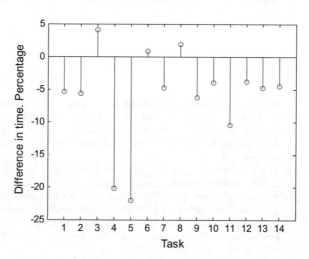

Fig. 32.5 Time improvement for 'cursive' students. Age 11–12

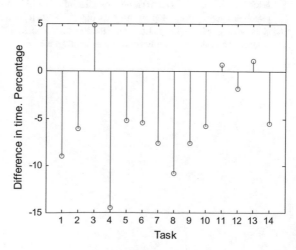

32.4   Conclusion

Time differences observed are small but significant enough. They support the conclusion that in this experimental framework cursive handwriting learning enables better ability to draw and write.

It is not yet verified if this better ability vanishes as students grow old, or if it is permanent. Also, students who learn only print-script may acquire other valuable skills during the long time devoted by other students to cursive handwriting.

Acknowledgements and Privacy Statement This work has been funded by Ministerio de economía y competitividad, Reference: TEC2016-77791-C4-2-R.

This study was approved by ethics inspection officer of the public schools involved according to Catalan government rules for the management of learning activities involving children. Full information about the study was provided to the schools' direction board who supervised the data acquisition, and to the participants in the study, and consent forms were acquired before handwriting and drawing assessments were carried out. All the material collected was anonymized for privacy concerns.

References

1. Bourne, L.: Cursive vs. Printing: Is One Better Than the Other? http://www.blog.montessoriforeveryone.com/cursive-vs-printing-is-one-better-than-the-other.html
2. Russell, H.: Signing Off: Finnish Schools Phase out Handwriting Classes. http://www.theguardian.com/world/2015/jul/31/finnish-schools-phase-out-handwriting-classes-keyboard-skills-finland
3. Van de Geyn, L.: The End of Cursive Writing in Schools? http://www.todaysparent.com/family/education/cursive-writing-in-schools/
4. Boone, J.: Cursive Handwriting Will No Longer Be Taught in Schools Because It's a Big, Old Waste of Time. http://www.eonline.com/news/481596/cursive-handwriting-will-no-longer-be-taught-in-schools-because-it-s-a-big-old-waste-of-time
5. Doverspike, J.: Ten Reasons People Still Need Cursive. http://thefederalist.com/2015/02/25/ten-reasons-people-still-need-cursive/
6. Klemm, W.R.: Why Writing by Hand Could Make You Smarter. https://www.psychologytoday.com/blog/memory-medic/201303/why-writing-hand-could-make-you-smarter
7. Konnikova, M.: What's Lost as Handwriting Fades. http://www.nytimes.com/2014/06/03/science/whats-lost-as-handwriting-fades.html?_r=3

Chapter 33
Italian General Elections of 2006 and 2008: Testing the Direct and Indirect Impact of Implicit and Explicit Attitudes on Voting Behaviour

Angiola Di Conza, Maria Di Marco, Francesca Tanganelli, Ida Sergi, Vincenzo Paolo Senese and Augusto Gnisci

Abstract Two studies were conducted during the Italian General Elections of 2006 ($N = 179$) and 2008 ($N = 607$), to investigate the relationships among implicit and explicit attitudes, voting intention and voting behaviour. Several structural equation models that included direct and indirect effect of implicit and explicit attitudes toward political objects (coalitions and leaders) on voting intention and behaviour were executed to test a prediction model of political preferences and voting behaviour. Notwithstanding some differences, the results of the two studies showed that (i) the implicit evaluations of political objects are more differentiated than the explicit ones; (ii) that implicit attitudes contribute in a specific and additive way to determine the voting intention and behaviour, and (iii) that the effect of the implicit attitude is also mediated by the explicit attitudes. Findings are discussed in the frame of dual cognition models and in the light of the peculiar political scenarios of the considered electoral process.

Keywords Implicit and explicit attitudes · Mediation of the intention · Voting behaviour · Dual cognition models

33.1 Introduction

The social psychological research have increasingly focused its attention on dual cognition models that assumes that social behaviour is determined by the joint activation of two process, known as automatic or deliberate, impulsive or reflective, aware or unaware, implicit or explicit [1–3]. Also the literature on political cognition has been influenced by these theoretical models leading to the consideration

A. Di Conza · M. Di Marco · F. Tanganelli · I. Sergi · V.P. Senese · A. Gnisci (✉)
Department of Psychology, University of Campania "Luigi Vanvitelli", Caserta, Italy
e-mail: augusto.gnisci@unicampania.it

A. Esposito et al. (eds.), *Multidisciplinary Approaches to Neural Computing*,
Smart Innovation, Systems and Technologies 69,
DOI 10.1007/978-3-319-56904-8_33

that even an high deliberative behaviour, as voting, is influenced by both processes [4, 5].

According to this approach, Lodge and colleagues [4, 6] showed that each socio-political object, once encountered and evaluated, becomes affectively charged, and that the associated affect is automatically activated at any time a political stimulus occurs, and influences the following decision making processes. Friese and colleagues [7] showed that implicit and explicit attitudes toward the five major German parties predicted the voting intention and the voting choice, and that implicit attitudes together with the intention improved the voting prediction. Arcuri and colleagues [8] showed that implicit attitudes predicted undecided voters' behaviour. Roccato and Zogmaister [9] showed that automatic evaluations of political objects slightly improve voting prediction over and above explicit attitudes.

Even though these studies have the merit of showing the relevance of the implicit attitudes on voting behaviour, they did not always differentiate the attitudes toward different political objects, as leaders and parties, whereas different studies [10, 11] showed that at the implicit level, the evaluation of different political objects are stored separately, while at a more reflective level, the associations are stronger, probably because of the intervention of cognitive processes of information integration. Moreover, they did not clarify which are the condition that strengthen or weaken the impact of the implicit evaluations.

Starting from the above mentioned literature, and according to the Reflective and Impulsive Model [3], the aims of the present contribution were twofold: (i) investigate if the attitudes toward political objects, that is leaders and coalitions, were influenced by the measurement method, that is implicit and explicit; (ii) to test a prediction model of voting decision, that is if the implicit influence the explicit attitudes [1, 5], and (iii) if the intention represents a mediator of both explicit and implicit attitudes' influence on the voting behaviour [1, 5, 7, 12]. To this aims, implicit and explicit attitudes, intention and vote were evaluated during Italian General Elections of 2006 (Study 1) and 2008 (Study 2), and the same models were tested by means of structural equation models.

33.2 Study 1

This study was carried-out during the Italian General Elections of 2006. For this election, parties formed two main coalitions: a centre-left one ("Unione"), and a centre-right one ("Casa delle Libertà"), under the leadership of Romano Prodi and Silvio Berlusconi, respectively.

33.2.1   Method

Sample. 392 Italian electors took part in this study. The ones having missing values and not voting for one of the two main coalitions were discarded, resulting in a final sample of 179 voters (91 men; age ranging from 18 to 40; $M = 22.7$, $SD = 2.9$).

Procedure. The experiment consisted into two phases. In the first phase, before the election, participants were administered an autobiographical questionnaire, the implicit and explicit attitude measures and asked to indicate their voting intention. Finally, they were informed that they will be contacted again after the election day for a very short follow up. In the second phase, after the election, they were contacted by telephone and asked to indicate the actual vote. All participants signed a written informed consent form which presented the research project, and explained how participants rights and the confidentiality of their data were guaranteed.

Implicit measures. Implicit attitudes toward the two main coalitions and their leaders were assessed by means of two Implicit Association Tests (IAT; αs > 0.78) [13]. The IAT scores were computed following the "D" improved algorithm; positive values indicate a positive evaluation of the centre-right coalition or leader, whereas negative values indicate a positive evaluation of the centre-left coalition or leader. The measures were administered in a counterbalanced order.

Explicit measures. Explicit attitudes were assessed by means of four semantic differential scales [14]. Respondents were asked to evaluate each target object by means of six bipolar adjectives on a seven-point scale. Mean values for centre-left and centre-right objects were subtracted each other to obtain a relative index of preference similar to the implicit score. Therefore, two aggregate of items for each object category were computed, for a total of four indexes: two for the leaders and two for the coalitions (α > 0.72).

Intention and Vote. Voting intention was asked at the end of the session. It was coded Unione $= -1$; no clear intention $= 0$; Casa delle Libertà $= 1$. The actual vote was required by phone after the election day and coded Unione $= 0$ and Casa delle Libertà $= 1$.

33.2.2   Data Analysis and Results

To investigate if the attitudes toward political objects were influences by the measurement method, three measurement models were compared: (i) a 4-factor model, with 4 latent dimensions related to centre-left coalition, centre-right coalition, centre-left leader and centre-right leader respectively; (ii) a 3-factor model, with 3 latent dimension, two for the implicit measures and one for the four explicit measures; and (iii) a 2-factor model, with 2 latent dimensions, one for the implicit measures and one for the explicit one. Results showed that the 3-factor model was

Table 33.1 Comparison between measurement models of implicit and explicit political attitudes

Model	χ^2	df	χ^2_{diff}	RMSEA	NFI	CFI
4-factor model	61.07***	14		0.14	0.94	0.97
3-factor model	74.62***	17	13.51**	0.14	0.95	0.96
2-factor model	131.54***	19	56.92***	0.18	0.92	0.93

Note χ^2 = Chi square; df = degree of freedom; χ^2_{diff} = change in Chi square; *RMSEA* = Root Mean Square Error of Approssimation; *NFI* = Normed Fit Index; *CFI* = Comparison Fit Index; $p < 0.05$; ** $p < 0.01$; *** $p < 0.001$

Table 33.2 Comparison between structural models ($N = 179$)

Model	χ^2	df	χ^2_{diff}	RMSEA	CFI	R^2_{Int}	R^2_{Voto}
1. Baseline model	172.96***	32		0.16	0.95	0.67	0.81
2. Add implicit att. toward leaders on intention	156.31***	31	16.65***	0.15	0.96	0.70	0.81
3. Add implicit att. toward coalitions on intention	145.36***	30	10.95***	0.15	0.96	0.71	0.81
3.1 Add explicit att. toward leaders on vote^	140.59***	29	4.77*	0.15	0.96	0.74	0.85
3.2 Add implicit att. toward leaders on vote^	146.60***	29	4.76*	0.15	0.96	0.72	0.83
3.3 Add implicit att. toward coalitions on vote^	137.92***	29	7.44**	0.14	0.96	0.69	0.80

Note Starting from the full mediation model (baseline model) and adding the direct effect of the implicit attitudes (leaders and coalitions, this order) on the intention; the direct effect of the explicit attitudes on the vote (same order) and finally the direct effect of the implicit attitudes on the vote (same order); ^ the reference model is the Model 3; χ^2 = Chi square; χ^2_{diff} = change in Chi square; *RMSEA* = Root Mean Square Error of Approssimation; *CFI* = Comparison Fit Index; * $p < 0.05$; ** $p < 0.01$; *** $p < 0.001$

the best fitting one, thus suggesting a specific implicit attitude related to the political object and a more general evaluation at the explicit level (see Table 33.1).

To test the prediction model of political preferences and actual voting behaviour, that is if the implicit attitudes influence the explicit ones, and if the intention represents a mediator of attitudes' influence on the voting behaviour, three structural models were compared: (i) a basic model, with implicit attitudes, explicit attitudes and intention considered as same-level predictors; (ii) a full mediation model, with the total mediation of implicit attitudes by the explicit ones and of explicit attitudes by the intention; and (iii) a partial mediation model, with the direct effects of implicit attitudes on the intention and on the behaviour, and of explicit attitudes on the behaviour.

Results showed that the partial mediation model was the best fitting one (see Table 33.2), thus suggesting that implicit attitude determine the explicit one, that both attitudes contribute in a specific and additive way to determine the voting intention, and that the intention mediate the effect of explicit and implicit attitudes on voting behavior (see Fig. 33.1).

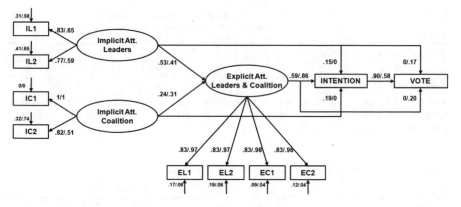

Fig. 33.1 Best fitting measurement and structural model for Study 1/Study 2. Only significant betas are printed

33.3   Study 2

This study was executed to replicate and extend the previous one, it was carried-out during the Italian General Elections of 2008. It is important to notice, that this elections were established and carried out in a very short time (2 months), giving to the electors a sense of urgency and a short time for the elaborations of the new political objects. Indeed, two main parties were proposed instead of the old coalitions, that represented new political aggregations of the old parties: a centre-left one ("Partito Democratico"), and a centre-right one ("Popolo delle Libertà"), under the leadership of Walter Veltroni and Silvio Berlusconi, respectively.

33.3.1   Method

Sample. 1036 Italian electors took part in this study. The ones having missing values and not voting for one of the two main coalitions were discarded, leaving a sample of 607 voters (244 men; age ranging from 18 to 69; $M = 25.0$, $SD = 8.5$).

Procedure and measures. The experiment procedure and measures were identical to the Study 1. All the measures were adapted according to the new coalitions and leaders.

33.3.2   Data Analysis and Results

The same measurement models used into the Study 1 were compared to investigate if the attitudes toward political objects were influenced by the measurement

Table 33.3 Comparison between measurement models of implicit and explicit political attitudes

Model	χ^2	df	χ^2_{diff}	RMSEA	NFI	CFI
4-factor model	108.14***	14		0.10	0.97	0.97
3-factor model	114.08***	17	5.94	0.10	0.96	0.96
2-factor model	144.07***	19	29.99***	0.10	0.95	0.96

Note χ^2 = Chi square; df = degree of freedom; χ^2_{diff} = change in Chi square; *RMSEA* = Root Mean Square Error of Approssimation; *NFI* = Normed Fit Index; *CFI* = Comparison Fit Index; * $p < 0.05$; ** $p < 0.01$; *** $p < 0.001$

Table 33.4 Comparison between structural models ($N = 607$)

Model	χ^2	df	χ^2_{diff}	RMSEA	CFI	R^2_{Int}	R^2_{Voto}
1. Baseline model	58.69***	32		0.07	0.98	0.53	0.64
2. Add implicit att. toward leaders on intention	58.62**	31	0.07	0.07	0.98	0.54	0.64
3. Add implicit att. toward coalitions on intention	58.62**	31	0.07	0.07	0.98	0.51	0.64
3.1. Add explicit att. toward leaders on vote	45.35*	31	13.34***	0.05	0.99	0.53	0.67
3.2. Add implicit att. toward leaders on vote	41.59	30	3.76*	0.05	0.99	0.53	0.68
3.3. Add implicit att. toward coalitions on vote	41.38	29	0.21	0.05	0.99	0.53	0.68

Note Starting from the full mediation model (baseline model) and adding the direct effect of the implicit attitudes (leaders and coalitions, this order) on the intention; the direct effect of the explicit attitudes on the vote (same order) and finally the direct effect of the implicit attitudes on the vote (same order); ^ the reference model is the Model 3; χ^2 = Chi square; χ^2_{diff} = change in Chi square; *RMSEA* = Root Mean Square Error of Approssimation; *CFI* = Comparison Fit Index; * $p < 0.05$; ** $p < 0.01$; *** $p < 0.001$

method. Results confirmed that the 3-factor model was the best fitting one (see Table 33.3).

The same structural models tested into the Study 1 were compared to test the prediction model of voting intention and behaviour (see Table 33.4).

Also in this study, results showed that the partial mediation model was the best fitting one, with some differences (see Fig. 33.1). The implicit attitude toward the leaders and the explicit attitude showed a direct effect on the voting behaviour, and the relation between the voting intention and behaviour was weaker.

33.4 Discussion and Conclusions

The two studies presented here were aimed to test the measurement model of implicit and explicit political attitudes, and to test their role in the prediction of voting behaviour across two different Italian general electoral rounds.

Overall, the results confirmed [10, 11] that at the implicit level attitudes toward coalitions and leaders are differentiated, thus should be considered separately; whereas explicitly they are less differentiated, that is they can be considered as the expression of a general political attitude. As regards the prediction of the voting behaviour, in line with theoretical models [1–3, 12], both studies showed that the impact of the implicit attitudes is partially mediated by the explicit one, and that the intention represents the closest determinant of the behaviour, even if when the electors have few knowledge or experience with the political coalition (Study 2) the intention has a weaker relation with the behaviour and the final decision is related to the general explicit evaluations and the implicit association to the leader.

Future studies should better investigate when and why the implicit attitudes exert their influence on the intention rather than on the vote and vice versa. In our studies, we hypothesize that, in a political scenario connoted by the presence of newly-constituted political aggregations (Study 2), and under the pressure of a new electoral call, that is in absence of a strong automated evaluations, the behaviour, more than the intention, is affected by both deliberate and automatic evaluations. As expected [1, 5], the influence of the implicit attitudes on the explicit attitudes is stronger when individuals do not try to control the implicit evaluations, and the effect of the implicit process extends to the behaviour. Future studies should test the hypothesis that people tend to control the expression of their implicit attitudes more in presence of less frequently evaluated objects under more controlled conditions. Notwithstanding the differences emerged, however, the two studies lead to the conclusion that, even if the specific role of each analysed variable can be different in different political rounds, because of contingent differences in the political scenarios, the involved cognitive processes are similar. Of course, the effects of the implicit component can be more relevant for some electors than for others, depending on their specific individual characteristics; e.g., electors with different characteristics can refer to their implicit attitudes more than others. Future studies should deepen the issue whether the observed effects are shared in a similar way by all the electors' categories or if the effect of the implicit evaluations is moderated by some variables; e.g., variables that able to modify the trustworthiness of the automatic evaluations [1, 5].

References

1. Gawronski, B., Bodenhausen, G.V.: Associative and propositional processes in evaluation: an integrative review of implicit and explicit change. Psychol. Bull. **132**, 692–731 (2006)
2. Smith, E.R., DeCoster, J.: Dual process models in social and cognitive psychology: conceptual integration and links to underlying memory systems. Pers. Soc. Psychol. Rev. **4**, 108–131 (2000)
3. Strack, F., Deutsch, R.: Reflective and impulsive determinants of social behavior. Pers. Soc. Psychol. Rev. **8**, 220–247 (2004)

4. Morris, J.P., Squires, N.K., Taber, C.S., Lodge, M.: Activation of political attitudes: A psychophysiological examination of the hot cognition hypothesis. Polit. Psychol. **24**, 727–745 (2003)
5. Gawronski, B., Galdi, S., Arcuri, L.: What can political psychology learn from implicit measures? Empirical evidence and new directions. Polit. Psychol. **36**(1), 1–17 (2015)
6. Burdein, I., Lodge, M., Taber, C.: Experiments on the automaticity of political belief and attitudes. Polit. Psychol. **27**, 359–371 (2006)
7. Friese, M., Bluemke, M., Wänke, M.: Predicting voting behavior with implicit attitude measures: the 2002 German parliamentary election. Exp. Psychol. **54**, 247–255 (2007)
8. Arcuri, L., Castelli, L., Galdi, S., Zogmaister, C., Amadori, A.: Predicting the vote: implicit attitudes as predictors of the future behaviour of decided and undecided voters. Polit. Psychol. **29**(3), 369–387 (2008)
9. Roccato, M., Zogmaister, C.: Predicting the vote through implicit and explicit attitudes: a field research. Polit. Psychol. **31**(2), 249–274 (2010)
10. Di Conza, A., Gnisci, A., Perugini, M., Senese, V.P.: L'influenza dell'atteggiamento implicito ed esplicito verso leader e partito sul voto delle elezioni europee del 2004 in Italia e delle elezioni politiche del 2005 in Inghilterra. Psicologia Sociale, 2, pp. 301–329 (2010)
11. Gnisci, A., Di Conza, A., Senese, V.P., Perugini, M.: Negativismo politico, voto e atteggiamento. uno studio su un campione di studenti universitari alla vigilia delle elezioni europee del 12–13 giugno 2004. Rassegna di Psicologia **26**, 57–82 (2009)
12. Ajzen, I., Fishbein, M.: The influence of attitudes on behavior. In: Albarracín, D., Johnson, B. T., Zanna, M.P. (eds.) The Handbook of Attitudes, pp. 173–221. Erlbaum, Mahwah, NJ (2005)
13. Greenwald, A.G., McGhee, D., Schwartz, J.L.K.: Measuring individual differences in implicit cognition: the implicit association task. J. Pers. Soc. Psychol. **74**, 1469–1480 (1998)
14. Osgood, C.E., Suci, G., Tannenbaum, P.: The Measurement of Meaning. University of Illinois Press, Urbana, IL (1957)

Chapter 34
Are the Gambler's Fallacy or the Hot-Hand Fallacy due to an Erroneous Probability Estimate?

Olimpia Matarazzo, Michele Carpentieri, Claudia Greco and Barbara Pizzini

Abstract Through two experiments we investigated, in a laboratory setting, whether a series of identical outcomes in a supposed random game would induce the gambler's fallacy or the hot-hand fallacy. By using two indices of fallacy, the choice of a card on which to bet and the probability estimate of the occurrence of a given outcome, we tested explicitly the widely accepted hypothesis that the two fallacies were based on erroneous probability estimates. Moreover, we investigated whether fallacies increase the proneness to bet. Our results support the occurrence of the gambler's fallacy rather than the hot-hand fallacy but suggest that choice and probability estimates are two reciprocally independent processes. Finally, probability estimates predict the amount bet.

Keywords Gambler's fallacy · Hot-hand fallacy · Probability estimates · Decision-making

34.1 Introduction

Humans frequently make errors of judgment when they have to estimate the probability of events and make choices. For instance, when people are faced with outcomes of random sequences, they tend to produce negative or positive autocorrelations between consecutive outcomes that are actually not present. The first case refers to the so-called gambler's fallacy (or fallacy of Monte Carlo), by which people tend to believe that, in a random sequence, a run of identical events (e.g. four consecutive heads in a coin toss) will be followed by the opposite event (e.g. tails). Consequently, as numerous studies have demonstrated [1–4], the gambler's fallacy (GF) can bias decision-making, by leading people to predict that in a random sequence, a future event will interrupt a long enough series of the same event.

O. Matarazzo (✉) · M. Carpentieri · C. Greco · B. Pizzini
Department of Psychology, University of Campania Luigi Vanvitelli, Caserta, Italy
e-mail: olimpia.matarazzo@unicampania.it

© Springer International Publishing AG 2018
A. Esposito et al. (eds.), *Multidisciplinary Approaches to Neural Computing*,
Smart Innovation, Systems and Technologies 69,
DOI 10.1007/978-3-319-56904-8_34

353

A phenomenon opposed to the GF is the hot-hand fallacy (H-HF), also defined as positive recency [5, 6], by which people tend to believe that a random sequence will continue in the wake of recent outcomes [7, 8]. This phenomenon was first investigated by Gilovich et al. [9], in which participants overestimated the likelihood that the basketball players who had just scored would score again, just because they had the "hot" hand.

Both fallacies suggest that people seem to believe that previous events affect the ones that follow, when actually they are reciprocally independent. This belief has been related to the law of small numbers [1] and to the representativeness heuristic [2], according to which people expect that even short random sequences would be representative of larger random sequences. So, if a sequence seems not to conform to the naïve representation of randomness—such as a series of the same event—people tend to believe that it will be balanced by future events, which are opposite to the previous ones and, consequently, tend to incur in the GF. On the other hand, a series of events not conforming to the randomness representation would no longer be seen as random but would be attributed to the agents' ability, i.e. to their "hot hand".

However, since the two fallacies entail opposite predictions in decision-making from the same type of events, many studies have attempted to specify, in addition to or in contrast with the representativeness heuristic, other putative factors leading to the occurrence of one fallacy rather than the other. Ayton and Fischer [6] proposed that prior expectations about the nature of the events could discriminate between the two phenomena. Events somewhat linked to human performance, such as sport scores, would be perceived as more controllable and less random than inanimate events, such as the outcomes of a roulette wheel. Therefore, the first type of event would be likely to elicit the H-HF, whereas the latter would elicit the GF. The results of their studies corroborated this hypothesis: Participants tended to predict red or blue at the roulette wheel on the basis of the GF, while they tended to predict wins and losses related to human performance on the basis of H-HF. Analogous results were obtained by other studies (e.g. [10]).

However, results emerging both from laboratory and field studies suggest that the occurrence of the GF versus H-HF can also be predicted by previous wins or losses in random games. For example, in a laboratory setting, Boynton [11] found that, irrespective of explicit attributions of randomness versus non-randomness of a binary sequence, successful predictions of outcomes were linked to the positive recency effect, while failures were associated to GF. The studies conducted by Croson and Sundali in casino setting [7, 12] showed that gamblers incurred in both fallacies, also at an individual level. They were more likely to incur in GF after a run of the same outcome (especially if they were losing rather than winning) but they were also more likely to bet on numbers that had recently won, in line with H-HF. Moreover, after previous wins, gamblers were more prone to bet and to stake higher. Also Suetens et al. [13], in a study of the Lotto, found that gamblers are influenced by both fallacies. They bet less on numbers drawn in previous weeks, in line with the GF, and bet more on numbers drawn more often in the recent past, in line with the H-HF.

A different interpretation, linking the H-HF to the availability heuristic and the GF to the representativeness heuristic, has been advanced by Braga et al. [14]. According to them, the availability heuristic, requiring less cognitive resources than representativeness heuristic, would underpin H-HF, which entails a simpler cognitive process than GH.

Although the processes underpinning the two fallacies are still blurred, there is a general agreement that both lead to overestimating the probability of the target event (i.e. the event expected to occur). The repercussions of this tendency on decision-making in financial markets [15–17], gambling (see for reviews, [18, 19]), sport [20] or legal contexts [21] have been largely documented. To explicitly test this hypothesis in laboratory, however, the relationship between the prediction of the future event, the probability estimate of such an event, and/or the decision of betting on this event should be examined.

Nevertheless, to our knowledge, no study has so far jointly considered these three indices. Most studies asked participants to predict the next outcome after varying the length of a run of identical outcomes and, in some studies, to indicate the confidence level in their predictions as well (e.g. [3, 11, 14, 18, 22]). A few studies explicitly asked participants to estimate the probability of a given event (e.g. [10, 15, 23]). For example, in a large survey performed with a representative sample of the population living in Germany, Dohmen et al. [15] invited participants to imagine a sequence of eight coin tosses and then asked them to indicate the probability with which events opposite to the last one would occur in the target trial. Results showed that about 21% of the participants fell in the GF, and about 9% gave probability estimates of less than 50% (in line with H-HF), while about 60% gave the correct answer. In a laboratory setting, Navarrete and Santamaria [23] asked participants to estimate the percentage probability of winning when betting on a colour, in conditions in which the number of sets in a roulette problem was varied. They found that increasing the extensional complexity of the problem reduced the occurrence of GF and improved the probability estimate. As already stated, Burns and Corpus [10] investigated both prediction and probability estimates of future outcomes but the two indices were analyzed separately.

Finally, a certain number of studies [4, 24–26] asked participants to predict the future event, after varying the length of a run of the same outcome, and to decide how much to bet on its occurrence. With a few exceptions [25], these studies found that higher bets were positively correlated to the GF.

34.2   Overview of the Study

The present study aimed to investigate, in a laboratory setting, GF and H-HF and their relationship with probability estimates and wager sizes in a fictional gambling task. More precisely, our first aim was to investigate whether a run of the same outcome in a supposed random game would induce GF or H-HF. Moreover, we aimed at investigating whether the induced fallacy would be linked to an increase of the probability of the outcome in conformity with the fallacy, and/or with an increase of the wager on such an outcome.

To this end, we created two experiments, with three between-subject conditions, where participants were invited to play in a drawing card game, in which cards were subdivided in high versus low cards.[1] In both experiments, participants were first presented with 12 supposed computer-generated random drawings, each of them entailing the possibility of winning or losing a certain amount of the available budget. In experiment 1, at the 13th draw, participants had to choose a card (high or low) and how much to bet on it, and to indicate the probability that the chosen card would be drawn. In experiment 2 the order of questions was inverted: Participants had first to indicate the draw probability of a given card and then to choose the card on which to bet and how much to bet. Actually, in each experiment, the 12 drawings were not random but were built in such a way as to present, in two conditions, a final run with four cards of the same type (high or low), while in the third condition (control) only the last card was high (half of the times) or low (the other half). In all three conditions, the last 4 cards drawn always involved a payout. Moreover, in all three conditions, after 12 draws, the initial budget was slightly increased. In this way, we created a procedure that equated the likelihood of occurrence of the two fallacies. In fact, in accordance with some of the studies presented above [6, 11, 12], on the one hand the alleged random sequence should favour the occurrence of the GF; on the other, the winning series should favour the occurrence of the H-HF.

The hypotheses underpinning the study were the following:

If the GF was induced, then in the two conditions with a final run of four of the same cards (high or low), participants should choose a card with the opposite value than the one in the run. If the H-HF was induced, then participants should choose a card with the same value as the one in the run. If no fallacy was induced (as well as in the control condition), then the participants' choices would be random.

In addition, if the card choice was based on, or associated with, the probability estimate about its future drawing, there should be a causal or correlational link between the two variables. More precisely, two results should occur: In the two fallacy-inducing conditions the probabilities of the card conforming to the fallacy should be overestimated, and a positive correlation should occur between the chosen card and the probability estimate of its future drawing.

Finally, if the amount of the wager chosen was based on the probability estimate of winning, and if in the fallacy-inducing conditions the drawing probability of the card on which to bet were to be overestimated, then a causal or correlational link should exist also between these three variables (condition, probability, and wager).

---

[1]As specified in the Materials and Procedure Section of the Experiment 1, high cards were the cards with numbers from 6 to 10 and low cards were those with numbers from 1 to 5. Note that in the deck of cards there were four types of cards, each of them having a number from 1 to 10.

Note that by using two separate indices—choice of the card on which to bet and probability estimate about the outcome of the future draw—to investigate the putative fallacies, we aimed to establish whether they were based on a cognitive heuristic, leading to erroneous beliefs on probability (of which individuals can be aware), or on a simpler type of heuristic, such as a preference heuristic, of which individuals are not aware. In the first case, we expected a relation between the two indices; in the second case, no relation was expected.

The rationale underlying the decision to carry out two experiments whose difference consisted in inverting the order of the questions about choice of the card and probability estimate was the following: In experiment 1, in line with the dual process theories (see for review, [27]), the first question concerned the choice of the card on which to bet because we were interested in bringing to light the outcome of an immediate process of decision-making, possibly preceding a more reflexive judgment process about the probability evaluation, brought to light by the question on probability estimate. Had both indices of GF or H-HF been found, it would have been possible to test their supposed relationship but without establishing a causal link between them, since participants could have chosen the card on which to bet on the basis of the probability estimate of its being drawn, or could have estimated this probability after their choice. This was the first limitation of experiment 1. The second limitation concerned the possibility that the question about the probability estimate that the chosen card would be drawn may have been misunderstood as a question about the confidence of winning. Since in experiment 1 no relation was found either between experimental conditions and probability estimate or between choice of card and probability, we speculated that participants could have actually indicated how lucky they felt rather than provide an "objective" evaluation of probability. In experiment 2, by inverting the order of questions, we tended to overcome these two limitations.

Both experiments were built through E-prime 2.0 software and were carried out with a laptop.

34.3   Experiment 1

34.3.1   Participants

One hundred thirty-five undergraduates (74 male and 61 female) of the Universities of Naples, Caserta, and Salerno participated in this experiment as unpaid volunteers, after signing the informed consent. They were aged from 19 to 28 years ($M = 22.61$; $SD = 2.16$) and were randomly assigned to one of the three conditions ($n = 45$ for each condition).

34.3.2 Materials and Procedure

Participants were informed that they had a budget of 20 tokens to take part in a card game consisting in 13 random drawings from a deck of 40 cards. In the deck there were four types of cards, each of them having a number from 1 to 10. The cards were subdivided into high cards (with numbers from 6 to 10) and low cards (with numbers from 1 to 5). Before each of the first 12 draws, the dealer would establish the amount of the stake and which type of card (high versus low) would win. So, at each draw, participants could win or lose the stake. It was specified that after each draw, the drawn card would be put back into the deck. At the 13th drawing, participants were invited to choose a card (low or high) and to specify how many tokens (from 1 to 10) they wanted to bet on it, by typing their choices on the laptop keyboard. Then, they were asked to indicate, on a scale from 1 to 100, the probability that the card on which they had bet would be drawn in the following draw. After that, the experiment ended and participants were debriefed and thanked.

As said above, the first 12 drawings were not random but had been built to create three conditions: Two fallacy-inducing conditions, and one control condition. In the two fallacy-inducing conditions, the sequence of drawings was built in such a way that the last four drawn cards were always low cards (condition with low card series —LCS) or high cards (condition with high card series—HCS), respectively. The two different series were built in order to distinguish, in the subsequent bet, the GF or the H-HF from a possible preference effect for high or low cards. In both conditions, the last series was always a winning one. In the control condition, the sequence of drawings was built so as to seem representative of a randomly generated series, by avoiding that cards of the same type followed one another more than twice. In order to counterbalance a possible "recency effect", half of the participants saw a high card as the last card, while the other half saw a low card. In all three conditions, after the first 12 drawings, participants reached a budget of 26 tokens.

34.3.3 Results

Table 34.1 shows the distribution of the cards on which participants decided to bet in the three conditions. The chi-square revealed a significant relationship between the experimental conditions and the choice of the card, $\chi^2 = 12.34$, df $= 2$; $p < 0.01$, in line with the GF. In the HCS condition, participants preferred to bet on low cards rather than on high cards, while the contrary occurred in the LCS condition. No significant difference was found in the control condition. However, although a one-way ANOVA conducted on the proportion of participants who

Table 34.1 Experiment 1—Distribution of the chosen card in the three experimental conditions

Condition	Card on which to bet		Total
	High card (%)	Low card (%)	(% = 100)
Control	26 (57.8)	19 (42.2)	45
High card series	14 (31.1)	31 (68.9)	45
Low card series	30 (66.7)	15 (33.3)	45
Total	70 (51.9)	65 (48.1)	135

chose a high card[2] showed a significant effect due to the condition, $F_{2,132} = 6.64$; $p < 0.01$; partial $\eta^2 = 0.091$, LSD pairwise comparisons revealed no significant difference between the control and the LCS conditions ($p = 0.383$). In both conditions participants tended to choose high cards (57.8% and 66.7%, respectively). On the contrary, the condition with HCS was significantly different from the control ($p. < 0.01$) and the LCS ($p. < 0.001$) conditions: The choice of the high card was lower (31.1%) than in the other two conditions. This finding suggests that participants preferred to bet on high cards, but when this tendency was in contrast with GF, the choice did no longer conform to the basic preference and was reversed in the one conforming to the GF.

In order to investigate the relationship between the manipulation of the last series of cards, the choice of the card on which to bet, and the probability estimate of its drawing, we first recoded the probability estimate in a categorical variable with three values: Less than 50%, equal to 50%, and more than 50%. The value of 50% was the correct estimate, whereas the other two values corresponded to subjective probabilities. The value of more than 50% corresponded to the fallacious overestimate, while that of less than 50% signaled incongruence between the choice of the card and the expected outcome. In Table 34.2 the distribution of the three-value probability estimate as a function of the experimental conditions and the choice of the card on which to bet was reported. Neither the chi-square conducted to assess the relationship between the experimental conditions and the probability estimate ($\chi^2 = 4.23$, df = 4; $p = 0.376$), nor the Phi coefficient used to assess the association between the choice of the card and the probability estimate ($\phi^2 = 0.158$; $p = 0.186$) were significant.[3] In the latter case we used a measure of association since the direction of the relationship between the two variables could not be clearly determined. Participants may have chosen the card on which to bet on the basis of

---

[2]At the request of a reviewer, we specified the following: Since the dependent variable was dichotomous (high versus low card), it was recoded as a single dummy variable: 1 = high card; 0 = low card. Then, the one-way ANOVA was performed on the proportion of high cards chosen by the participants (note that the proportion of low cards can be obtained through subtraction).

[3]Note that no significant effect emerged even from the analyses performed by treating the probability estimate as a continuous variable. The linear regression carried out to test the effect of experimental conditions (coded as two dummy variables) on probability estimate showed no relationship between the variables. Even the point-biserial correlation coefficient did not show any significant relationship between the choice of the card (coded as dummy variable) and the probability of its drawing.

Table 34.2 Experiment 1—Distribution of the probability estimate (recoded as a nominal scale) of card drawing in function of the experimental conditions and the choice of the card on which to bet

Condition	Probability estimate of drawing the chosen card				Total (% = 100)
	Choice of the card	Less than 50 (%)	50 (%)	More than 50 (%)	
Control	High card chosen	5 (19.2)	15 (57.7)	6 (23.1)	26
	Low card chosen	6 (31.6)	7 (36.8)	6 (31.6)	19
	Total Control	11 (24.4)	22 (48.9)	12 (26.7)	45
High card series	High card chosen	4 (28.6)	5 (35.7)	5 (35.7)	14
	Low card chosen	7 (22.6)	9 (29)	15 (48.4)	31
	Total HCS	11 (24.4)	14 (31.1)	20 (44.4)	45
Low card series	High card chosen	8 (26.7)	12 (40)	10 (33.3)	30
	Low card chosen	5 (33.3)	4 (26.7)	6 (40)	15
	Total LCS	13 (28.9)	16 (35.6)	16 (35.6)	45
Total HCC		17 (24.3)	32 (45.7)	21 (30)	70
Total LCC		18 (27.7)	20 (30.8)	27 (41.5)	65
Overall Total		35 (25.9)	52 (38.5)	48 (35.6)	135

Legenda: HCS = High card series; LCS = Low card series; HCC = High card chosen; LCC = Low card chosen

the probability estimate of its drawing or may have estimated this probability after their choice. However, the results suggest that choice and probability estimates are reciprocally independent.

Finally, to examine whether experimental conditions affected wager amount, we carried out a one-way ANOVA, which did not show any significant effect, $F_{2,132} = 0.52$; $p = 0.989$; partial $\eta^2 = 0.000$: in all the three conditions the average wager was about 5 tokens. On the contrary, the correlation between probability estimate and wager amount was positively significant, $r = 0.343$; p (2-tailed) < 0.001.

34.3.4   Discussion

The results of experiment 1 are quite puzzling. They showed evidence in line with the GF, revealing that such a fallacy was strong enough to counteract and overcome the preference for choosing high cards, which is manifested in the control condition, whereas in the condition with the low card series the choice was congruent with the GF and the preference for high cards. Only one third of participants made choices in conformity with the H-HF. However, our results suggest that choice and probability estimates are based on two reciprocally independent processes. An alternative hypothesis is that participants could have misunderstood the question about the probability estimate that the chosen card would be drawn, as a question about the confidence level in their choice, or about the subjective probability of a good outcome. The order of the two questions could have favored such misinterpretation. Even the findings that about a quarter of participants underestimated the probability of card drawing across all the conditions and that in the control condition, the correct estimate was only provided by about half of participants may support the hypothesis of misinterpretation. Experiment 2 was set up to compare the two hypotheses.

34.4 Experiment 2

This experiment was carried out to establish whether the choice of cards congruent with GF or H-HF depended or not on erroneous probability estimate. To this end, we maintained the procedure of experiment 1 but inverted the order of the final instructions: We first asked participants to evaluate the probability that a card opposite to the last they had seen would be drawn, and then to choose the card on which to bet and how many tokens to bet. In this way, we asked participants to evaluate the probability estimate independently of the choice and, therefore, we aimed to prevent a misinterpretation of the question as subjective belief in individual good luck.

It should also be noted that from the perspective of the dual process theories, such a procedure, by asking participants to first provide a judgment of probability and then to choose, would facilitate a decision-making process based on System 2 rather than on System 1. Therefore, this procedure would favor congruence between the two indices of fallacy we investigated: A judgment-based index and a behavioral index. If this congruence was not found, it would be inferred that GF and/or H-HF lay on two separate processes and that the explicative hypotheses in term of erroneous probability estimate should be revised and integrated.

34.4.1 Participants

One hundred ninety undergraduates (85 male and 105 female) of the Universities of Naples, Caserta, and Salerno participated in this experiment as unpaid volunteers, after signing the informed consent. They were aged from 19 to 33 (M = 22.99; SD = 2.79) and were randomly assigned to one of the three conditions (60 to each fallacy-inducing condition and 70 to the control condition).

34.4.2   Materials and Procedure

This experiment was in response to Experiment 1: Therefore, the materials and procedures were the same except for the order of the final questions aimed at investigating the GF and the H-HF. As above, at the thirteenth draw, participants were asked to indicate: (1) the probability that a card with a value opposite to the last one would be drawn; (2) the type of card on which they wanted to bet; (3) how many tokens they wanted to bet. More precisely, participants having seen a high card at the 12th draw were asked to indicate the probability of a low card, while participants having seen a low card at the 12th draw were asked to indicate the probability of a high card.

34.4.3   Results

Two causal models were tested: A model in which the experimental conditions affected the choice of the card via probability estimate, and a model in which the experimental conditions affected the wager amount via probability estimate.

The first model was tested initially by recoding the probability estimate as a three-value categorical variable. Recoding the probability estimate in a categorical variable aimed to distinguish the correct estimate (50%) from those congruent with the two fallacies. Since we asked participants to estimate the probability of the type of card opposite to the last drawn, estimates of more than 50% would be congruent with GF and those of less than 50% would be congruent with the H-HF. Note that in the control condition, half of participants estimated the probability of the high card and the remaining half the probability of the low card; in the high card series condition, participants estimated the probability of the low card, while the opposite was requested in the low card series condition.

In Table 34.3 the distribution of the choice of the card on which to bet as a function of the experimental conditions and the probability estimate of its being drawn was reported.

Since in the control condition half of participants were asked to estimate the probability of low card, and the other half the probability of high card, we performed the statistical analyses by splitting the sample based on the last card drawn. In this way, we compared the condition with low card series to the control condition with last-drawn low card, and the condition with high card series to the control condition with last-drawn high card. The two chi-squares performed to examine whether experimental conditions affected probability estimates did not reveal any significant effect: $\chi^2 = 2.51$, df = 2; p = 0.285, for conditions with last-drawn high card(s); $\chi^2 = 1.56$, df = 2; p = 0.459, for conditions with last-drawn low card(s). Irrespective of the experimental conditions, the participants' answers did not significantly differ between correct estimate, over- and underestimation of the probability. The two chi-squares carried out to examine whether

Table 34.3 Experiment 2—Distribution of the choice of the card on which to bet in function of the experimental conditions and the probability estimate (recoded as a nominal scale) of its drawing

Condition	Card on which to bet			
	Probability estimate (%)	High card	Low card	Total (%)
Control with last-drawn high card	<50	7	4	11 (31.4)
	50	10	8	18 (51.4)
	>50	4	2	6 (17.1)
	Total (%)	21 (60)	14 (40)	35 (100)
Control with last-drawn low card	<50	4	5	9 (25.7)
	50	12	4	16 (45.7)
	>50	6	4	10 (28.6)
	Total (%)	22 (62.9)	13 (37.1)	35 (100)
High card series	<50	12	12	24 (40)
	50	6	15	21 (35)
	>50	3	12	15 (25)
	Total (%)	21 (35)	39 (65)	60 (100)
Low card series	<50	6	11	17 (28.3)
	50	16	4	20 (33.3)
	>50	19	4	23 (38.3)
	Total (%)	41 (68.3)	19 (31.7)	60 (100)
Overall total (%)		105 (55.3)	85 (44.7)	190 (100)

N.B. In the conditions with last high card, the probability referred to the low card drawing; in those with last low card, it referred to the high card drawing

experimental conditions affected the choice of the card did not reveal any significant effect for conditions with last-drawn low card(s), $\chi^2 = 0.29$, df $= 1$; p $= 0.586$, and a significant effect for conditions with last high card(s), $\chi^2 = 5.60$, df $= 1$; p < 0.01. Participants in the condition with high card series chose more low cards than participants in the control condition (with only the last-drawn high card), in line with the GF. On the contrary, when the last card(s) were low, participants in both conditions tended to choose high cards. The two chi-squares conducted to examine the influence of probability estimate on the choice of the card on which to bet revealed a significant effect of probability when the last-drawn card(s) were low, $\chi^2 = 12.46$, df $= 2$; p < 0.01, but did not show any effect when the last-drawn card (s) were high, $\chi^2 = 2.61$, df $= 2$; p $= 0.271$. When the last-drawn card(s) were low and participants were asked to estimate the probability of a high card, those who gave correct or overestimated probabilities tended to choose high cards, whereas those who gave underestimated probabilities tended to choose both types of cards. In the conditions where last-drawn card(s) where high and it was requested to estimate the probability of a low card, the participants' choices were not affected by their estimates. Even those who gave correct or overestimated probabilities on low cards tended to bet on both types of cards.

Although the results of the chi-square tests were clear enough, we also tested a causal model of the relationship between experimental conditions and choice of the card, via probability estimate treated as continuous variable. To this end, we conducted two mediation analyses using the INDIRECT macro developed by Preacher and Hayes [28]. The first analysis was performed by selecting the conditions with last-drawn high card(s). The second analysis was performed by selecting the conditions with last-drawn low card(s).

In both analyses they were coded as a dummy variable ($1 =$ inducing-fallacy condition; $0 =$ control condition) and introduced in the regression analyses as predictor; probability estimate was introduced as mediator, and the chosen card, coded as dummy variable ($1 =$ high card, $0 =$ low card), was the criterion variable. In line with the results of the chi-square tests, the results of the first analysis showed only a total effect of the condition on the choice of the card, $B = -1.024$; S. E. $= 0.438$; $p = 0.019$; Wald $= 5.458$, analogous to the one already reported. The condition did not affect the probability estimate and the latter did not affect the choice of the card. The results of the second analysis showed only a direct effect of the probability estimate on the card choice, $B = 0.043$; S.E. $= 0.017$; $p = 0.010$; Wald $= 6.573$: the choice of the high card increased as probability estimate increased. The condition did not affect the probability estimate or the choice of the card.

To test the causal model according to which the experimental conditions affected the wager amount via probability estimate, we conducted two mediation analyses using the MEDIATE macro [29], following a procedure analogous to the above-mentioned one. However, since in this case the dependent variable was continuous, the mediation analyses were performed through linear regressions rather than logistic regressions. The first analysis, with the conditions with last-drawn high card(s) as predictor, did not reveal any significant effect. The second analysis, with the conditions with last-drawn low card(s) as predictor, showed only a significant effect of the probability estimate on the wager amount, $B = 0.063$; S.E. $= 0.025$; $t = 2.475$; $p = 0.015$: The more the probability estimate increased, the more the wager amount increased. No other effect was found.

34.5 Discussion and Conclusion

This study had three goals: (1) to establish whether a winning run would be more likely to elicit the GF or the H-HF in a fictional gambling game; (2) to investigate whether fallacies increase the proneness to bet; (3) to test the hypothesis, widely accepted to date, that the GF or the H-HF are based on an incorrect estimate of the probability of a given event that, in turn, lays on judgment heuristics, such as the representativeness heuristic. The third goal was the most relevant from a theoretical point of view, and we aimed to achieve it by using two different indices of fallacies: A behavioral index (the choice of the card on which to bet in the gambling game) and a cognitive index (the probability estimate of drawing a given card). Through two different experimental procedures, we varied the order with which these indices were investigated to prevent possible misinterpretations of the instructions. The results of the two experiments are congruent and allow the formulation of a single explicative hypothesis. They suggest that deciding on which card to bet was a process based mainly on the preference for high cards, which in turn could be based on the simple heuristic that "high is better than low". Only the occurrence of a long series of high cards seems able to contrast and overcome this tendency by reversing the preference for high cards in favor of low cards. The choices consonant with the GF clearly emerged only in these conditions, since in the condition with low card series, choosing a high card is consonant with both the preference heuristic and the GF. Thus, the two processes are indistinguishable.

Is the GF based on overestimating the probability of the target outcome? Our results suggest a negative answer. First of all, in both experiments, the card manipulation did not affect the probability estimates, which were almost equally distributed between correct, overestimated, and underestimated answers, with a slight, but not significant, tendency toward correct answers in the control condition. If the probability estimate were the only index of fallacies, this finding could have been interpreted as an evidence that, regardless of the experimental manipulation, almost two third of the participants spontaneously incurred in the GF, in the H-HF or, at least, in negative or positive recency. However, by using both a behavioral and a cognitive index of fallacy, we were able to establish whether the participants' beliefs about the probability resulted or not in a congruent choice. In experiment 1, no correlation was found between the two variables. In experiment 2, we found that probability estimate affected choice in the conditions with last-drawn low card(s). Nevertheless, this finding seems to be an effect of the preference for high card rather than a judgment outcome. Participants who correctly estimated or overestimated the probability of a high card tended to bet on a high card, apparently conforming their behavior to previous estimates. However, participants who underestimated the probability of high card distributed their choices between the two types of card, instead of focusing them on low card, as they should have done if they had been consistent with their estimates. Moreover, when the probability of low card had to be estimated, participants did not adapt their choices to previous estimates: Regardless of the values attributed to the probability of drawing a low card, they betted almost equally on both types of cards. These findings further corroborate the idea that the incidence of probability on choice emerges only when there is congruence between preference-based choices and probability-based choices. When there is incongruence between the two types of choice, the participants' behavior seems to follow two opposing decision criteria: The congruence between judgment and choice, and the matching between preference and choice.

Although these results need further studies to be substantiated, at present they corroborate the idea, firstly advanced by Zajonc [30], that affect heuristic (such as the preference for a giving stimulus) is the basic and automatic process informing decision-making. The judgment heuristics, such as the representativeness heuristic thought to underlie the GF and/or the H-HF, are somewhat more complex cognitive strategies. Hence, only in some circumstances are they able to overcome the influence of affect heuristic on choices.

As to the second goal of our study, whether fallacies increased the proneness to bet, we found that wager size increased in function of probability estimate. However, since the experimental manipulation did not affect the probability estimate, but only affected the choice of the card on which to bet, no relationship could be established between the fallacy-inducing conditions and the wager amount. To attempt to understand this finding, we need to consider that in control conditions of both experiments, in which the sequences were built according to the randomness representation, less than half of participants estimated probability correctly. The others seemed to formulate their judgments on the basis of positive or negative recency: The probability of a future outcome was over- or under-estimated in function of the last outcome observed. Maybe the experimental manipulation did not affect the probability estimate since participants "naturally" followed the principles underlying the H-HF (positive recency) and the GF (negative recency). Again, further studies are needed.

In conclusion, our results do not corroborate any of the two causal models tested: The occurrence of a relationship between experimental manipulation and choice mediated by probability estimates, the occurrence of a relationship between experimental manipulation and proneness to bet via probability estimates. They suggest that choice and probability estimates are two reciprocally independent processes and that the choice is largely dependent on affect (preference) heuristic.

34.6   Ethics Approval

This study was approved by the Ethics Committee of the Department of Psychology of the University of Campania Luigi Vanvitelli (previously, Second University of Naples)—approval number: 23/2016.

References

1. Tversky, A., Kahneman, D.: Belief in the law of small numbers. Psychol. Bull. **76**, 105–110 (1971)
2. Kahneman, D., Tversky, A.: Subjective probability: a judgment of representativeness. Cognit. Psychol. **3**, 430–454 (1972)
3. Xue, G., He, Q., Lei, X., Chen, C., Liu, Y., Chen, C., Lu, Z.L., Dong, Q., Bechara, A.: The gambler's fallacy is associated with weak affective decision making but strong cognitive ability. PLoS ONE **7**, 1–5 (2012)
4. Studer, B., Limbrick-Oldfield, E.H., Clark, L.: 'Put your money where your mouth is!': effects of streaks on confidence and betting in a binary choice task. J. Behav. Dec. Mak. **28**, 239–249 (2015)

5. Jarvik, M.E.: Probability learning and a negative recency effect in the serial anticipation of alternative symbols. J. Exp. Psychol. **41**, 291–297 (1951)
6. Ayton, P., Fischer, I.: The hot hand and the gambler's fallacy: two faces of subjective randomness? Mem. Cognit. **32**, 1369–1378 (2004)
7. Croson, R., Sundali, J.: The gambler's fallacy and the hot hand: empirical data from casinos. J. Risk. Uncertainty **30**, 195–209 (2005)
8. Xu, J., Harvey, N.: Carry on winning: the gamblers' fallacy creates hot hand effects in online gambling. Cognition **131**, 173–180 (2014)
9. Gilovich, T., Vallone, R., Tversky, A.: The hot hand in basketball: on the misperception of random sequences. Cognit. Psychol. **17**, 295–314 (1985)
10. Burns, B.D., Corpus, B.: Randomness and inductions from streaks: "gambler's fallacy" versus "hot hand". Psycho. B. Rev. **11**, 179–184 (2004)
11. Boynton, D.M.: Superstitious responding and frequency matching in the positive bias and gambler's fallacy effects. Organ. Behav. Hum. Dec. **91**, 119–127 (2003)
12. Sundali, J., Croson, R.: Biases in casino betting: The hot hand and the gambler's fallacy. Judg. Dec. Mak. **1**, 1–12 (2006)
13. Suetens, S., Galbo-Jørgensen, C.B., Tyran, J.R.: Predicting lotto numbers: a natural experiment on the gambler's fallacy and the hot hand fallacy. J. Eur. Econ. Assoc. https://www.eeassoc.org/doc/upload/Suetens-Galbo-Tyran20150511042958.pdf. (2015)
14. Braga, J., Ferreira, M.B., Sherman, S.J.: Disentangling availability from representativeness: gambler's fallacy under pressure. In: Andrade, C., Garcia, J., Fernandes, S., Palma, T., Silva, V.H., Castro, P. (eds.) Research directions in social and organizational psychology, pp. 109–124. Edições Sílabo, Lisboa (2013)
15. Dohmen, T., Falk, A., Huffman, D., Marklein, F., Sunde, U.: Biased probability judgment: evidence of incidence and relationship to economic outcomes from a representative sample. J. Econ. Behav. Organ. **72**, 903–915 (2009)
16. Rabin, M.: Inference by believers in the law of small numbers. Q. J. Econ. **117**(3), 775–816 (2002)
17. Stöckl, T., Huber, J., Kirchler, M., Lindner, F.: Hot hand and gambler's fallacy in teams: evidence from investment experiments. J. Econ. Behav. Organ. **117**, 327–339 (2015)
18. Clark, L.: Decision-making during gambling: an integration of cognitive and psychobiological approaches. Phil. Trans. R. Soc. B **365**, 319–330 (2010)
19. Fortune, E.E., Goodie, A.S: Cognitive distortions as a component and treatment focus of pathological gambling: a review. Psychol. Addict. Behav. **26**, 298–310 (2012)
20. Koehler, J.J., Conley. C.A.: The "hot hand" myth in professional basketball. J. Sport Exerc. Psychol. **25**, 253–259 (2003)
21. Chen, D.L., Moskowitz, T.J., Shue, K.S.: Decision-making under the gambler's fallacy: evidence from asylum judges, loan officers, and baseball umpires. Fama-Miller Working Paper. http://ssrn.com/abstract=2538147 or http://dx.doi.org/10.2139/ssrn.2538147. (2016). Accessed SSRN
22. Clark, L., Studer, B., Bruss, J., Tranel, D., Bechara, A.: Damage to insula abolishes cognitive distortions during simulated gambling. PNAS **111**, 6098–6103 (2014)
23. Navarrete, G., Santamaría, C.: Adding possibilities can reduce the gambler's fallacy: a naïve-probability paradox. J. Cognit. Psychol. **24**, 306–312 (2011)
24. Roney, C.J., Trick, L.M.: Grouping and gambling: a gestalt approach to understanding the gambler's fallacy. Can. J. Exp. Psychol. **57**, 69–75 (2003)
25. Lyons, J., Weeks, D.J., Elliott, D.: The gambler's fallacy: a basic inhibitory process? Front. Psychol. **4**, 72 (2013)
26. Marmurek, H.H.C., Switzer, J., D'Alvise, J.: Impulsivity, gambling cognitions, and the gambler's fallacy in university students. J. Gambl. Stud. **31**, 197–210 (2015)
27. Evans, J.S.B., Stanovich, K.E.: Dual-process theories of higher cognition: advancing the debate. Perspect. Psychol. Sci. **8**, 223–241 (2013)

28. Preacher, K.J., Hayes, A.F.: Asymptotic and resampling strategies for assessing and comparing indirect effects in multiple mediator models. Behav. Res. Method **40**, 879–891 (2008)
29. Hayes, A.F., Preacher, K.J.: Statistical mediation analysis with a multicategorical independent variable. Brit. J. Math. Stat. Psychol. **67**, 451–470 (2014)
30. Zajonc, R.B.: Feeling and thinking: preferences need no inferences. Am. Psychol. **35**, 151–175 (1980)

Chapter 35
When Intuitive Decisions Making, Based on Expertise, May Deliver Better Results than a Rational, Deliberate Approach

Mauro Maldonato, Silvia Dell'Orco and Raffaele Sperandeo

Abstract In the last 30 years, the systematic analysis of human thought has provided new evidences on intuition's nature. It has been observed in experimental level that in front of decision-making problems, most people unknowingly adopt adaptive solutions that are different from logical inferences of normative rationality. To cope with the temporal and cognitive limitations, humans always use heuristic strategies that allow them to gather quickly useful information for survival. Naturally formal logic can lead to adequate choices, but its processes are slow and cognitively expensive. In this paper we intend to show how, in specific situations and contexts, the paths of formal logic and of natural logic (heuristics, intuitions and so on) diverge dramatically.

Keywords Intuition · Decision-making · Heuristics · Natural logic · Rationality

35.1 Introduction

Classical theories of decision, such as the *Expected Utility Hypothesis* [1], defined the decision behavior as an integration process of information in which, through the *Weighted Additive Strategies* (WADD), it was always possible to maximize our own expected utility [2]. Indeed, because of the cognitive-computational and environmental limits that characterize most of everyday situations, individuals often

M. Maldonato (✉) · S. Dell'Orco · R. Sperandeo
Department of Human Sciences DISU, University of Basilicata, Potenza, Italy
e-mail: mauro.maldonato@unibas.it

S. Dell'Orco
e-mail: silvia.dellorco@unibas.it

R. Sperandeo
e-mail: raffaele.sperandeo@gmail.com

© Springer International Publishing AG 2018
A. Esposito et al. (eds.), *Multidisciplinary Approaches to Neural Computing*,
Smart Innovation, Systems and Technologies 69,
DOI 10.1007/978-3-319-56904-8_35

rely, rather than on logical-formal strategies [3], on an automatic cognitive processing: particularly on intuition, a form of instinctive and unconscious knowledge [4, 5] that allows us to look at things in new and often decisive ways. Intuition can be understood as an expression of our ecological rationality [6, 7] that takes over when we are not able to assess the available data. Thanks to it, we can process, quickly and without great effort [8], a lot of information sedimented in our memory, urging an immediate and often reliable recognition of the present situation. The efficacy of this process is proportional to the experience level we have in that particular domain [9]. Just think of a chess champion who, after a quick look, performs the best move possible in that situation; or a doctor who, in an emergency situation, recognize the forthcoming risk of life in a patient [10]; or, still, a manager who can predict the consumers' response to a new marketing initiative [11, 12]. The ability to distinguish between thousands of different situations and objects is one of the fundamental tools of the expert, and also the main source of his insights. In the scope of "Naturalistic Decision-Making" [13]—a research program that studies how experts decide in situations characterized by various constraints as time pressure, incomplete knowledge of the alternatives, emotional tension, uncertainty, ill-defined goals or high stakes—a bunch of experiments have shown that experts, unlike the beginners decide without evaluating analytically the pros and cons of each option [13, 14]. In fact, they can overcome attentional and procedural memory limits through the internalization of cognitive processes that have shown to be useful in the past [9] and through a process of recognition of the situation that occurs comparing the alternatives and the potential courses of action according to certain criteria of acceptability. In particular, they consider situations holistically [15]: they identify intuitively the objectives to be pursued, the most important clues to observe and monitor, the possible evolutions of the situation and action plans to be followed [16].

When an expert makes a decision, he 'photographs' the present situation: the association between detected clues previous experience allows him to define quickly a possible plan of action [13]. The cognitive processing focuses entirely on the operation of that choice regarding the perceived environment, allowing also a higher control on the course of the chosen action. So, if the systematic comparison between different solutions creates an ideal solution, the investment on a reasonable solution leaves the opened possibility that it may be modified. An eloquent example is the so called "evidential paradigm of the medical semiotics" [17]. Which is not so based on analytical reasoning, but on an intuitive noesis that allows the expert doctor to make a diagnosis based on the emergence of superficial symptoms that would say nothing to inexperienced people. No doctor makes diagnosis based only on standardized diagnostic protocols. It is a kind if knowledge that is neither formalized nor communicable that, in addition to individual inclinations, also requires imponderables elements such as flair, a good glance and so on [18]. Some elements unhide only to careful and experienced observer.

35.2 Instinctively Understand: The Role of Intuition

But what is an intuition? Is it creativity, tacit knowledge, implicit learning and memory, sixth sense, heuristics, emotional intelligence? Intuition has characteristics in common with these and other definitions [19, 20]. It takes place almost instantaneously and consists of emotional and somatic processes, without any role (at least apparently) of the conscious and logical-deductive thinking [21]. An intuition, in fact, almost always presents a somatic correlate: a gut feeling that comes suddenly freeing us from our doubts. Although based on a limited amount of information, especially in stressful situations or limited time conditions, intuition links our actions with the experiences stored in our long-term memory [22]. Of course, intuitive thinking without reasoning can be inaccurate and misleading. But we do not make mistakes only when we rely on our intuition. We also make mistakes when we reflect too much on what to do, because we 'smother' our visceral sensations, depriving us of their 'wisdom'. This phenomenon, called *choking*, manifests itself particularly in the sensory-motor skills of experts, for example of an artist or a professional athlete [23]. In fact, if beginners need to focus on every technical detail of their performances, the same is not true for professionals, because controlling an automated task could be counterproductive for them [24]. Indeed, when executing a performance in conditions of psychological pressure (like an important debut, audition or exam), professionals will tend to focus more on technical details usually performed automatically, since they are acquired by experience. This increased focus on execution of actions step by step damages the automated abilities, producing a change in routine that often results in sub-optimal performance [25]. In that way an actor becomes insecure on the acting of his lines, a dancer loses fluidity in his movement, a baseball player has more difficulty recovering the ball and so on. In short, the margin for error increases greatly, there's a lack in the natural flow of the performance and the grace of talent disappears [26]. That's why at certain levels of experience we can stop thinking.

35.3 Non Compensatory Strategies and Dual Process Theories of Reasoning

The increased interest in the intuitive decision-making strategies aroused a parallel interest for preparation of alternative decision-making strategies to the normative ones. Among them there is the Consistency-Maximizing Strategy (CMS) according to which the decision-making process consists of three phases [27]: (1) in front of a decision-making situation, people would activate immediately a sort of temporary network of important information that are available in the memory to form a first mental representation of the situation; (2) automatic cognitive processes reduce the incoherences between network information by creating a coherent representation of

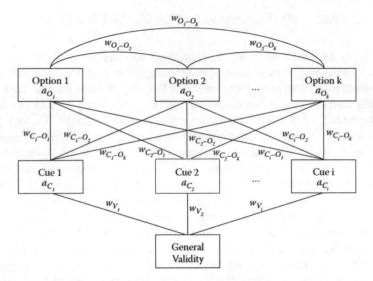

Fig. 35.1 A general model for basic probabilistic inferences. *Boxes* represent nodes; *lines* represent links, which are all bidirectional. Connection weights can range from −1 to +1 and are labelled **w**. Using the iterative updating algorithm, coherence is produced in the network by changing activations (*a*). The special node "General Validity" has a constant activation of +1 and is used to supply the network with energy. The indexes **o** and **c** refer to option and cue n des, the index **v** refers to connections with the general validity node (27, pp. 218)

the situation; (3) people will represent to themselves those problems in which a decisive option prevails (Fig. 35.1).

Current studies on intuition have roots in the research program *Heuristics and Biases Approach* of Kahneman and Tversky that, since the seventies of the twentieth century, has experimentally investigated the deviations of human decisions from normative models of rationality [28, 29]. These investigations have shown that, in many tasks that typically require a type of analytical reasoning, people rely on intuitive inferences [30]. One possible explanation is that the concept of normative rationality does not coincide with that of adaptive rationality [31], or that the utility maximization for the individual is different from the utility maximization for the species. From heuristics research on decision-making, research has seen a proliferation of approaches that go beyond the formal cognitive processes deliberated by neoclassical rationality. Research lines incorporating intuition into decision making processes are those defined *dual-process theories* [32–37], which since 1970 have become common in various areas of cognitive and social psychology. One example, among others, is the distinction between automatic and controlled processes [38], heuristic versus systematic thinking [39], conscious versus unconscious cognition [40], or affect versus cognition.

Traditionally, analytical thinking and intuition were considered alternative cognitive processes: the first one characterized by precise formal reasoning and compensatory strategies; the second one is fast, not formal, indescribable and often unconscious in producing answers to problems [41]. The dual-process theories of reasoning have grown in recent years [42]. For example, the *Heuristic Analytic Theory* of Evans [32], unlike the theory of Tversky and Kahneman [43], suggests that the preconscious heuristics would have the function of selecting the representations of a decision problem. For their part, the rules of rationality theories do not provide a satisfactory evaluation method of reasoning and often lead to irrational conclusions. Evans and Over [44] propose two distinct forms of rationality: the first (System 1) is fast, implied, associative, pragmatic, based on previous experiences and not aware. Its function is domain-specific (the knowledge gained by this system are developed in highly specialized sectors) and its domain-general mechanism can be compared to a neural network in which knowledge derives from the activation of particular units of the network and not from rules of specific-content. System 2 is unitary, analytical, slow, conscious, sequential and explicit. The biases would ensue from the fact that in the first heuristic processing phase the information logically relevant can be neglected, or even omitted. Sloman is the author of another *dual-process theory,* [37] according to which cognitive processes are characterized by two systems: associative system and rule-based system. The first one is based on temporal contiguity levels and relations of similarity acquired by personal experience: a system that is fast, automatic, concrete and of which we only know the final outcome. The second, the rule-based system, draws inferences from analytical-formal processes and is aware. According to Sloman, these two reasoning systems, even though produce opposite solutions, are not separated, but act simultaneously: this simultaneity is defined as "Criterion S".

Stanovich and West [45] combine the *Cognitive Experiential Self-Theory* [35] with the subsequent theories and present a unified theory based on two distinct reasoning systems also called here "System 1" (experiential) and "System 2" (cognitive). The first one provides intuitive responses to problems, while the second one monitors and, if necessary, correct these answers. The primacy of System 1 prevents individuals to think about an issue according to its logical properties. Stanovich [46] defines this tendency to contextualize automatically problems as "fundamental computational bias". According to the authors, these differences in cognitive abilities would correspond to two different kinds of intelligence: analytic intelligence (measured by psychometric tests) and interactional intelligence (social pragmatic intelligence). Among the three dual-process theories, that of Stanovich and West [45] provides the most detailed explanation on how the two reasoning systems develop from the evolutionary point of view. System 1 is at service of evolutionary rationality and is designed to monitor and identify the regularities of the environment, while System 2 at its turn is functional to the instrumental rationality. In addition, the variability in problem-solving and decision-making tasks and the discrepancy between normative and descriptive models cannot be explained only by the presence of performance errors and cognitive computational limitations, but it's necessary also to consider the many individual differences—as the needing to conclude the task as soon as possible, the tendency to think deeply, the propensity to confirm our own hypotheses and so on.

Fig. 35.2 Dual systems and
information processing (42,
p. 92)

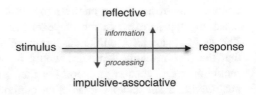

In line with the dual-system models, the *Reflective Impulsive Model* [42] explains the behavior as the result of the joint function of two processing systems each one of them operate according to different principles (Fig. 35.2): the *reflective system* (RS) and the *impulsive-associative system* (IS). While the first produces judgments and decisions following intentional patterns, the second is activated unknowingly by previously learned associations and by motivational orientation.

The *multimodal associative memory* of IS, which includes behavioral patterns and motivational orientation, has a key role in the decision-making process because it generates different emotional responses and behavioral trends to the problems posed by the environment [47]. These *internal cubes* are received by RS and included in the judgment's formation. The RS operations are slower than the IS functions, are based on a few, distinct symbolic representations and put together strategic plans of action to achieve a goal, inhibiting the instinctive responses (e.g., impulses or habits) [48]. At this point, the RS usually selects those cues that (a) meet best the goal and (b) are compatible with the cognitive capacity and the available environmental resources.

35.4 Conclusions

Over the last 30 years, systematic analysis of human decisions has come up with new and surprising evidence. In particular, experimentally, it has been shown that in their behaviour the great majority of people follow spontaneous intuitions, unconsciously adopting adaptive solutions which are incompatible with normative rationality [49, 50]. In particular, it was seen that to decide people would activate quickly a temporary network of important information available in the memory for a first context representation; that automatic cognitive processes reduce the incoherences between network information by creating a coherent environment representation; that people will represent to themselves those problems in which a decisive option prevails.

This recognition, derived from study and analysis of the mind carried out throughout the 20th century, should constitute the basis for a further conceptual revolution that will have to redefine the role of external constraints to human action (of resources and information available to the subject), giving back importance to internal constraints, (in computing capacity [51, 52], memory and more). It's a matter of inserting these evidences into the framework of a *natural logic* founded on the theoretical and experimental system of cognitive neurosciences. Soon it should be possible to clarify that each individual adopts conduct rules that allow him, on one hand, to simplify his decision-making models and, on the other hand, to coordinate himself with other agents to resize the uncertainty proper to every complex system.

References

1. Von Neumann, J., Morgenstern, O.: Theory of Games and Economic Behavior. Princeton, Princeton University Press (1947)
2. Newell, B.R., Lagnado, D.A., Shanks, D.R.: Straight Choices: The Psychology Of Decision Making. Psychology Press, Hove (2015)
3. Simon, H.A.: Reason in Human Affairs. Stanford University Press, Stanford, CA (1983)
4. Newell, B.R., Shanks, D.R.: Unconscious influences on decision making: a critical review. Behav. Brain. Sci. **38**(01), 1–19 (2014)
5. Epstein, S.: Intuition from the perspective of cognitive-experiential self-theory. In: Plessner, H., Betsch, C., Betsch, T. (eds.) Intuition in Judgment and Decision Making, pp. 23–37. Lawrence Erlbaum Associates, New York (2008)
6. Todd, P.M., Gigerenzer, G.: Ecological Rationality: Intelligence in the World. Oxford University Press, Oxford (2012)
7. Dell'orco, S.: Intuition, decision and ecological rationality: the toolbox of evolution. Hum. Evol. **28**(1–2), 65–77 (2013)
8. Kahneman, D.: Thinking, Fast and Slow. Farrar Straus & Giroux, New York (2011)
9. Moxley, J.H., Ericsson, K.A., Charness, N., Krampe, R.T.: The role of intuition and deliberative thinking in experts' superior tactical decision-making. Cognition **124**(1), 72–78 (2012)
10. Calder, L.A., Forster, A.J., Stiell, I.G., Carr, L.K., Brehaut, J.C., Perry, J.J., Vaillancour, C., Croskerry, P.: Experiential and rational decision making: a survey to determine how emergency physicians make clinical decisions. Emerg. Med. J. **29**(10), 811–816 (2011)
11. Bauer, J.C., Schmitt, P., Morwitz, V.G., Winer, R.S.: Managerial decision making in customer management: adaptive, fast and frugal? J. Acad. Mark. Sci. **41**(4), 436–455 (2012)
12. Basel, J.S., Brühl, R.: Rationality and dual process models of reasoning in managerial cognition and decision making. Eur. Manag. J. **31**(6), 745–754 (2013)
13. Klein, G.: A naturalistic decision making perspective on studying intuitive decision making. J. Appl. Res. Mem. Cogn. **4**(3), 164–168 (2015)
14. Gore, J., Flin, R., Stanton, N., Wong, B.L.W.: Applications for naturalistic decision-making. J. Occup. Organ. Psychol. **88**(2), 223–230 (2015)
15. Epstein, S.: Integration of the cognitive and the psychodynamic unconscious. Am. Psychol. **49**(8), 709–724 (1994)
16. Klein, G.A., Crandall, B.W.: The role of mental simulation in naturalistic decision making. In: Hancock, P., Flach, J., Caird, J., Vicente, K. (eds.) Local Applications of the Ecological Approach to Human-Machine Systems. Lawrence Erlbaum Associates, Hillsdale (1995)
17. Schwartz, S., Griffin, T.: Medical Thinking: The Psychology of Medical Judgment and Decision Making. Springer, New York (2012)
18. Maldonato, M., Dell'orco, S.: Toward an evolutionary theory of rationality. World Futures **66**(2), 103–123 (2010)
19. Dörfler, V., Ackermann, F.: Understanding intuition: the case for two forms of intuition. Manag. Learn. **43**(5), 545–564 (2012)
20. Sinclair, M.: Handbook of Intuition Research. Edward Elgar Publishing Limited, Cheltenham (2011)

21. Hayles, N.K.: Cognition everywhere: the rise of the cognitive nonconscious and the costs of consciousness. N. Lit. Hist. **45**(2), 199–220 (2014)
22. Gigerenzer, G., Murray, D.J.: Cognition as Intuitive Statistics. Lawrence Erlbaum Associates, Hillsdale, NJ (2015)
23. Goldin-Meadow, S., Beilock, S.L.: Action's influence on thought: the case of gesture. Perspect. Psychol. Sci. **5**(6), 664–674 (2010)
24. Gigerenzer, G.: Simply Rational: Decision Making in the Real World. Oxford University Press, Oxford (2015)
25. Maldonato, M., Dell'Orco, S.: Natural Logic: Exploring Decision and Intuition. Sussex Academic Press, Brighton (2011)
26. Maldonato, M., Dell'Orco, S.: The predictive brain. World Futures **68**(6), 381–389 (2012)
27. Glöckner, A., Betsch, T.: Modeling option and strategy choices with connectionist networks: towards an integrative model of automatic and deliberate decision making. Judgm. Decis. Mak. **3**(3), 215–228 (2008)
28. Hogart, R.: Educating Intuition. The University of Chicago Press, Chicago (2001)
29. Maldonato, M., Dell'Orco, S.: Making decision under uncertainty: emotions, risk and biases. Smart Innovation Syst. Technol. (Springer) **37**, 293–302 (2015)
30. Maldonato, M.: Decision Making: Towards an Evolutionary Theory of Rationality. Sussex Academic Press, Brighton (2010)
31. Volchik, V., Zotova, T.: Adaptive rationality in the evolutionary and behavioral perspectives. Middle East J. Sci. Res. **17**(10), 1488–1497 (2013)
32. Evans, J.S.: Intuition and reasoning: a dual-process perspective. Psychol. Inq. **21**(4), 313–326 (2010)
33. Evans, J.S.: Dual-process theories of reasoning: Contemporary issues and developmental applications. Dev. Rev. **31**(2–3), 86–102 (2011)
34. Chaiken, S., Trope, Y.: Dual-process Theories in Social Psychology. Guilford Press, New York (1999)
35. Epstein, S.: Integration of the cognitive and the psychodynamic unconscious. Am. Psychol. **49**, 709–724 (1994)
36. Hammond, K.R.: Human Judgment and Social Policy: Incredible Uncertainty, Inevitable Error, Unavoidable Justice. Oxford University Press, New York (1996)
37. Sloman, S.A.: The empirical case for two systems of reasoning. Psychol. Bull. **119**, 3–22 (1996)
38. Schneider, W., Shiffrin, R.M.: Controlled and automatic human information processing: detection, search, and attention. Psychol. Rev. **84**, 1–66 (1977)
39. Chaiken, S.: A theory of heuristic and systematic information processing. Handbook of Theories of Social Psychology, vol.1, pp. 246–166; vol. 39, pp. 752–766 (2011)
40. Greenwald, A.G.: New look 3: reclaiming unconscious cognition. Am. Psychol. **47**, 766–779 (1992)
41. Osman, M.: An evaluation of dual-process theories of reasoning. Psychon. Bull. Rev. **11**(6), 988–1010 (2004)
42. Strack, F., Deutsch, R.: The reflective-impulsive model. In: Sherman, J.W., Gawronski, B., Trope, Y. (eds.) Dual-process Theories of the Social Mind. Guilford Publications, New York (2014)
43. Tversky, A., Kahneman, D.: Judgment under uncertainty: heuristics and biases. Science **185** (4157), 1124–1131 (1974)
44. Evans, J., Over, D.: Rationality in reasoning: the problem of deductive competence. Cah. psychol. cogn. **16**, 102–106 (1997)
45. Stanovich, K.E., West, R.F.: Individual differences in reasoning: implications for the rationality debate. In: Gilovich, T., Griffin, D., Kahneman, D. (eds.) Heuristics and Biases: The Psychology of Intuitive Judgment. Cambridge University Press, New York (2002)
46. Stanovich, K.E.: Who is rational?: Studies of Individual Differences in Reasoning. Lawrence Erlbaum Associates, Hillsdale (1999)

47. Smith, E.R., DeCoster, J.: Dual process models in social and cognitive psychology: conceptual integration and links to underlying memory systems. Pers. Soc. Psychol. Rev. **4**, 108–131 (2000)
48. Hofmann, W., Friese, M., Strack, F.: Impulse and self-control from a dual-systems perspective. Perspect. Psychol. Sci. **4**(2), 162–176 (2009)
49. Maldonato, M., Dell'Orco, S.: The natural logic of action. World Futures J. Gen. Evol. (Routledge) **69**(3), 174–183 (2013)
50. Maldonato, M., Montuori, A., Dell'Orco, S.: The Exploring Mind Natural Logic and Intelligence of the Unconscious. Kindle Edition (2013)
51. Esposito, A., Jain, L.: Toward robotic socially believable behaving systems volume II—"Modeling Social Signals". Springer, Berlin, Heidelberg (2016)
52. Esposito, A., Jain, L.C.: Modeling emotions in robotic socially believable behaving systems. In: Esposito, A., Jain, L.C. (eds.) Toward Robotic Socially Believable Behaving Systems-Volume I, pp. 9–14. Springer, Berlin, Heidelberg (2016)

Chapter 36
Artificial Entities or Moral Agents? How AI is Changing Human Evolution

Mauro Maldonato and Paolo Valerio

Abstract A large amount of theoretical and experimental research—from dynamic systems to computational neurosciences, from statistical learning to psychobiology of development—indicates that the encounter between Humans and very powerful AI will lead, in the near future, to organisms capable of going over the simulation of brain functions: hybrids that will learn from their internal states, will interpret the facts of reality, establish their goals, talk with humans and, especially, will decide according to their own 'system of values'. Soon the traditional symbolic-formal domain could be overcome by the construction of systems with central control functions, with cognition similar to the biological brain. This requires a clarification on how they will act and, mostly, how they will decide. But what do we know today about decision-making processes and what is their relationship with the emotional spheres? If, traditionally, emotions were considered separate from the logical and rational thought, in recent years we have begun to understand that emotions have deep influence on human decisions. In this paper, we intend to show how emotions are crucial in moral decisions and that their understanding may help us to avoid mistakes in the construction of hybrid organisms capable of autonomous behavior.

Keywords Emotions · Decision-making · Moral dilemma · Natural logic · Rationality · Artificial intelligence · Brain · Mind

M. Maldonato (✉)
Department of Human Sciences DISU, University of Basilicata, Potenza, Italy
e-mail: mauro.maldonato@unibas.it

P. Valerio
Department of Neurosciences, University of Naples Federico II, Naples, Italy
e-mail: paolo.valerio@unina.it

© Springer International Publishing AG 2018
A. Esposito et al. (eds.), *Multidisciplinary Approaches to Neural Computing*,
Smart Innovation, Systems and Technologies 69,
DOI 10.1007/978-3-319-56904-8_36

36.1   Introduction

For a long time philosophy and then science argued on one hand that our mind is equipped with logical-deductive tools to reach valid conclusions regardless of the premises and to obtain the maximum benefit from their decisions [1] and, on the other hand, that the emotions overshadow our rationality [2]. This is a long tradition of thought that includes, for example, Plato, who considered the passions and other feelings "barbaric slush" that hinder the thought [3]; Descartes [4] who believed that the mind is entirely separated from the body: the first one entirely reflected in the logic, the second one with emotions and feelings; Hegel [5] who hopes in the removal of anything that does not correspond to the *Idea*. Other famous examples could help us out. However, at least until the birth of cognitive science (in the first half of the twentieth century), the decision model resembled a sort of "moral algebra" that allowed to know in advance the effects of our conduct [6]. This model of rational choice, associated with the *Bayes' theorem* [7] has been an essential tool for those who recognized, in the logical-mathematical process, the essence of reasoning. Around the middle of the twentieth century, von Neumann and Morgenstern [8] highlighted an element overlooked by equilibrium theorists: the influence of social interactions on individual decisions. Economic processes, they argue, do not derive from actions of agents capable of perfect predictions, but from multiple interaction games that unpredictably affect individual decisions [9]. In the game of economy, in which an individual has partial information, only a probabilistic calculation can allow, although with margins of error, a successful strategy. An economic science that is rigorous must have a 'normative' foundation and coherent formal criteria [10] that allow rational decisions facing risky alternatives, reducing most of the subjective constraints.

Humans won their developmental challenge without relying too much on the formal-logical reasoning. For a long time almost everything—from social organization to individual lifestyles—has been functional to survival. The thought was the body and the body was the thought, and this reciprocity has shaped the architecture of the mind [11]. Environmental pressures, primarily the scarcity and the non-reproducibility of natural resources, have committed our ancestors in challenges of all kind [12]. Only high emotional and intuitive index decisions would guarantee their survival [13]. Today we know that thoughts and emotions, concepts and feelings, permeate each other [14, 15] and, especially, that without sensory and motor skills (therefore bodily) many aspects of human thought and knowledge would be unexplained [16]. Therefore underestimating role of the body in human thought means to devalue the role in the analysis of contexts, in problem solving, in the readiness of answer, in the rapidity of action, in achieving the purpose [17]. On the other hand, a thought is not independent from those who think [18].

36.2 Brains that Decide

Among the abilities of the body that contribute to the activities of mind there are also the sensory and motor ones [19]. Emotions—that are the oldest expression of the living body—favor quick decisions and answers worthy of the environmental challenges [20]; unlike logical reasoning which is slow and almost always delayed compared to the action [21]. Most of our reasoning is often based on non-deductive and unaware inferences and on simplified schemes often influenced by representations and distorted perceptions of risk: those variables make an optimal response unlikely [22]. In this sense, a *natural logic* (heuristics, informal adaptive strategies, and more) is wider than a formal logic [23, 24].

In the second half of the twentieth century, a large amount of experimental research has shown not only that unpredictability and uncertainty, because of partiality or insufficient information, determines risk situations that are constantly present in human action [12], but above all, that extra-cognitive factors like emotional assessment of risk, perseverance, fear for the consequences of an action, frustration tolerance, courage and self-esteem intervene in our decisions. Nevertheless, if few doubt the fact that the brain has shaped human thought, it defined the characteristics of our rationality and forged our decisions, very few doubt that it is able to control the instinct and emotionality, to evaluate situations objectively and to choose the most advantageous among the various alternatives. Kahneman [25] has shown that this conjecture is fallacious and that, instead, we are constantly exposed to influences that undermine our judgments and our actions [26]. So, if it is true that the organization of our thought is efficient and productive and allows us to develop skills and abilities, making us capable to easily perform complex operations, it is also exposed to systematic errors when intuition is influenced by stereotypes [27].

At a neurobiological level, the decision making process is related to the orbitofrontal cortex [28], to the ventromedial prefrontal cortex [29–32], to the amygdala [33–36], to the anterior cingulate cortex and to the hippocampus. Interferences in the integrated functioning of these structures affect our decisions [37]. The relationship between these areas is not linear and is characterized by interactions that determine its course [38]. That's why decision-making processes cannot be considered purely rational, but as an expression of a dynamic balance of emotions, memories and adaptive habits deeply rooted in the brain structures [19]. After all, most of our behaviors are supported by unaware automatisms. Emotions tie the information encoded by the brain [39]. Some authors distinguished between "hot" and "cold" reasoning [40]: the former is supported by the ventromedial prefrontal cortex [41], which is active in the decision-making process that involves emotions; the latter is supported by the dorsolateral prefrontal cortex located in the front higher part of prefrontal cortex.

36.3 Deciding Like Humans Do

It is now shared opinion that in the near future, organisms will appear which are capable of exceeding simulations of brain functions: hybrids able to process the abstract-symbolic functions similar to those of a biological brain; they will learn from their internal states, will interpret the facts of reality, establish their goals and, most importantly, will decide on the basis of its own 'system of values' [42]. For now, research programs [43] are engaged in the projection of communication forms with strong emotional-affective index, forms of communication based on (a) the creation of new learning algorithms that jointly analyze multimodal channels of information [44]; (b) new assessment techniques of frustration, of stress and of mood through the natural interaction and conversation [45]; (c) methods that improve the self-awareness, the awareness of their own emotional state and of communication with the others [46]; (d) the assessment of individual health [47]; (e) the examination of the ethical implications in the field of affective informatics.

It won't be easy to reproduce all this in hybrid agents, because these may acquire different skills from those of humans [48]. It is not paradoxical to believe that their rational behaviors could be even more efficient than the human ones, because of their ability to evaluate, before complex situations, a wider range of options without emotional conditioning. We can just consider ethically sensitive fields such as medical or military ones [49], in which there are already advanced forms of cooperation between humans and robots (software agents) to carry out critical missions where human life is at stake: rescue missions where computational cognitive models include moral values and human concerns [50]; emergency situations that could be compromised by slow human decision-making processes; or even war situations where hybrid agents may have more measured behaviors than the human soldiers, because hybrids agents are more respectful of the rules of engagement [51].

But what do we mean by moral decision in hybrid organisms and agents? Years ago, Picard [52] listed some reasons in favor of an assignment of emotions to hybrid agents: in fact, they would render the emulation of human action more credible, they would improve the relationship between humans and machines and, finally, through their modeling they would even help understand human emotions. Beyond utilitarian analysis limits, those are not unreasonable arguments. Are they sufficient, however, to face the questions posed by agents capable of moral decisions? For example, how would these agents act in situations of potential conflict between opposing moral instances as saving lives of a group or of an external member? Would they unconditionally obey the authority or would they do what they 'consider' right? Could recognizing and understanding the emotions and concerns of the others agents influence their own moral decisions?

Finally, if it's true that moral behaviors of humans are defined by the interaction among biological, cultural and social elements, which values and hierarchies of values will inspire their behaviors? Would an advanced autonomy suggest an "out of control" freedom? In conclusion, due to the increasing power and interconnections of the web, will the skills and the acquired prerogatives become the bases of an universal hybrid moral grammar [53, 54], or the first step towards the free will of machines, without any human control?

Before any discussion, we must be clear on some essential points. First of all, a moral should cover basic rules such as do not kill, do not steal, do not fool, be honest, loyal and unselfish, trust in man's ability to learn and commit to a shared system of moral rules [55]. The social reciprocity, besides, is an important resource because it promotes virtuous behaviors, punishes those that aren't virtuous, pushes to the deferment of actions over time and so on [56]. In this sense, the possibility of hybrid agents to take moral decisions opens the field to series of questions. Can these agents be considered holders of rights and, vice versa, can they be considered responsible for their actions? Could they be convicted because of the violation of law? If, as someone has argued [57], in a few decades they will be more intelligent than humans, can we be sure that they will still be 'friends' of humans? And, if we cannot be sure, should we perhaps abandon the research? Understandably, the intellectual excitement for such questions is proportional to the importance of the ethical, philosophical and practical issues at stake [58]. In front of a large freedom of those agents, we will need high moral standards [59]. Although there are points of views that tend to limit their role in improving the quality of human life, the ethical concerns about their behavior cannot be easily eluded [60, 61].

36.4 Moral Decisions Agent Out of Control?

We may consider the idea of applying the famous experiment, called "the railway carriage dilemma" to hybrids [62, 63]. A railway carriage without brakes is running fast towards five people working on the rails. In order to save these people, you have to press a button that will divert the race on an alternative rail, where another worker is doing his job. If you press the button, five people will be saved, but one person will die (Fig. 36.1). When asked the question: "Would you divert the carriage to save five people but sacrificing the life of another?", most respondents consider plausible to push the button. Even if it is horrible, sacrificing one person instead of five has its own rationality. There is also a second experiment, not too different from the first one. The scene is the same: an out of control train and five people working on the rails, unaware of what is going on. Compared to the first experiment, there is only one difference: on an overpass, next to the experimenter, there is a burly stranger. The overpass is halfway between the carriage and the five workers working on the rails below. The risk of a massacre is very high. The only way to avoid it is to push off the stranger. The generous proportions of his body certainly will stop the carriage. The poor man will die, but his sacrifice will save five workers. Here is the question: "Would you push the stranger to death in order to save the five workers?" At this point, things get more complicated. In both cases, you sacrifice one man in order to save five of them. From the utilitarian point of view, both choices meet the criterion of good for the greatest number of people. However, if in the first case most of the respondents would push the button, in the

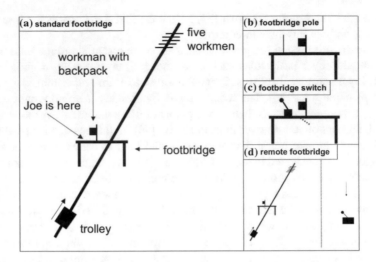

Fig. 36.1 Diagrams for the **a** standard footbridge dilemma (physical contact, spatial proximity, and personal force), **b** footbridge pole dilemma (spatial proximity and personal force), **c** footbridge switch dilemma (spatial proximity), and **d** remote footbridge dilemma. (Panels b-d depict details of diagrams presented to subjects with labels and some pictorial elements removed for clarity.) (cf. 63: 8)

second case, even if they are not able to explain it, most of them would refuse to push the stranger down.

Now, how will an artificial organism decide when facing: the personal or impersonal conflict, the not always justifiable prevalence of means over ends, the fact that actions always have an own story, so they must be analyzed for their real intentions, that one thing is to save the largest number of people, and another thing is not to hurt an unsuspecting and innocent man? In short, would this organism maximize the utility, regardless any other consideration? The difficulties of these research programs, and even more the construction of advanced hybrids, are completely confirmed [64]. If man has always solved the issues that tormented him with the indirect help of evolution, now the advanced development of hybrid agents causes accelerations and transformations that will fundamentally change human relationships and the relationship that humans have always had with the environment. Besides, the combined implementation of genetic methods and nanotechnology will result in formidable opportunities and adaptive challenges that, at least for now, we can guess but we cannot predict [57]. We are within a change that will last a long time. Many people believe that technology will sooner or later prevail over humans, others trust the way that such processes will be governed: among the latter, a part of the productive world considers any parameter and rigid constraint for the activity of hybrid agents as an obstacle to progress and innovation [65]. It is a technological unpredictable evolution, but it is evident that progress in this area could help the development of agents and organisms able to engage in reasoning and moral decisions.

36.5   Conclusions

A moral decision deals with emotions, awareness, elaborations of the semantic content of the information, and social skills [66]. Conceiving the presence of such beyod rational sphere in hybrids with decision-making skills imposes extremely relevant considerations: not only for the fundamental role of emotions [67, 68], but mainly because morality presupposes an ontology. A moral decision, in fact, arises from agents plunged in their environment, in their own culture, in relationship with other individuals, each one of them with their own goals, values and desires [69]. Moreover, in social situations what is moral is not necessarily predetermined. In fact, the appropriateness of certain actions often derives from real situations and from precise interactions between individuals [70]. In this sense, one thing is to build decision architectures in hybrid organisms or internal representations of environments that predict potential actions, and another thing is to predict human actions.

Until today, decisions and moral judgments have been guaranteed by the experience accumulated during the evolution and by just as many ecological, rational constructions [6]. It is plausible to consider that moral spheres in hybrid agents must arise from the interaction of both these levels, that on one hand allow dynamism and flexibility to morality compared to external inputs; and on the other hand the rational assessment of choices and actions that correspond to moral principles. In summary, by combining emotional-affective instances with rational-utilitarian ones [61].

Before us there is a chance to reach the construction of systems that can integrate skills that facilitate the development of higher-order faculties [71]. The focus, limited until today to the implementation of operating skills in hybrid organisms, inevitably will open to spheres that go beyond the ability to reason and decide, especially for the implications that concern self-consciousness and free will. They are inescapable supra-rational individual and social issues that will determine the success or the failure of a path on which the fate of humanity as we have known until now largely depends.

References

1. Manktelow, K.I.: Thinking and Reasoning: An Introduction to the Psychology of Reason, Judgment and Decision Making. Psychology Press, Hove, East Sussex (2012)
2. Russell, B.: History of Western Philosophy. Routledge: Collectors Edition (2013)
3. Plato. Republic. London: Macmillan (1866)
4. Descartes, R.: Discourse on Method; Translated with an Introd. Liberal Arts Press, New York (1950)
5. Hegel, G.W.F.: Phänomenologie des geistes. BoD–Books on Demand (2015)
6. Todd, P.M., & Gigerenzer, G.: Ecological Rationality: Intelligence in the World. Oxford University Press (2012)

7. Bayes, T.: Studies in the history of probability and statistics. Biometrika **45**(3/4), 293–315 (1958)
8. Neumann, J.Von, Morgenstern, O.: Theory of Games and Economic Behavior. Princeton University Press, Princeton (1947)
9. McFall, J.P.: Rational, normative, descriptive, prescriptive, or choice behavior? The search for integrative metatheory of decision making. Behav. Dev. Bull. **20**(1), 45 (2015)
10. Chater, N., & Oaksford, M.: Normative systems: Logic, probability, and rational choice (pp. 11–21). The Oxford handbook of thinking and reasoning (2012)
11. Jeannerod, M.: Motor Cognition: What Actions Tell the Self. Oxford University Press, Oxford (2006)
12. Maldonato, M., Dell'Orco, S.: How to make decisions in an uncertain world: Heuristics, biases, and risk perception. World Futur. **67**(8), 569–577 (2011)
13. LeDoux, J.: Rethinking the emotional brain. Neuron **73**(4), 653–676 (2012)
14. Damasio, A., Carvalho, G.B.: The nature of feelings: evolutionary and neurobiological origins. Nat. Rev. Neurosci. **14**(2), 143–152 (2013)
15. Shiv, B., Loewenstein, G., Bechara, A., Damasio, H., Damasio, A.R.: The dark side of emotion in decision-making: when individuals with decreased emotional reactions make more advantageous decisions. Cognitive Brain Res. **23**(1), 85–92 (2005)
16. Varela, F.J., Thompson, E., Rosch, E.: The Embodied Mind: Cognitive Science and Human Experience. MIT Press, Cambridge, MA (1991)
17. Gigerenzer, G., Gaissmaier, W.: Heuristic decision making. Annu. Rev. Psychol. **62**(1), 451–482 (2011)
18. Maldonato, M.: The Predictive Brain. Sussex Academic Press, Brighton, Chicago, Toronto (2014)
19. Oliverio, A.: Cervello. Bollati Boringhieri, Torino (2012)
20. LeDoux, J.E.: Rethinking the emotional brain. Neuron **73**(4), 653–676 (2012)
21. LeDoux, J.E.: Anxious: using the Brain to Understand and Treat Fear and Anxiety. Viking, New York (2016)
22. Gigerenzer, G., Gaissmaier, W.: Heuristic decision making. Ann. Rev. Psychol. **62**(1), 451–482 (2011)
23. Cellucci, C.: Rethinking Logic: Logic in Relation to Mathematics, Evolution, and Method. Springer, Berlin (2013)
24. Maldonato, M., Dell'Orco, S.: Natural Logic: Exploring Decision and Intuition. Sussex Academic Press, Brighton (2011)
25. Kahneman, D.: Thinking, Fast and Slow. Farrar Straus & Giroux, New York (2011)
26. Kahneman, D., & Tversky, A.: Prospect theory: an analysis of decision under risk. Econom.: J. Econom. Soci., 263–291 (1979)
27. Newell, B.R., Lagnado, D.A., & Shanks, D.R.: Straight Choices: The Psychology of Decision Making. Psychology Press (2015)
28. Montague, R.: Why Choose this Book?: How we Make Decisions. EP Dutton (2006)
29. Bechara, A., Damasio, H., Damasio, A.R.: Emotion, decision making and the orbitofrontal lcortex. Cereb. Cortex **10**(3), 295–307 (2000)
30. Paulus, M.P., Frank, L.R.: Ventromedial prefrontal cortex activation is critical for preference judgments. NeuroReport **14**(10), 1311–1315 (2003)
31. Clark, L., Bechara, A., Damasio, H., Aitken, M.R.F., Sahakian, B.J., Robbins, T.W.: Differential effects of insular and ventromedial prefrontal cortex lesions on risky decision-making. Brain **131**(5), 1311–1322 (2008)
32. Zald, D.H., Mattson, D.L., Pardo, J.V.: Brain activity in ventromedial prefrontal cortex correlates with individual differences in negative affect. Proc. Nat. Acad. Sci. **99**(4), 2450–2454 (2002)
33. Sokol-Hessner, P., Camerer, C.F., & Phelps, E.A.: Emotion regulation reduces loss aversion and decreases amygdala responses to losses. Soc. Cognit. Affect. Neurosci., nss002 (2012)
34. Gupta, R., Koscik, T.R., Bechara, A., Tranel, D.: The amygdala and decision-making. Neuropsychologia **49**(4), 760–766 (2011)

35. Urry, H.L., Van Reekum, C.M., Johnstone, T., Kalin, N.H., Thurow, M.E., Schaefer, H.S., Davidson, R.J.: Amygdala and ventromedial prefrontal cortex are inversely coupled during regulation of negative affect and predict the diurnal pattern of cortisol secretion among older adults. J. Neurosci. **26**(16), 4415–4425 (2006)
36. Bechara, A., Damasio, H., Damasio, A.R.: Role of the amygdala in decision-making. Ann. N. Y. Acad. Sci. **985**(1), 356–369 (2003)
37. Daw, N.D., O'Doherty, J.P., Dayan, P., Seymour, B., Dolan, R.J.: Cortical substrates for exploratory decisions in humans. Nature **441**(7095), 876–879 (2006)
38. Maldonato, M.: Decision making: towards an evolutionary psychology of rationality. Sussex Academic Press (2010)
39. Esposito, A., Jain, L.C.: Modeling emotions in robotic socially believable behaving systems. In: Esposito, A., Jain, L.C. (eds.) Toward Robotic Socially Believable Behaving Systems—Volume I: 9–14. Springer Verlag, Berlin, Heidelberg (2016)
40. Goel, V., Dolan, R.J.: Explaining modulation of reasoning by belief. Cognition **87**(1), B11–B22 (2003)
41. Kunda, Z.: The case for motivated reasoning. Psychol. Bull. **108**(3), 480 (1990)
42. Kurzweil, R.: How to Create a Mind: The Secret of Human Thought Revealed. Penguin (2012)
43. Fortunati, L., & Vincent, J.: Introduction. In: Vincent, J., Fortunati, L. (eds.), Electronic Emotion: The Mediation of Emotion Via Information and Communication Technologies, (pp. 1–31). Oxford: Peter Lang (2009)
44. Martinez, A., Du, S.: A model of the perception of facial expressions of emotion by humans: research overview and perspectives. J. Mach. Learn. Res. **13**(1), 1589–1608 (2012)
45. Zisook, M., Taylor, S., Sano, A., and Picard, R.W.: SNAPSHOT Expose: stage based and social theory based applications to reduce stress and improve wellbeing. In: CHI 2016 Computing and Mental Health Workshop (2016)
46. Scherer, K.R., Bänziger, T., & Roesch, E.: A Blueprint for Affective Computing: A Sourcebook and Manual. Oxford University Press (2010)
47. Luxton, D.D., June, J.D., Sano, A., & Bickmore, T.: Intelligent mobile, wearable, and ambient technologies for behavioral health care. Artif. Intell. Behav. Ment. Health Care, 137 (2015)
48. Esposito, A., Jain, L.: Toward Robotic Socially Believable Behaving Systems Volume II —"Modeling Social Signals". Berlin, Heidelberg: Springer Verlag (2016)
49. Anderson, M., Anderson, S.L.: Machine ethics: creating an ethical intelligent agent. AI Mag. **2804**, 15–26 (2007)
50. Kamm, F.M.: Moral status and personal identity: clones, embryos, and future generations. Soc. Philos. Policy **22**(02), 283–307 (2005)
51. Arkin, R.C..: Governing lethal behavior: Embedding ethics in a hybrid deliberative/reactive robot architecture part I: Motivation and philosophy. In: Proceedings of the 3rd International Conference on Human Robot Interaction—HRI 2008, pp. 121–128 (2008)
52. Picard, R.W.: Affective computing: challenges. Int. J. Hum. Comput. Stud. **59**(1), 55–64 (2003)
53. Hauser, M.D.: Moral Minds: How Nature Designed Our Universal Sense of Right and Wrong. Ecco, New York (2006)
54. Mikhail, J.: Universal moral grammar: theory, evidence and the future. Trends Cognit. Sci. **11** (4), 143–152 (2007)
55. Prinz, J.J.: The Emotional Construction of Morals. Oxford University Press, New York (2008)
56. Rawls, J.: A Theory of Justice. Harvard University Press (2009)
57. Kurzweil, R.: The Singularity is Near: When Humans Transcend Biology. Penguin, New York (2005)
58. Wallach, W.: Artificial morality: bounded rationality, bounded morality and emotions. In: Smit, I., Lasker, G., Wallach, W. (eds.), Proceedings of the Intersymp. Workshop on Cognitive, Emotive and Ethical Aspects of Decision Making in Humans and in Artificial Intelligence, pp. 1–6, Baden-Baden, Germany, IIAS, Windsor, Ontario (2004)
59. Picard, R.W.: Affective Computing. MIT Press, Cambridge, MA (1997)

60. Anderson, M., Anderson, S.L.: Toward ensuring ethical behavior from autonomous systems: a case-supported principle-based paradigm. Ind. Robot: An Int. J. **42**(4), 324–331 (2015)
61. Wallach, W., Franklin, S., Allen, C.: A conceptual and computational model of moral decision making in human and artificial agents. Topics Cognit. Sci. **2**(3), 454–485 (2010)
62. Foot, P.: Virtues and vices and other essays in moral philosophy. Cambridge University Press, Cambridge (2002)
63. Greene, J.D., Cushman, F.A., Stewart, L.E., Lowenberg, K., Nystrom, L.E., Cohen, J.D.: Pushing moral buttons: the interaction between personal force and intention in moral judgment. Cognition **111**(3), 364–371 (2009)
64. Webb, B.: Can robots make good models of biological behaviour? Behav. Brain Sci. **24**(06), 1033–1050 (2001)
65. Bostrom, N.: Superintelligence: Paths, Dangers, Strategies. OUP Oxford, Oxford (2014)
66. Mason, H.E.: Moral Dilemmas and Moral Theory. Oxford University Press, New York (1996)
67. Damasio, A.R.: Descartes' Error: Emotion, Reason, and the Human Brain. Putnam, New York (1994)
68. Salovey, P., Mayer, J.D.: Emotional intelligence. Imagin. Cogn. Pers. **9**(3), 185–211 (1990)
69. Breazeal, C.: Emotion and sociable humanoid robots. Int. J. Hum.-Comput. Stud. **59**(1–2), 119–155 (2003)
70. Nozick, R.: Moral complications and moral structures. Nat. Law Forum **13**, 1–50 (1968)
71. Scassellati, B.M.: Foundations for a Theory of Mind for a Humanoid Robot. Massachusetts Institute of Technology, Massachusetts (2001)

Printed in the United States
By Bookmasters